Semiconductor
X-Ray Detectors

Series in Sensors

Series Editors: Barry Jones and Haiying Huang

Other recent books in the series:

Semiconductor X-Ray Detectors

B. G. Lowe
R. A. Sareen

CRC Press
Taylor & Francis Group
Boca Raton London New York

CRC Press is an imprint of the
Taylor & Francis Group, an **informa** business

CRC Press
Taylor & Francis Group
6000 Broken Sound Parkway NW, Suite 300
Boca Raton, FL 33487-2742

First issued in paperback 2017

© 2014 by Taylor & Francis Group, LLC
CRC Press is an imprint of Taylor & Francis Group, an Informa business

No claim to original U.S. Government works

ISBN-13: 978-1-4665-5400-9 (hbk)
ISBN-13: 978-1-138-03385-6 (pbk)

Library of Congress Cataloging-in-Publication Data

Lowe, B. G., author.
   Semiconductor x-ray detectors / B.G. Lowe, R.A. Sareen.
      pages cm -- (Series in sensors)
   Includes bibliographical references and index.
   ISBN 978-1-4665-5400-9 (hardcover : alk. paper)
   1. Detectors. 2. X-rays--Measurement. 3. Semiconductors. I. Sareen, R. A., author. II. Title. III. Series: Sensors series.

   TK7871.85.L69 2014
   621.381--dc23   2013038219

Visit the Taylor & Francis Web site at
http://www.taylorandfrancis.com

and the CRC Press Web site at
http://www.crcpress.com

Contents

Chapter 2: Detector Response Function 129

Preface

The authors waited more than 5 years before recording this detailed account of semiconductor x-ray detectors. During their careers, they witnessed the evolution of the Si(Li) as it gradually replaced early detectors such as scintillators, gas counters, and surface barrier detectors. As will be shown, history has repeated itself and the Si(Li) has now largely been replaced by the silicon drift detector over a period of more than 25 years. It became a welcome relief for many users to relinquish the need to regularly fill their cryostats with liquid nitrogen and to conveniently use detectors using electronic cooling. The one remaining advantage of the Si(Li), its sensitive depth, was not seen sufficiently important compared to the convenience of electronic cooling. Before the knowledge was finally lost, the authors wished to record the science surrounding the Si(Li) and have drawn together this information from industry, government laboratories, and universities. The Si(Li) is restricted to the few special instances where its large sensitive depth is essential. We hope this record will be of interest and an important supplement to the knowledge now required for achieving full understanding of the workings of its replacement, the silicon drift detector.

Acknowledgments

We have both been very fortunate in having the support and patience shown by our families for the many hours we have sat either at a computer or in a library or communicating with former colleagues and associates trying to gather together the information needed for this book.

Many talented people have supported us throughout our careers. In the early days, Rob had a young assistant, Heather Usher, who painstakingly processed the semiconductors and manufactured the early range of detectors needed for Nuclear Physics research at Liverpool and Manchester Universities. Again, while working at Ortec in Oak Ridge, Tennessee, in the early 1970s, Rob appreciated the technical help from the detector R&D group and also the hours spent in lively discussions with Charlie Inskeep, Manny Elad, and John Walter. This pattern of collaboration between physicists and technical specialists became the blueprint for success when transferred into new ventures.

We both enjoyed long careers in industry establishing Link Systems as a leading supplier of detectors, and this pattern was continued later at Gresham Scientific Instruments. During the mid 1980s, we benefitted by acquiring outstanding companies such as Tennelec, The Nucleus, X-tech, and Camscan, and we both enjoyed the collaboration and services of working with talented people from a wide range of disciplines. They were involved not only in the successful development and manufacture of detectors and their associated electronics but also in their promotion into the world markets. This list is understandably long, and we apologise if we have left out any names, but over this period of 40 years there are a number of former colleagues we would like to thank in writing for all their kind efforts and innovations.

In detector manufacturing, we are grateful to Irene Sears and her team for the manufacturing of the Si(Li)s and the HPGe detectors, and in FET manufacturing and assembly we are indebted to Michelle Smith. Claire Whitelock contributed enormously both in system testing and later in cryostat assembly. Ted Whitehouse led our mechanical design and was a tremendous asset to the business. When Ted retired, Chris Tyrell very successfully continued his work. Our company strategy was always to be the technical leader in our chosen industry and with the help of an enormously talented technical team that included Graham White, Tawfic Nashashibi, Bob Daniel, Chris Cox, Steve Bush, Dave Sareen, and Cos Antoniou, we were at the forefront with innovative products.

Barrie is indebted to the help and assistance given by Stuart Tyrrell while at Link Systems and later Oxford Analytical, and to Irene Sears at Oxford Instruments and Gresham Scientific Instruments, and later to Fang Xiang at Gresham Scientific Instruments. We both benefitted from our work with Greg Bale, Mark Gray, Steve Bush, and all the staff in detector test at Gresham Scientific Instruments.

Acquisitions can often lead to destructive competitive disadvantages and major personality conflicts that ultimately destroy the benefits perceived for the acquisition. Because we were technically led, our acquisitions did not suffer so much from these problems, and we are grateful to Pat Sangsingkeow, Larry Darken, Bruce Coyne, and their colleagues at Tennelec and The Nucleus. It was an enormous pleasure to work with these talented individuals.

Oxford Instruments acquired Link Systems and it was a pleasure to work with Peter Williams, Paul Winson, John Pilcher, and John Woodgate, and it is comforting to see that the former Link Systems microanalysis products are still world leading products.

Gresham Scientific Instruments was established as a division of Gresham Power Electronics, a specialist power supply company and the purchaser of Camscan. Again, this gave both Rob and Barrie the opportunity to work with new people who brought new insights to development and manufacturing. Rob is indebted to the help and wisdom from Mervyn Hobden and the support from Shada Kazami. It was also a pleasure to work with Tony Bosley, Peter Smith, Dave Sareen, Tawfic Nashashibi, Bob Daniel, Greg Bale, Irene Sears, Claire Whitelock, Michelle Smith, and Graham Ensell, and all the other staff who supported this growing business.

Finally, the team efforts from all the other departments in a company are just as vital to its success. We could not have made the advances at Link without our financial team, led by Mac Seaton and then by Danny Weidenbaum. The sales department took on the difficult role of combining end user sales with OEM sales using a direct sales force and agents in many countries. These were ably led by Ron Jones and supported enthusiastically by Roger Bowring, Tom Sheehan, Eric Samuel, Neil McCormick, Frank Brown, Don Grimes, Pat Campos, and many others. The purchasing department under Anne Hart was a constant help to the challenges of a growing business, and the service department under Robin Hart made sure our customers were happy with their installations. Just as a company is no use without a good sales department, it is also extremely difficult to sell a detector without the means of handling the signals generated by the detector. This soon extrapolates into the concept of making a system that can display the histograms and with the appropriate software analyze the complex spectra. For this reason, we have been most fortunate to work alongside expert digital designers such as Brian Sharp, Colin Robinson, and Zaffa Ali, and talented analytical specialists and software designers such as Chris Millward, Peter Statham, Brian Cross, and Will Clark.

To all these people and to many others, we are most appreciative of their contributions to our understanding of the complex physics associated with radiation detection and the challenges in making large numbers of state-of-the art reproducible products that have allowed other scientists to expand their horizons. This probably summarises the difference between studying one device at a university on a research program and having to repeatedly make similar devices in large numbers for a wide range of uses. The challenges are very different, and both are important reservoirs of knowledge. In our own small way, we hope we have documented some of the things we have learned during our time in industry, and we acknowledge that this collection of information is incomplete and biased toward our own experiences and limited understanding of how these devices work.

Authors

School and Universities

 **B. G. Lowe** was born in Chester, UK, and attended Chester City Grammar School and Liverpool University. He graduated with a degree in Physics with distinction in 1962 and with Honours in 1963 and completed his Ph.D. in Nuclear Physics in 1968. He then took a Research Fellowship in Nuclear Physics at Southampton University. His main interest was acoustic spark-chamber particle detectors. He then became a Commonwealth Education Officer based at Colombo University, Sri Lanka, and later took a Lectureship at the Science University in Penang, Malaysia. His research interest was Environmental Radioactivity using low background Ge(Li) detectors. He returned to the UK in 1977 as a lecturer in Physics at the N.E. Wales College of Higher Education.

Industry

Barrie joined Link Systems Ltd UK as chief physicist in 1978, working on the development of lithium-drifted silicon and high purity germanium detectors under Rob Sareen. He became Physics director at Oxford Instruments Microanalysis Group in 1983 as well as head of Development of the Industrial Analysis Group in 1991. The work included the development of a Peltier cooled p–i–n diode and Stirling engine cooled Si(Li) detector for industrial x-ray fluorescence (XRF). From 1993 to 2000, he was senior scientist for the Oxford Instruments

Analytical Systems Division that included the companies X-Tech (x-ray tubes) and Nuclear Measurements Group (large volume HPGe detectors), both based in the United States.

In 1994 he gave the course 'X-ray Detectors' at the IEEE Nuclear Science Symposium in Virginia, USA. During the 1990s, he was also involved with the UK Government–sponsored 'IMPACT' project investigating novel semiconductor x-ray detectors such as silicon pixel arrays and CCDs. Collaborators included Leicester University, e2v plc, and the Rutherford Appleton Laboratory. He also worked on CdZnTe detectors with Greg Bale (then at Leicester University). From 2000 up to his retirement in 2006, he was senior scientist for e2v Scientific UK, mainly concerned with Si(Li) and HPGe detectors.

Publications and Patents

Barrie was an author on more than 20 publications on radiation detectors and the 'conditioner' patent for Si(Li) detectors.

He is married with two sons and two daughters and lives in an Oxfordshire village.

School and Universities

R. A. Sareen was born in Colwyn Bay, North Wales, and attended Colwyn Bay Grammar School. He graduated with a degree in Physics with Honours at Liverpool University in 1963 and then completed a 1-year Graduate Apprenticeship at Rank Bush Murphy in Welwyn Garden City before returning to Liverpool University as a research assistant. Initially, he worked on the design and construction of a Von Ardenne duoplasmatron and then on Silicon Surface Barrier detectors to measure the ions selected by a magnetic spectrometer attached to this ion source. The success of this program led to the establishment of a Detector Group to investigate a wide range of detectors to support the researchers at the Accelerators at both Liverpool and Manchester Universities. Detectors included silicon surface barrier detectors for particle physics, lithium-drifted germanium detectors for detecting gamma rays, and lithium-drifted silicon detectors for x-ray measurements. This work led to a Masters Degree in Physics and a consultancy with the pioneering detector company, Ortec Inc., in the United States. Later, Rob returned to Manchester University and completed his Ph.D. in Nuclear Physics.

Industry

In 1968, Rob left with his family to join Ortec as a research scientist in Oak Ridge, Tennessee. He worked in the company's Detector Group, designing both linear and radial position-sensitive particle detectors with some of the pioneers on silicon and germanium detectors including John Walter, Rex Trammell, Emmanuel Elad, and Charlie Inskeep. Rob also investigated new front contacts for silicon particle and x-ray sensors.

After a period of 4 years, he returned to the UK to establish a detector company initially called Nuclan Ltd. Factors instrumental to making this decision included the offer of a government grant from the Department of Industry and the knowledge of the exciting work on time variant pulse processing by Wrangy Kandiah's Group at UKAEA Harwell. Also, Kevex, a California-based company, had just released an optical restoring x-ray spectrometer and Ortec had decided it did not want to pursue this technology, but instead focus its efforts on the growing opportunities in germanium gamma ray detectors.

Nuclan quickly merged with Link Systems Ltd., a recently formed analytical company making a computer-controlled energy dispersive analyzer to work with imported Kevex x-ray detectors. Rob, in his role as technical director, introduced a range of silicon x-ray spectrometers based on the principles of lithium drifting to form deep structures for absorbing x-rays up to 30 keV. The company grew rapidly during the 1970s with successes in Europe, the USSR, and the United States.

In 1980 the company was acquired by a leading UK plc, UEI, and Rob became managing director of the Link Scientific Group and a main board director of UEI. The Link Group grew rapidly during this period and also acquired X-Tech in California, a company specialising in X-ray Tubes, and then The Nucleus in Tennessee, the owners of Tennelec, a nucleonics company.

Carlton Communications purchased UEI in the late 80s and then sold the Link Scientific Group to Oxford Instruments. Rob joined the Oxford Group as an executive director responsible for the Link Group of Companies and the XRF Analytical business that Oxford had grown.

In 1992, Rob left industry and returned to academia at Manchester University for 2 years and then returned into industry having purchased Gresham Power Electronics in Salisbury and Camscan in Cambridge. Gresham introduced a range of sensors to satisfy the needs of a growing number of specialist analytical companies that had the computer and software skills but needed an appropriate sensor to make an analytical instrument. Gresham was acquired by e2v in 2006 and is now firmly established as a successful supplier of a wide range of x-ray sensors.

Awards and Publications

Rob was awarded an MBE in 1991 for Services to Science and became a Fellow of the Institute of Physics in 1986. He is also a Fellow of the Royal Microscopical Society.

Link Systems was awarded Queens Awards for both Technology and Export and an R&D 100 award from the United States. In 1999, he was awarded the Presidential Science Award for 'Outstanding Contribution to the Theory and Practice of Microbeam Analysis' by the Microbeam Analysis Society.

During his working career, he liaised with several government departments including the Security Services and met with the Prime Minister, Margaret Thatcher, several times to discuss the challenges of growing UK businesses in an international market. He has also participated as a committee member on topics such as Nuclear Strategy and Home Land Security. He is married and has two sons living in Marlow, Buckinghamshire.

Rob authored more than 15 publications on radiation detectors and holds two patents.

Acronyms

a-C	Amorphous Carbon
AEC	Atomic Energy Commission (USA or Australia)
AEI	Associated Electrical Industries
AEM	Analytical EM
AERE	Atomic Energy Research Establishment, UK
a-Ge	Amorphous germanium
ANU	Australian National University
APD	Avalanche photodiode
APS	American Physical Society
a-Si	Amorphous silicon
ASIC	Application-specific IC
ASM	Associated Semiconductor Manufacturers Ltd., UK
ASTM	American Society for Testing and Materials
AT&T	American Telephone and Telegraph (became Bell Telephone Co.)
ATW	Atmospheric pressure–supporting thin window
AWRE	Atomic Weapons Research Establishment, UK
BNL	Brookhaven National Laboratory, USA
BTH	British Thomson-Houston Co.
CCD	Charge coupled device
CCE	Charge collection efficiency
CDS	Correlated double sampling
CEA	Nuclear Energy Research Centre, Saclay, France
CEN	Nuclear Energy Research Centre, Saclay, France
CEO	Chief executive officer
CERN	European Organisation for Nuclear Research, Geneva
CES	French Atomic Energy
CFTH	French Thomson-Houston Co.
CMOS	Complementary MOS
CNRS	Centre National de la Recherché Scientifique, France

CP4	Silicon or germanium etchant
CPPM	Centre de Physique des Particles de Marseille, France
CRO	Cathode ray oscilloscope
CSF	Compagnie Generale de Telegraphie sans Fils, France
CTE	Charge transfer efficiency (CCDs)
C–V	Capacitance–voltage (plot)
CVD	Chemical vapour deposition
DEPFET	Depletion-mode FET
DI	Deionise (high resistivity) water
DoE/D	Department of Energy/Defence (USA)
e2v	Originally English Electric Valve (EEV) Co., UK
EDX	Energy dispersive x-ray spectrometry
EDXMA	Energy dispersive x-ray microanalysis
EDXRF	Energy dispersive x-ray fluorescence
EDXRS	Energy dispersive x-ray spectrometry
EG&G	Edgerton, Germeshausen and Grier Co., USA
EM	Electron microscope
EMP	Electron microprobe
EMRS	European Materials Research Society
ENC (enc)	Equivalent noise charge
EPA	Electron probe analysis
epi-Si	Epitaxial grown Si
EPMA	Electron probe microanalysis
ESA	European Space Agency
ESCA	Electron spectroscopy for chemical analysis
ESTEC	European Space Research and Technology Centre, Netherlands
ETH	Eidgenossische Technische Hochschule, Zürich, Switzerland
EXAFS	Extended x-ray absorption edge fine structure
FBC	Feedback capacitor
FET	Field effect transistor
FWHM	Full-width half-maximum
FZ	Float zone
GE	General Electric, USA

GEC	General Electric Company, UK
Ge(Li)	Lithium-drifted germanium
GKSS	Helmholtz Research Centre, Germany
G-R	Guard ring
HEP	High energy physics
HNU	Manufacturer of Photo-Ionisation Detectors, USA
HPB	High-pressure Bridgman (crystal growth)
HPGe	High-purity Ge
HPSi	High-purity Si
HTO	High take-off (angle)
HV	High voltage
IAEA	International atomic energy authority
IC	Integrated circuit
ICC	Incomplete charge collection
IEEE	Institute of Electrical and Electronics Engineers
IMO	Inverted mode operation (same as MPP)
INR	Institute of Nuclear Research, Poland
IOP	Institute of Physics, UK
IR	Infrared
IRE	Institute of Radio Engineers (became IEEE)
JEOL	Japanese Electron-Optics Labs Ltd.
JFET	Junction FET
JPL	Jet Propulsion Laboratory, USA
KAPL	Knolls Atomic Power Laboratory, USA
KEK	Japanese synchrotron center
LALC	Large area, low capacitance
LBL	Lawrence–Berkeley National Laboratory, USA
LEC	Liquid encapsulated Czochralski (crystal growth)
LED	Light-emitting diode
LHC	Large Hadron Collider, CERN
LLNL	Lawrence Livermore National Laboratory, USA
LN	Liquid nitrogen
LPB	Low pressure Bridgman (crystal growth)
LTT	Compagnie Lignes Telegraphiques et Telephoniques, France

MAS	Microbeam Analysis Society
MBE	Molecular beam epitaxy
MCA	Multichannel analyzer
MDL	Minimum detectable limit
MIT	Massachusetts Institute of Technology, USA
MOS	Metal oxide semiconductor (technology)
MOSFET	Metal oxide semiconductor FET
MPI	Max Planck Institute, Germany
MPP	Multipinned phase (same as IMO)
MRS	Materials Science Society, USA
MUX	Multiplexing
NAA	Neutron activation analysis
NAS	National Academy of Sciences, USA
NASA	National Aeronautical and Space Administration, USA
NBS	National Bureau of Standards, USA
NE	Nuclear Enterprises Ltd., UK
NEC	Nuclear Equipment Corporation, USA
NIM	Nuclear Instrumentation Module
NIM	*Nuclear Instruments & Methods in Physics Research* (journal)
NRC	National Research Council, USA
NRD	Nuclear Radiation Development, Canada
NRDL	Naval Radiological Defence Laboratory, USA
N-S	IEEE Nuclear Science Symposium
NSI	Nuclear Semiconductor Inc., USA
NTD	Nuclear Transmutation Doping
NYO	New York Operation Office (USAEC)
OI	Oxford Instruments plc. UK
ORNL	Oak Ridge National Laboratory, USA
PAD	Pixel array detectors
P/B	Peak-to-background ratio
PCB	Printed Circuit Board
PGT	Princeton Gamma Technology Inc., USA
PHA	Pulse height analysis (or analyzer)
p–i–n	p-intrinsic–n type silicon diode structure

PIXE	Particle-induced x-ray emission
pn-CCD	p–n junction CCD
poly-Si	Polycrystalline Si
PSD	Position-sensitive detector
PSG	Phosphosilicate glass
PTFE	Teflon
PTIS	Photothermal ionisation spectroscopy
PVC	Polyvinyl chloride
PXD	Pixel (x-ray) detectors
QE	Quantum efficiency
R	Rads (Roentgen x-ray dose)
RAL	Rutherford-Appleton Laboratory, UK
RC	Resistance Capacitance product (time constant)
RCA	Radio Corporation of America, USA
RF	Radio frequency
RIDL	Radiation Instrument Development Laboratory, USA
RMD	Radiation Monitoring Devices, USA
rms	Root mean square deviation
RT	Room temperature
RTC	La Radiotechnique, France
RW	Ramo-Wooldridge Co., USA
SBIR	Small business innovative research
SCD	Swept charge device
SDD	Silicon drift detector
SEM	Scanning EM
SERL	Services Electronics Research Laboratory, UK
Si(Li)	Lithium-drifted silicon
SOG	Spin-on glass
SPIE	Society of Photographic Instrumentation Engineers, USA
SSR	Solid State Radiations Inc., USA
T	Absolute temperature in Kelvin
TCE	Trichloroethylene
TEC	Thermo-Electric (Peltier) cooler
TEM	Transmission EM
THM	Travelling heater method (crystal purification)

TI	Texas Instruments Inc., USA
TMC	Technical Measurements Corporation, USA
TUM	Technical University Munich, Garching, Germany
TW	Thin (x-ray) window
TXRF	Total reflection XRF
UCRL	University of California Lawrence Radiation Laboratory (report)
UEI	United Engineering Industries, UK
UHV	Ultrahigh vacuum
UKAEA	United Kingdom Atomic Energy Authority
URS	United Research Services, USA
USAEC	United States Atomic Energy Commision
UTW	Ultrathin (x-ray) window
VG	Vacuum Generators Ltd., UK
VLSI	Very large silicon IC
VP	Vice president
WBGS	Wide bandgap semiconductors
WDS	Wavelength dispersive spectrometry
WDXRF	Wavelength dispersive XRF
WW2	World War 2
XAFS	X-ray absorption edge fine structure
XMM	X-ray multi mirror satellite
XRF	X-ray fluorescence
XRS	X-ray spectrometry (also a journal)

1

Introduction

Since 1950, we have witnessed the development of a range of radiation detectors exploiting the properties of semiconductor materials. The progress has been documented in many textbooks, some of which are listed in chronological order in Appendix 1A. We have also added a list of books that cover certain aspects of semiconductor x-ray detectors where the main topic is x-ray analysis.

It may legitimately be asked, 'Why do we need another book on the subject?' To answer that, we note that there has been rapid progress recently in devices such as silicon drift detectors (SDDs) and charge-coupled devices (CCDs) and renewed interest in alternative materials such as CdZnTe and diamond. This makes the choice of detector for particular applications rather bewildering for potential users and it has become necessary to compare their characteristics with the lithium-drifted silicon 'Si(Li)' x-ray detectors that have been the mainstay of the x-ray scientists using semiconductor detectors. Added to this, the practical construction and historical development of these detectors has not, to date, been given its just exposure in detail. The high-purity germanium (HPGe) detector has been shown to have superior qualities to the Si(Li) detector in many of its characteristics and the reasons for its present lack of application and availability need to be aired.

Going back even before the dawn of semiconductor detectors, Serge Korff [1] wrote in 1946 in the preface to his book, *Electron and Nuclear Counters* (published by Van Nostrand),

It is the purpose of this book to gather together and summarise the facts regarding the theory of the discharge mechanism and the practical operation of various types of counters. Although they have been known for about forty years, they are even today surrounded by an atmosphere of mystery and their construction and operation are claimed by many competent scientists to involve 'magic'. Various laboratories have developed special procedures for their manufacture and use, often without knowing why particular techniques appear to be successful.

That book concerned *gas proportional detectors*, but Korff's remarks could apply equally well today when looking back over the 40-odd years since the 'maturity' of the Si(Li) x-ray detector in the mid to late 1960s. Gas detectors are still manufactured and the technology is still difficult, as is that for semiconductor x-ray detectors. Indeed, even as early as 1970, Fred Goulding [2] of the Lawrence Radiation Laboratory wrote:

Work on ultra-high resolution spectrometer systems using semiconductor detectors has been excessively time-consuming, and, for the most part frustrating. To my knowledge this remark has been true (if we include ionisation chambers) for the past 25 years although the details have changed.

It is interesting to note that CdZnTe and SDD detectors are now reaching their 'maturity' in terms of applications, but the gestation time again has been in the order of 20 and 30 years, respectively. Even in the case of Si(Li) detectors, some areas are still not well understood after nearly 50 years of 'maturity'. It is our hope that this book will explain the painful evolution of semiconductor x-ray detectors and will remove at least some of the 'magic' involved in their construction.

It should be explained at the outset that this book is not a textbook on semiconductor physics or electronics and will give only as much of the theory as is required to describe the principles of the detectors. If further theoretical details are required, references are made to where this can be found. Instead, this book will concentrate on the principles of the semiconductor transducers themselves. Some attention will be given to the field effect transistor (FET) that developed historically with the transducer and shared many of its manufacturing problems (material, surface states, noise) and ended, in practice, closely coupled (indeed sometimes integrated) with it.

Because both are usually cooled, this union is often referred as a 'cryo-head'.

Semiconductor detectors were first used in nuclear physics research as particle detectors for electrons, protons, alpha particles, and—with a suitable converter—neutrons. They were also almost immediately used to try to measure gamma rays and x-rays. Information on the developments of associated materials and measuring electronics is abundant in the scientific literature. In this book, we attempt to bring some of this information together and, where possible, credit the innovators. Our focus will be on how these developments paved the way for the modern x-ray sensor.

Our perspective as participators in the industries created by these detector developments enables us to bring a working view of how both the science and the companies involved progressed. We will focus on x-ray detectors made from silicon in the main as used in the x-ray microanalysis and XRF (x-ray fluorescence) industry, but we will also cover the use of germanium and other semiconductors.

Our searches through the scientific journals and the various technical books has shown how history often repeats and also how recognition of creative thought can be confused and misinterpreted. Both government scientists and industrial scientists are sometimes limited in what they can publish, and many universities and institutions have allegiances with industry that can produce similar restrictions. These constraints tend to dissolve with time, and we have been fortunate enough to examine many technical journals; some of these are now available on the Internet. Some information, as for example on the development of FETs, was held in the National Archives in the UK and was subjected to restrictions that have now been released under the 30-year rule. Other information has been assembled from both company published articles and by conversations with some of the people involved in the history of the development, the manufacturing, and the marketing of these devices. Putting this all together has taught us how important it is for young engineers and scientists to examine the early work and to make use of the excellent online search techniques available for reviewing literature and patents. In particular, and where appropriate, we reproduce text and drawings from the patents and papers to give a meaningful time-base to these developments. Many hours spent on reading these patents have demonstrated what a fertile source of information they are regarding both the science and the claimed invention.

Our working knowledge of the x-ray spectroscopy industry while these developments were progressing has been supplemented with information from conversations with other similarly placed scientists and engineers. Our researches have shown how the development of semiconductor radiation detectors tracked the development of semiconductor materials and semiconductor devices such as diodes, transistors and integrated circuits (ICs). The technical challenges that had to be overcome to produce the detectors in current use depended on the parallel development of FETs, as noted earlier, as well as on x-ray transmission windows and signal processing electronics. In the following, we use 'FET' to define the family that includes junction field effect transistors (JFETs) and metal oxide semiconductor FETs (MOSFETs). Most of the x-ray detectors needed cryostats constructed from the correct materials with efficient thermal design and freedom from vibration effects. It is our intention to review the development of all the key components that were required for today's x-ray spectrometers and to show how the various sensors evolved. The overview that follows will describe the essentials without entering the details that will, in many cases, be discussed in later chapters. It is hoped that this Introduction and the details that will be discussed next (Chapters 2 through 10) will help to set the history (Chapter 11) in perspective. Chapter 11 can be omitted, but it does shed light on why the detectors evolved, rightly or wrongly, in the way they did.

1.1 Detector and Charge Sensitive Preamplifier: A System Overview

To give an overview, it is instructive to consider the electronic circuit representing the basis of most semiconductor x-ray detectors (Figure 1.1).

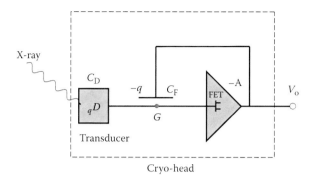

Figure 1.1 X-ray sensor front end or cryo-head.

Here, '–A' represents the preamplifier open loop-gain that, by negative feedback through the capacitor C_F, maintains point G at a virtual ground potential. The first stage of the preamplifier is the FET. If a charge q has accumulated at G (by ionisation due to an x-ray passing into the transducer D or by leakage currents into G), then the virtual ground is maintained by the preamplifier reacting to pump an equal and opposite charge $–q$ onto C_F. This results in an output voltage

$$\text{Vo} = -q/\text{CF}. \qquad (1.1)$$

This is a simple negative feedback circuit by way of the capacitor, the feedback capacitor (FBC), C_F, to one electrode of the transducer D and the gate of the FET. Its simplicity, however, belies the physical and engineering difficulties in realising such an arrangement capable of resolving x-rays of energies of the order of 100 eV (or about 30 electrons in silicon) up to more than 20 keV. The FET gate G represents a very high impedance point in the circuit, and any charge that accumulates there has virtually no leakage to true ground. In the absence of some charge reset mechanism, the FET would be driven off its linear excursion and introduce signal distortion. The practical methods of providing such a mechanism proved to be just one of the technological challenges.

Considering the absorption of x-rays in matter, we expect the fraction of intensity absorbed $–dI/I$ to be proportional to the incremental penetration depth, dx

$$dI/I = -dx/L,$$

where L is a constant for any given material and x-ray energy and has dimensions of length. The result of integrating this is an equation known as Beer's law,

$$I = I_0 \exp(-x/L),$$

where I_0 is the x-ray intensity at surface $x = 0$, and L is the reciprocal of the 'linear absorption coefficient' of the absorbing medium. As the absorption in matter is a strong function of its density ρ, it is most often written as

$$I = I_0 \exp(-\mu_a \rho x), \qquad (1.2)$$

where μ_a is the mass absorption coefficient and generally decreases with increasing x-ray energy. It will be noticed that the most probable depth for absorption in a uniform medium is always near $x = 0$ (i.e., at the surface). As an example, Figure 1.2 shows the attenuation of 20 keV x-rays in silicon and 50 keV x-rays in germanium using the published values of μ_a.

It is seen that approximately 3 mm depth of silicon or equivalent is required for reasonable efficiency at 20 keV. Germanium at this depth is efficient up to ~100 keV. Incidentally, GaAs being a compound of elements either side of germanium in atomic number, has a similar attenuation to it. As a result of the quantisation of the electron energy levels in atoms, the mass absorption coefficients have step discontinuities, known as absorption edges. For example, as the x-ray energy moves from below to above the binding energy of the K-shell (see Section 1.3.2) electron (1.84 keV in silicon and 11.1 keV in germanium), there becomes enough energy to release the electron and the absorption coefficient steps up in magnitude at these energies. These discontinuities are not shown in Figure 1.2.

The detector is based on the principle that the charge generated by the absorption of the x-ray in the transducer is proportional to the energy of the x-ray. It is important to realise that the exponential decay in intensity of low-energy

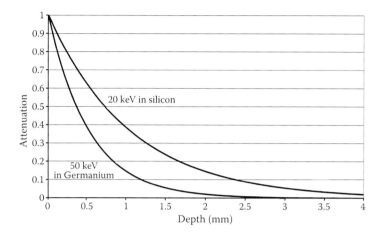

Figure 1.2 Attenuation of x-rays in silicon and germanium.

x-rays described by Beer's law above (with the absorption edge discontinuities) does not in any way affect the *energy* of the x-rays. This is unlike charged particles that lose energy quasi-continuously along their path. Thus, from a spectroscopic viewpoint, it is more important that the charge generated is fully accounted for than that x-rays are not absorbed in any 'dead layers' (including any x-ray window). In other words, that all such charge is contained in the active volume of the transducer and there are negligible contributions from absorption from outside it.

To achieve the low x-ray energy threshold sensitivity, the noise must be reduced and the transducer and the FET must be cooled, but not necessarily to the same temperature. For sensitivity using silicon, the transducer must be 3 mm or more in diameter and so the active volume is ~30 mm^3 or more. This is in complete contrast to the microelectronic development that has taken place in silicon technology driving the dimensions smaller and smaller. The cryo-head must be screened from extraneous electrical, magnetic, radiation, and vibrational interference, and the virtual ground (G in Figure 1.1) must present the lowest possible capacitance as it is connected to the gate of the FET and the FET noise depends on its capacitive loading. This capacitive loading includes the transducer capacitance, C_D, the gate capacitance of the FET, C_F, and any stray capacitance, C_S. In practice, this means that the physical size of the conductor at G (the transducer electrode and FET gate contact) must be restricted, and there must be virtually no changes in C_F and C_S (both of the order of a fraction of a pF) due to vibration (microphony). Another source of noise is the dielectric used to hold these components together, and care must be taken in selecting materials. Added to these impositions, in electron microscope (EM) applications, the cryo-head needs to be on the end of a long probe (typically 30–50 cm) and typically 12–16 mm in diameter. We will discuss how such transducers and FETs can be manufactured, and how the other conditions can satisfactorily be engineered.

1.2 Transducers

The transducer converts the energy E of the x-ray entering it into an electrical signal, usually in the form of charge q induced onto the electrode connected to G in Figure 1.1. The major part of this book will be devoted to the transducer, its physical properties, manufacture, and evolution. If we are to measure E in terms of q, it is clear that q should represent *all* of energy E, which should in turn be completely absorbed in the transducer. It would also be advantageous if they were directly proportional. Or, in terms of the number n of *electronic* charges e, for example

$$q = ne$$

$$E = \omega n, \tag{1.3}$$

where ω is a constant. This constant ω (measured in eV) represents the x-ray energy required to induce one electronic charge onto G and is a characteristic of the absorbing medium.

We can picture the process using the gravitational analogy of dropping a coin into a glass of beer (say), as depicted in Figure 1.3.

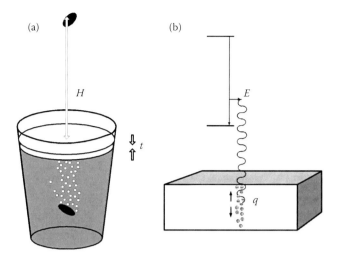

Figure 1.3 Coin dropped into a liquid. (a) Pebble in a liquid and (b) x-ray absorbed in matter.

X-ray detectors rely on the linearity between energy absorbed and charge collected. The number of bubbles liberated (say, thickness t of froth) is proportional to the potential energy (height) of the coin. The x-ray emitted from the atomic level transition is fully absorbed in the solid and by ionisation, generates charge q on the electrode.

The absorption of x-rays is stronger, in general, in solids than in liquids or gases. The specific absorption (mass absorption coefficients, μ_a) and density of the solid has to be a consideration for the more penetrating higher energy x-rays (or gamma rays) because spectroscopy relies on the total absorption of the incoming x-ray energy. At the energies under consideration, photoelectric absorption (where the x-ray energy is transferred to an atomic electron producing a 'photoelectron') is dominant, but scattering (Compton scattering) of the x-ray may be significant in the higher energy range. Compton scattering of an x-ray is the 'billiard ball'-like collision of the photon by an atomic electron. The probability of scattering is proportional to the number of electrons available (atomic number Z) and is very low in silicon at the energies we are mainly concerned with (below 20 keV). In photoelectric absorption, the atom of the absorbing medium is left in an excited state and can relax by the emission of a fluorescence x-ray, but this is usually also absorbed back into the medium. The probability of photoelectric absorption varies as a high power (~4.5) of the atomic number Z of the atom involved, which explains the higher absorption in germanium relative to silicon of the same energy x-ray (Figure 1.2). Higher Z materials have advantages of high efficiency in the case of higher energy (more penetrating x-rays and gamma rays). The reduced energy of the Compton scattered x-ray and the energy of the recoil electron are also usually absorbed. Charged particles, such as electrons, have shorter ranges and are all usually stopped in thick detectors.

The energy of the photoelectron is equal to the energy of the absorbed x-ray minus the atomic binding energy of the absorbing electron. The emission of the photoelectron leaves behind a hole. This 'hole' is filled by an electron from a higher atomic level, followed by the emission of a fluorescence x-ray or an Auger ('O-J') electron (Figure 1.4). The photoelectron (and Auger electron, if emitted) is usually reabsorbed back into the transducer. It should be pointed out that with Compton scattering through wide angles and subsequent photoelectric absorption, even a thin detector totally absorbs some high-energy events. The higher density solids are preferred for the higher energy range of x-ray spectrometry to

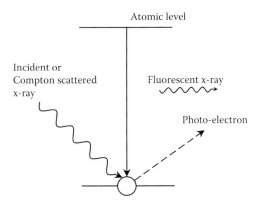

Figure 1.4 Radiation loss processes.

improve the detectors efficiency, and in this respect solid-state detectors are preferred to gas proportional detectors in most cases. If we are to measure the free charge generated in a solid volume, it is clearly advantageous also if the volume does not already contain free charge. Using our gravitational 'glass of beer' detector analogy, it would not be very effective if the beer were full of bubbles to begin with (cool beer is best!). This immediately rules out metals. In fact, it is the free charge at the surface of metals that gives the metal surface its reflective lustre. Insulators normally have a dull surface and semiconductors (such as graphite, silicon, and germanium) only a partial lustre. It would seem from this point of view that insulators or 'semi-insulators' should be best and, in fact, these were the first to be used (crystal detectors) and are still used (materials such as mercuric iodide, diamond) when cooling is not possible, convenient, or necessary.

The fast-moving photoelectrons and Auger electrons ionise the solid in a cascade of charge generation, and the transducer can be regarded as a 'charge amplifier'. Looking at Equation 1.3, it is seen that the inverse of ω is a measure of the sensitivity or 'intrinsic charge gain' of the transducer. The smaller ω becomes, the greater is the charge signal q for a given value of E. With modern FET amplifiers, this is not a significant issue (substituting $C_F = 0.01$ pF into Equation 1.1 gives $V_o \sim 6$ μV even for just a million electrons). However, the *statistical spread* of the signal *is* important. This is given by the statistical spread in the number n, and in Poisson statistics this would be $\sqrt{n}$. So, the relative width (or 'resolution') of the distribution in n might be expected to be $\sqrt{n}/n = 1/\sqrt{n}$. We will see later that because all of the energy E is not converted to charge, even for complete absorption, the spread in n is actually given by $\sqrt{(Fn)}$ rather than $\sqrt{n}$. Here, F is the celebrated Fano factor that, apart from ω, is the next most important parameter for the absorbing medium. It is quite surprising, however, that the value of F (first calculated for gases by Ugo Fano [3] in 1946 and is ~0.1) does not vary much from one absorbing medium to another, even from solids to gases.

1.3 Why Semiconductors?

Semiconductors have higher intrinsic gains because bound electrons are more easily broken free, for example, by radiation or heat, and have ω measured in just a few eVs. The rare gases have values of a few 10s of eV. The 'Cooper pairs' of bound electrons found at very low temperatures (liquid helium ~2 K) in superconductors have binding energies of the order of a few meV and give much higher sensitivity and resolution detectors. Apart from the very low temperature superconducting and bolometer detectors, which have their own cost and technology problems and are not the topic of this book, semiconductors therefore appear to be the most attractive.

The values of ω (or ionisation energy) for a number of gases, semiconductors, semi-insulators, and insulators are given in Table 1.1 (see also Table 10.1).

TABLE 1.1
Value of ω (or Ionisation Energy) for Specified Materials

Medium	Formula	Atomic Number Z (Average)	Density (gm/cm³)	Average Energy per Electron–Hole Pair (eV)
Neon gas	Ne	10	0.9×10^{-3}	36.2
Argon gas	Ar	18	1.7×10^{-3}	26.2
Xenon gas	Xe	54	5.9×10^{-3}	21.5
Silicon dioxide (quartz)	SiO_2	15.3	2.2	17
Silicon	Si	14	2.3	3.65
Germanium	Ge	32	5.3	2.96
Gallium arsenide	GaAs	31.3	5.4	4.3
Mercuric iodide	HgI_2	80.53	6.4	4.2
Lead iodide	PbI_2	82.5	6.2	7.68
Cadmium telluride	CdTe	48.5	6.1	4.4
Silicon carbide	SiC	10	3.12	7.8
Diamond	C	6	3.5	13.25

There is, however, the problem of free quiescent charge in the semiconductor bulk. Even if the temperature is lowered to liquid nitrogen temperature (LNT) (77 K), as was done with the early crystal detectors, there is sufficient free bulk charge to add to the statistical broadening (noise). It took some time for users to realise that semiconductors had another unique advantage, however. They can be fabricated in such a way as to *deplete* the bulk of its free charge, thereby revealing a highly x-ray–sensitive layer. This was achieved by forming a reverse bias p–n junction *diode*. To explain this, we need to look at the properties of semiconductors that make them so special.

1.3.1 Properties of Semiconductors

From the earliest investigations of electrical conduction, solids fell into the well-separated categories of insulators and conductors. This turned out to be extremely convenient practically for the development of electricity as the free charge conduction in one conductor (e.g., copper) could be isolated (e.g., by ceramic or a plastic coating) from that in another. Then, in the late 1800s, it became clear that there was a category of solids, 'semiconductors', which fell between the two. In fact, unlike metals, their conductivity increased with temperature and illumination. Then in the 1970s, it was discovered that even some plastics (conjugated polymers) could conduct electricity. The range is illustrated in Figure 1.5.

1.3.2 Energy Gap

The explanation for these differences lies in the 'electronic band structure' of the materials. Electrons can only occupy specific energy levels and in solids these levels form bands, each of which can only accommodate a certain number of electrons because of the quantum mechanical exclusion principle for electrons. This stipulates that only one pair of electrons (one spin up and one spin down) can occupy any energy level in a quantum mechanical system such as an atom. In metals, the highest band is only partially occupied and is known as the conduction band. In it, the electrons are free to

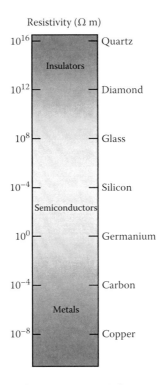

Figure 1.5 Resistivities of various materials.

move under the influence of an applied electric or magnetic field. In insulators, the conduction band is empty and the highest occupied band is the valence band. The valence band is occupied by electrons that are bound and localised (to a greater or lesser extent) to the atom. All states are occupied and because of the exclusion principle, no electron is free to move as the surrounding states are already occupied. The energy gap (E_g) between them is so great ($E_g > 10$ eV) that electrons are energetically incapable of leaping this gap at normal temperatures and electric fields, and no conduction takes place.

The same situation also exists in semiconductors, but here the energy gap E_g is less (a few eV) and electrons can, under several circumstances, leap from the valence band into the conduction band. This is illustrated schematically in Figure 1.6.

In metals, the highest occupied bands are only partly filled, and they can conduct electric charge. In insulators, the

Band structure in solids

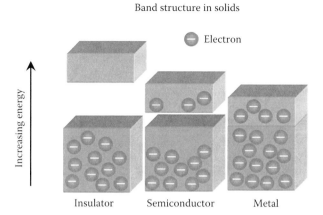

Figure 1.6 Electrons available for conduction in materials with a small, medium, or no bandgap.

highest occupied bands are filled and are separated from the next empty band by an energy gap. If the energy gap is small enough, electrons stimulated by heat or radiation can jump from the valence band to the conduction band and the material is a semiconductor. To see how the energy gap arises, we turn to the structure of atoms and the periodic table of elements. As we increase the number of atomic electrons (atomic number Z) because of the exclusion principle, they fill the available levels in sequence. The first few levels and their maximum occupancy are 1s(2), 2s(2), 2p(6), 3s(2), and 3p(8), and because of the way their energies are distributed they form closed 'shells'—1s(2), 2s–2p(8), and 3s–3p(8). These are given notations (K, L, M, etc.). This gives the periodic table and the grouping of elements with similar chemical and physical properties as shown below.

The center of the table (Group 4 elements, highlighted red in Figure 1.7) represents atoms with an outer shell of four electrons and hence four empty levels in the outer valence shell. The shell of any such atom can be 'quasi-completed' by bringing four other atoms into close proximity so that electrons can be 'shared'. This sharing forms four 'covalent bonds'. This is shown schematically in two dimensions in Figure 1.8 for the case of silicon.

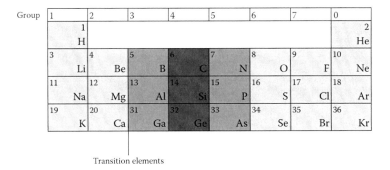

Transition elements

Figure 1.7 **(See colour insert.)** Part of the periodic table.

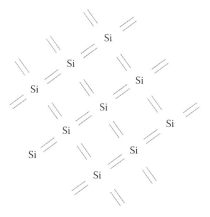

Figure 1.8 Two-dimensional representation of the silicon lattice.

When a number of silicon atoms are brought together, they form a diamond like silicon lattice; both the 3s level and the 3p level are broadened into bands that overlap and as the atoms approach their equilibrium position, these levels separate into two distinct bands with a gap between them. This band formation is illustrated in Figure 1.9, calculated for the case of nine such atoms. The nine allowed levels form quasi-continuous bands.

We refer to the lower band as the valence band and the upper band as the conduction band, and they are separated by the forbidden energy gap E_g as shown. This type of arrangement suggests that the silicon atom is already sharing its valence electrons throughout the lattice, a point we will make use of when describing conduction. Under certain conditions,

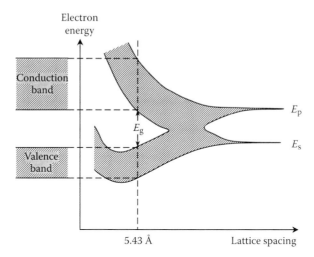

Figure 1.9 Formation of energy gap in a semiconductor as lattice atoms reach their equilibrium positions.

the valence electron can acquire sufficient energy to escape from this bond into the interstitial space becoming virtually free. It has now entered the conduction band, and the energy required to break the covalent bond is exactly equal to the bandgap energy of 1.1 eV in silicon. Thermal energy is an example of how this bond can be ruptured. Another example would be by photon absorption, where the energy of the incident photon is sufficient to break the bond. The excess energy is carried off by the subsequent photoelectron. By forming the energy gap, the quantum nature of every atom manifests itself throughout the semiconductor crystal and dominates its electrical characteristics.

The situation is, in fact, more complicated than described. We have described a *direct* bandgap. But in silicon (an indirect bandgap semiconductor), for the transition of an electron across the gap in either direction, momentum from a third source (usually lattice vibration, a 'phonon') is required (see Section 1.7.3). We see from the periodic table (Figure 1.7) that the crystalline Group 4 elements C (diamond), Si, and Ge are expected to be semiconductors. Figure 1.9 illustrates how the lattice spacing influences E_g. The closer the atomic packing, the larger the bandgap. The largest Group 4 bandgap, diamond (5.4 eV), has the closest atomic packing of all, resulting in its extreme hardness. Because of the screening effect of the more tightly bound electrons, the energy gap decreases with atomic number Z. For germanium, it is 0.67 eV. It happens that the combining of atoms from either side of the Group 4 semiconductors into the so-called **3–5** (**III–V**) compounds (marked orange in Figure 1.7) also produces semiconductors. They tend to be wide bandgap materials such as BN, AlN, GaAs, GaP, GaN, and InSb. Furthermore, **2–6** (**II–VI**) compounds can also give important wide bandgap semiconductors (WBGSs) such as CdTe. The lattice spacing (and hence, E_g and ω) can be influenced by both temperature and strain, and also the alloying of semiconductors, such as SiGe and CdZnTe. The latter is termed 'bandgap engineering' and will be further discussed in Chapter 10. The compounds involving elements closer to the edges of the periodic table (Figure 1.7) form weaker bonds by transferring electrons. They are thus more ionic in nature.

1.4 Fermi–Dirac Statistics

At very low temperatures, the free electrons in a metal fill the conduction band up to a characteristic energy called the Fermi energy, E_F. As the temperature rises, this abrupt falloff in the electron distribution, $n(E)$, is smoothed out according to the Fermi–Dirac distribution $F(E)$, which gives the probability that an electron quantum energy state is filled.

$$1/F(E) = \exp\{(E - E_F)/kT\} + 1 \qquad (1.4)$$

This is illustrated in Figure 1.10, which shows the Fermi–Dirac distribution (a) (b) (c), for three increasing temperatures.

As the temperature increases, the tail of the Fermi–Dirac distribution can extend above zero energy (the 'vacuum' energy), and electrons can be evaporated off the surface (thermionic emission). Note that the Fermi–Dirac distribution value always remains half of its maximum value at the Fermi energy E_F. The concentration of electrons $n(E)$ is the product of the density of states available and the Fermi–Dirac

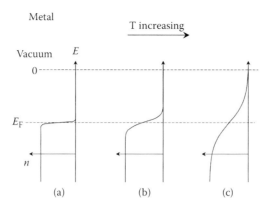

Figure 1.10 Fermi–Dirac distribution in metal. (a) Distribution at low temperature, (b) distribution at normal temperatures, and (c) distribution at high temperatures.

probability distribution of filling them. The diffusion of electrons will require that E_F remains constant throughout the metal if the metal has uniform properties.

At low temperatures and with no radiation of any sort present, semiconductors are, for all practical purposes, insulators and even at room temperature they never conduct as well as metals. When an electron makes the leap into the conduction band, two significant things happen. It upsets the charge neutrality of the site it came from (leaving an excess positive charge associated with the site), and it leaves an unoccupied electron state in the valence band as shown in Figure 1.11.

Thus, another electron from the valence band can take its place (itself creating a hole), and so effectively the 'hole' with its associated positive charge can move incrementally through the crystal. In an electric field, it moves in a way analogous to a positive carrier of charge and contributes to the conduction. This conduction is illustrated in Figure 1.12.

The holes and electrons are termed positive and negative 'charge carriers', and their densities in the semiconductor are represented by p and n. The motion of the charge carriers is characterised by their mobility μ in the semiconductor in terms of their drift velocity v in the presence of an electric field E. Thus,

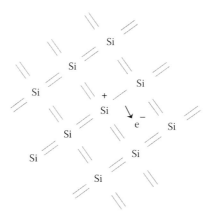

Figure 1.11 Electron moving into the conduction band.

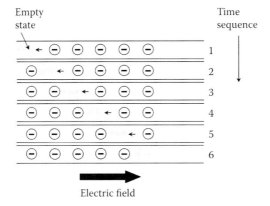

Figure 1.12 Movement of a hole in the presence of an electric field.

$$v = \mu E. \tag{1.5}$$

Because the mobility of carriers is dependent on scattering off the crystal lattice vibrations, termed 'phonon scattering' (and as we will see later, any impurities), it increases rapidly as the temperature is lowered. Figure 1.13 shows the variation of electron and hole mobility with temperature for silicon and germanium.

Unlike electrons, the holes cannot be regarded as point charges as they are associated with a distortion over many unit cells of the lattice and therefore not localised. As a result,

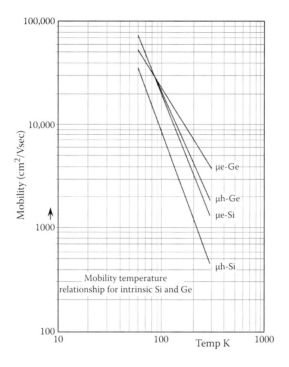

Figure 1.13 Mobility as a function of temperature.

their effective mass is higher and their mobility generally, lower and they are more prone to being trapped, for example, by dislocations or impurities in the lattice. As the field is increased, v increases linearly but reaches saturation velocity v_s when the energies become larger than the magnitude of the thermal energies. This occurs for electrons in silicon at room temperature at fields above ~10 kV/cm and at 100 K for silicon and germanium at fields above ~1 kV/cm, where $v_s \sim 10^7$ cm/s. The situation is not very different for holes. For cooled silicon and germanium detectors these fields are normally exceeded and we can assume near-saturation velocity for carriers. At fields another order of magnitude above (greater than ~100 kV/cm in silicon at room temperature), the electrons and holes can be accelerated sufficiently to cause 'impact ionisation'. At very high fields, an avalanche of carriers can be produced effectively increasing the charge gain considerably. This effect is used in avalanche photodiodes (APDs), and we will devote Section 7.2 to these photodiodes (used as x-ray detectors) later.

The mobility μ, involving the velocity of charges or—in other words—an electric current, can be related to electric resistance via Ohm's law

$$\Delta V = IR.$$

If we consider an element with area A through which the charge due to electrons moves dx in a time dt, Ohm's law gives

$$dV = (Ane\,dx/dt)(\rho\,dx/A), \qquad (1.6)$$

where ρ is the resistivity. In terms of the electric field, E, it follows from 1.5 that

$$E = -dV/dx = nev\rho.$$

Using Equation 1.5 for v,

$$\rho = 1/(ne\mu).$$

Considering holes as well as electrons, this is usually written as the conductivity σ (= $1/\rho$)

$$\sigma = e[n\mu(n) + p\mu(p)]. \qquad (1.7)$$

Because it is always physically the movement of electrons, the mobility of the hole μ(p) in the valence band is of the same order of magnitude as the mobility of electrons μ(n) in the conduction band. Indeed, the mobility of holes can even, in some circumstances, exceed that of electrons as it does in GaAs and Ge (at LNT). The type of conduction described above is known as 'intrinsic' conduction, because it is an intrinsic property of the semiconductor alone. Intrinsic conduction can be produced by energetic vibrations in the crystal lattice (thermal 'phonons' or in other words, heat) or by ionising radiation. A detector based on the latter would be termed an 'intrinsic detector', and an example is the HPGe detector. Note that, in these circumstances, for every electron promoted to the conduction band, there is a corresponding hole in the valence band. Thus, if n_i is the density of negative electrons and p_i is the density of positive holes in an *intrinsic* semiconductor then

$$n_i = p_i. \tag{1.8}$$

In this case, 1.7 reduces to

$$\sigma = en_i[\mu(n) + \mu(p)]. \tag{1.9}$$

The density of intrinsic carriers n_i is a strong function of temperature, as shown in Figure 1.14.

These intrinsic carriers are the 'quiescent bubbles in our beer analogy detector', and it illustrates why cooling (especially of germanium) is usually necessary. The Fermi–Dirac distribution for the valence band still applies for semiconductors but extends across the energy gap. The relevance of the Fermi energy E_F is now as a reference level at which the distribution falls to half maximum value, and so is the energy at

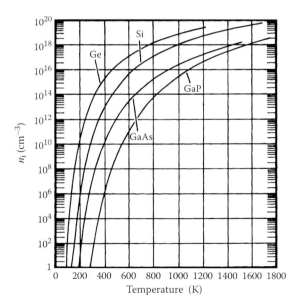

Figure 1.14 Carrier density as a function of temperature.

which the probability of finding an electron is equal to that of finding a hole. In intrinsic semiconductors, E_F is 'pinned' to mid-gap energy. As mentioned above, the concentration of electrons is the product of the density of available states and the Fermi–Dirac probability distribution. In an intrinsic semiconductor there are no such states in the energy gap. Figure 1.15 illustrates the band diagram for an intrinsic semiconductor at two different temperatures, (a) and (b). The mobile charges are depicted as the 'wheels' $\oplus$, $\ominus$ (fixed charges are depicted as symbols + and –).

In case (b), the temperature is high enough for electrons to leap the energy gap, producing mobile electrons in the conduction band and mobile holes in the valence band. The electrons quickly relax back down to the lowest levels in the conduction band, and the holes relax back (holes being the absence of electrons 'fall upward') to the highest levels in the valence band.

In thermal equilibrium, the intrinsic carrier concentration n_i in the conduction band can be shown [4] to be given by

$$n_i^2 = N_C N_V \exp(-E_g/kT),\qquad(1.10)$$

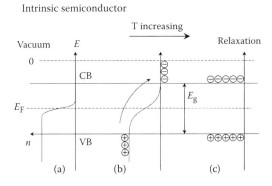

Figure 1.15 Intrinsic semiconductor bandgap and charge excitation. (a) At low temperatures the bands are free from charge, (b) as temperatures increase electrons cross the gap to populate many levels in the conduction band leaving holes in the valence band, and (c) as the excited electrons lose kinetic energy they relax back to the bottom of the conduction band and the holes move upwards to the top of the valence band.

where N_C and N_V are the effective density of states at the conduction and valence band edges, respectively. This leads to the relation

$$n_i = CT^{3/2} \exp(-E_g/2kT), \qquad (1.11)$$

where C is a constant.

At normal temperatures, the exponential term dominates and the electrical conductivity increases rapidly with increasing temperature (unlike metals) and is reduced considerably in wide bandgap materials such as GaAs (see Figure 1.14).

1.5 Doping of Semiconductors

To make practical semiconductor devices (diodes, transistors, ICs), the electrical conductivity is increased by several orders of magnitude by adding to (or 'doping') the semiconductor with small amounts of specific impurity atoms. These dopant atoms occupy sites normally occupied by the semiconductor atom itself and have the effect of either *donating* an extra electron into the conduction band or forming a hole in the valence band. This hole can *accept* an electron from the conduction band. The two types of dopants are known as *donors* and *acceptors*. Respective examples are phosphorus (P) and boron (B). This is illustrated in Figure 1.16.

The Coulomb force binding this electron to the P atom is reduced by the screening effect of other electrons, by the dielectric constant (~12 for silicon), and its effective orbit radius is increased by the same factor. This results in a lowering of the bond energy by a factor of 144 from 1.1 eV to a value of ~7.6 meV. The P atom can be regarded as a large 'hydrogen-like' atom extending over a great many lattice sites. The donor P atom is at first electrically neutral, but the electron being weakly bound will move away from the site under the influence of an electric field and/or thermal energy. This then exposes the fixed positive charge on the P nucleus (the donor site) that will act to terminate electric field lines and scatter carriers as they move through the crystal. This adds to the

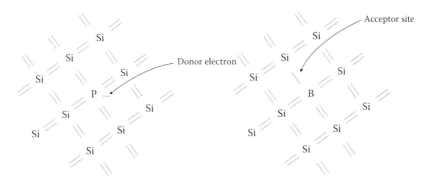

Figure 1.16 Illustrating a donor atom 'P' (resulting in a mobile electron) and an acceptor atom 'B' (resulting in a mobile hole) in the Si lattice.

thermal scattering of carriers discussed above and decreases their mobility considerably.

Likewise, the B atom in Figure 1.16 can lose its associated hole when an electron fills it (or 'combines' with it), charging it negatively and making it also a scattering site (the B acceptor site). The Si atom that supplied the electron to the acceptor site becomes positively charged, and in this way the hole is associated with a positive charge that moves with it through the crystal as described in Section 1.4. Because the conduction is now dominated by externally introduced impurities, the type of conduction caused by the motion of these mobile holes and electrons is termed 'extrinsic' conduction. The conduction in conjugate polymers is quite different and is attributable to the alternating double and single covalent bonds within the structure.

In the above example of doped semiconductors, the majority of charge carriers are negative in the donor case and positive in the acceptor case and they are referred to as n-type and p-type semiconductors. The carriers are referred to as the *majority carriers* (electrons in n-type and holes in p-type) and the intrinsic thermal carriers, or carriers generated by radiation, supply the *minority carriers*. If we dope the semiconductor with a density of N_D donors, we get an extra density of majority carriers $n = N_D$ (free electrons) without introducing any holes at all. Likewise, in a p-type semiconductor, we get an extra $p = N_A$. Normally, the densities of carriers introduced by doping far exceeds the density of intrinsic carriers, and we can equate $p \sim N_A$ and $n \sim N_D$. It is important to realise that the crystal in all these cases remains overall electrically neutral. This is known as the principle of neutrality. In terms of the band diagrams, the acceptor impurities give rise to energy levels within the bandgap close to the valence band and the donor impurities give rise to energy levels close to the conduction band. This has the effect of displacing the Fermi level E_F downward from mid-gap position in the first case (p-type semiconductor) and upward in the second (n-type semiconductor). This is illustrated in Figure 1.17.

Such 'band diagrams' must always be interpreted as showing the potential energy of *electrons*. The majority carriers introduced by these impurities dominate the conductivity of a 'doped' semiconductor. This dominates over the effect of any decrease in mobility caused by them, and Equation 1.7 becomes

$$\sigma = e[N_D\mu(n) + N_A\mu(p)].$$

Figure 1.18 shows the effect of doping on the resistivity of silicon at room temperature.

Because μ for holes is $\sim 1/3$ that for electrons, p-type materials have $\sim 3\times$ the resistivity of n-type for the same level of doping. Of course, a semiconductor can be doped with both donors and acceptors. Whether the material is n-type or p-type will depend on the sign of $(N_D - N_A)$. When the level of both types of dopants is very high [i.e., when $(n \sim p)$], the Fermi level is 'pinned' to near mid-gap. This situation is sometimes (somewhat misleadingly) termed 'intrinsic'.

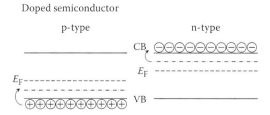

Figure 1.17 Movement of the Fermi level in a doped semiconductor.

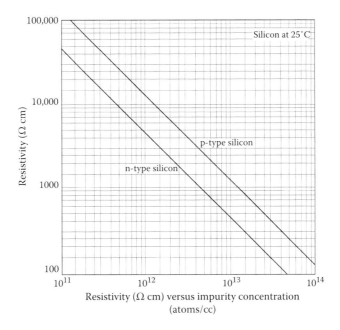

Figure 1.18 Resistivity vs. doping concentration at 25°C.

1.5.1 Law of Mass Action

Doping takes the carrier densities from n_i to n and p_i ($=n_i$) to p. As we have seen, the holes and electrons can recombine and the rate at which this happens will be proportional to the product of their densities np. At a given temperature, the thermal generation rate of holes and electrons in the doped semiconductor will only depend on the density of the valence bonds in the crystal, and this is always far higher than the density of impurity atoms. The generation rate will therefore be virtually the same as for the undoped intrinsic material, and the recombination rate for this, by Equation 1.8, is

$$n_i p_i = n_i^2.$$

However, by the law of mass action in thermal equilibrium, the recombination rate must be equal to the generation rate (in both intrinsic and extrinsic material) and we must have

$$np = n_\mathrm{i}^2. \tag{1.12}$$

This equation describes the equilibrium between recombination and generation. Thus, the effect of adding mobile electrons by adding donor impurities has the effect of *decreasing* the hole concentration p. Similarly for acceptor impurities, the electron concentration n is suppressed. For example, in n-type silicon at room temperature, with modest donor impurity concentration (say one atom in 4×10^{10}), its conductivity is dominated by electrons. The generation rate n_i^2 depends on the density of states at the bottom of the conduction band, the temperature T, and energy gap E_g of the semiconductor as shown by Equation 1.10. Because it does not depend on the Fermi level, it applies to doped as well as intrinsic semiconductors.

If charge generation dominates, we have nonequilibrium conditions, and

$$np \ll n_\mathrm{i}^2.$$

This is the case, for example, where we have a field (as in the depletion layer of a biased p–n junction diode) that separates the holes and electrons before they can recombine. However, we have a 'quasi-equilibrium' condition where

$$np \sim \text{constant.} \tag{1.13}$$

If the temperature is low enough, we might expect $p = 0$ for n-type material, but this is not true. There is always a hole density given by Equation 1.12

$$p = n_\mathrm{i}^2 / n = n_\mathrm{i}^2 / N_\mathrm{D}.$$

Another case of nonequilibrium is the injection of charge into the base region in transistor action. We might expect that the density of minority carriers could exceed the density of majority carriers! However, this is also not so as the system will always adjust toward charge neutrality. A corresponding opposite charge is drawn from the base contact in a very short time (dielectric relaxation time, $\tau = \varepsilon\varepsilon_0\rho$). Thus, the density of minority carriers never exceeds the density of majority carriers.

1.5.2 Depletion Layers and Band Bending

Consider an n-type semiconductor (right-hand side of Figure 1.19). The left-hand side is an isolated uniform material, which is occupied by electrons, but we will not specify the material just yet. It could, for example, be a metal, another crystalline semiconductor, or some amorphous material. The situation where they are not in contact is shown in Figure 1.19a. We will bring these two materials into intimate contact in what we will call a 'junction'. Assume that both have populations of electrons described by Fermi levels E_F in each. Their difference is shown as $\Delta\varphi$, but in isolation they do not have any bearing on each other.

If they are first connected together by a conducting wire, because of the difference in Fermi levels and the need to

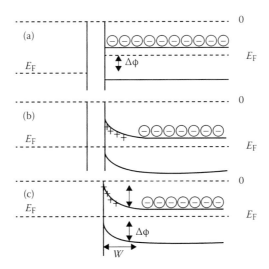

Figure 1.19 Illustration of how, in equilibrium, bands have to bend to allow the Fermi level to be continuous throughout the structure. (a) Unspecified material on the left hand side not in contact with the n-type semiconductor on the right hand side, (b) material now connected to the semiconductor with a wire, and (c) material in intimate contact with the semiconductor.

establish thermodynamic equilibrium, electrons will flow from the surface of the semiconductor through the wire to the lower energy states until the Fermi levels equalise (analogous to water, it finds its own level). This is shown in Figure 1.19b. The flow of mobile electrons from the semiconductor surface uncovers the fixed positive charge on the donor atoms and sets up a potential difference between the two materials that will oppose further flow. Note the fixed positive charges cause the energy bands to bend upward as shown. This is termed 'band bending'. The region drained of the mobile electrons is called a *depletion layer*. It acts as an insulating layer, and there is effectively a barrier for the flow of electrons in the conduction band from the left-hand side to the semiconductor. Its importance lies in the fact that such a layer presents an almost ideal mobile charge-free volume for the detection of ionising radiation. If we continue to move the materials into contact, the potential difference $\Delta\varphi$ between the materials will be taken up across the insulating depletion layer width (W) as shown in Figure 1.19c. The matching of Fermi levels is brought about by raising all energy levels to the left of the junction by $\Delta\varphi$ and because they are in intimate contact, also the levels *at the junction* and immediately to the right of the junction in the surface of the semiconductor. In the n-type semiconductor, this can only happen by all the energy levels (including the conduction and valence band levels) at the junction bending upward (or downward depending on the relative depths of the initial Fermi levels). The bending can be exaggerated by lifting the Fermi level on the left-hand side (by applying a negative potential across the junction). This will also increase the depth of the depletion layer.

In practice, there are a number of ways band bending can be achieved. For example, the left-hand side material in Figure 1.19 can be a metal deposited onto the semiconductor surface or it can be part of the same semiconductor but doped p-type. The first is known generally as a 'Schottky or surface barrier' (this is the method used for a Si(Li) detector junction) and the second as a 'p–n junction barrier' (which is the method used for a p-type HPSi or HPGe detector where a lithium-diffused junction is formed at the back of the detector). We will look at these two basic methods in a little more detail here and in greater depth in Chapter 4.

1.5.3 Schottky Barrier

Figure 1.20 shows the effect of bringing a metal into contact with an n-type semiconductor.

The metal 'work function', Φ_m, is the energy depth of the Fermi level and is the energy required to extract an electron from the Fermi level to vacuum level. It can be loosely regarded as the least energy (or electric potential) required to achieve electron emission from the metal surface. The work function of the semiconductor Φ_s is again the depth of the Fermi level as shown in Figure 1.20a, where we assume $\Phi_m > \Phi_s$. A closer description of the least energy required to remove an electron to vacuum in the semiconductor is χ_s, the 'electron affinity'. This is the depth of the bottom of the conduction band or lowest electron level in the conduction band. Note that it is independent of any doping of the semiconductor.

If we now bring the metal into contact with the semiconductor as shown in Figure 1.20b, all energy levels at the junction of the two must move up by the work function difference $(\Phi_m - \Phi_s)$ in order that the two Fermi levels will match. Thus, the valence band and the conduction band must bend upward by this amount at their junction. We see that a barrier for electrons in the metal of height Φ_B is formed, where

$$\Phi_B = (\Phi_m - \Phi_s) + (\Phi_s - \chi_s) = (\Phi_m - \chi_s). \qquad (1.14)$$

So, basically, the barrier height is expected to be the difference between the work function of the metal and the electron

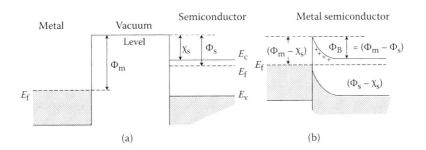

(a)   (b)

Figure 1.20 Metal semiconductor contact. (a) Metal on the left hand side separated from the n-type semiconductor by a vacuum and (b) metal in contact with the semiconductor.

affinity of the semiconductor (~4 eV for Si and Ge) and is calculated to be between 0.5 and 0.9 eV in most cases. However, it must be noted that, in practice, the barrier heights measured for metals on any semiconductor are almost independent of the metal work function Φ_m. The reason for this is that in practice there are interfacial states and often also an oxide layer, both of which will dominate what actually happens at the junction. These are referred to as 'surface barrier' contacts. This will be discussed in greater depth later.

Let us now look at what decides the depth of the depletion layer. The bands are bent upward by an energy $\Phi_m - \Phi_s$ (Figure 1.20). This is achieved by the mobile electrons in the conduction band receding away from the junction to uncover the fixed positive charge on the now uncompensated donor atoms. We thus have, in effect, a capacitor formed with the depletion layer as dielectric with positive charge in the semiconductor balanced by an equal and opposite charge at the surface of the metal. Because the donor concentration is always much less than the concentration of the electrons in a metal, the electrons occupy only a depth of ~0.05 nm, whereas to balance this a much greater depth (many μm) of donor atoms in the semiconductor is required to be uncovered. Thus, the purer the semiconductor, the greater the depletion depth. As stated previously, the barrier can be enhanced by applying a negative potential to the metal relative to the semiconductor, thereby requiring more positive charge and depleting the semiconductor to greater depth. In this way, depletion depths of many millimetres of silicon and many centimetres of germanium can be achieved with reasonable bias (<few kV) and the available material purity.

1.6 p–n Junction Barrier

Consider the case when the material on the left-hand side as shown in Figure 1.19 is also a semiconductor. Furthermore, suppose that this is achieved in a single crystal of the semiconductor by doping the left side with acceptors (p-type) and the right side with donors (n-type). This is similar to that shown in Figure 1.17, but brought together in a *single crystal*. Before equilibrium, the band diagram will be as shown in Figure 1.21a, which—as far as the Fermi levels are concerned—is similar to Figure 1.19a. As discussed above, this is not a stable configuration as they are now in intimate contact in the same crystal and the Fermi levels must equalise bending the bands on the right upward with respect to the bands on the left as shown in Figure 1.21b.

In the p–n junction, we have a concentration gradient of majority carriers that causes the *diffusion* of mobile holes from the p-type into the n-type and mobile electrons from the n-type into the p-type side of the semiconductor. This is a conventional current from left to right as shown in Figure 1.21a. The movement of these mobile carriers uncovers the fixed charges on dopant atoms in either side of the junction as shown in Figure 1.21b and sets up an opposing field

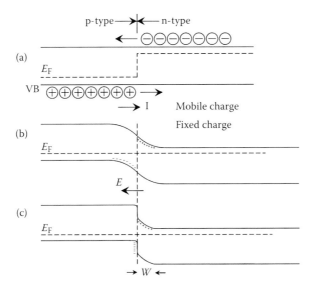

Figure 1.21 Band bending in a p–n junction. (a) p-type semiconductor on left hand side of a single crystal semiconductor with n-type semiconductor on the right hand side before equilibrium is established, (b) fermi levels equalise in equilibrium and band bending is established, and (c) p-type doping level high so that the majority of the depletion layer exists in the n-type material.

E that reduces the current to zero and establishes equilibrium. Like the Schottky barrier, we have in effect a capacitor formed with the depletion layer W as the dielectric with positive space charge in the n-type semiconductor balanced by an equal and opposite space charge in the p-type semiconductor. If the doping levels are similar on both sides of the junction, the depletions into both will be roughly equal and we have a *symmetric p–n junction*. Note from Figure 1.21b that the band bending upward implies positive space charge in the region and downward a negative space charge. If the doping level in the p-type side was much higher than in the n-type side, the amount of depletion depth to balance charge on the left hand would be much less than the depletion W on the right as shown in Figure 1.21c. This type is an *asymmetric p–n junction* (sometimes termed an 'abrupt' or 'high-low' doped junction), which is typical of p–n junctions formed by the diffusion or ion implanting of dopant atoms. It is also similar in form to the Schottky barrier shown in Figure 1.19c. The advantage of the p–n junction is that it is a 'buried' junction and therefore does not suffer from the surface vagaries of the Schottky barrier. For this reason, the development of a good surface barrier that supported a high bias voltage (and therefore deep depletion) lagged for several years behind the diffused junction.

1.6.1 Application of Reverse Bias

In the Schottky barrier, the metal acts rather like a very highly doped p-type semiconductor in an abrupt p–n junction. This would be the case, for example, of gold deposited on n-type high-purity silicon or germanium. For this reason, we say the gold has formed a 'p-type contact'. Even without any applied bias, we have a small depletion layer with a 'built-in' voltage across it because of the charge build-up in the layer. In silicon, this 'contact potential' or 'built-in bias' is about $V_o = 0.5$ V. Any additional reverse bias V_b we supply is effectively added to this. In reverse bias, we enhance the field E across the junction by applying positive potential to the n-type with respect to the p-type as shown in Figure 1.22.

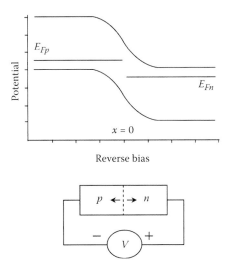

Figure 1.22 p–n junction.

Note that the barrier height for current flow has now been raised, and the depletion depth has increased as has the space charge either side of the barrier. This increases the band bending and enhances the barrier to majority carriers. The Fermi levels no longer equalise because we no longer have equilibrium conditions. Note, incidentally, that because the energy band diagram represents the potential energy of mobile *electrons*, its gradient is opposite to that of a conventional electrostatic potential representing that of positive charge. Carriers (electrons) are continuously being supplied by the external bias, and this constitutes a 'leakage current'. Under *forward* bias, the conductivity is governed by the majority carriers as in normal conduction, but in reverse bias, as in radiation detectors, the leakage current is due to the *minority* carriers that do not experience the barrier. We saw from Equation 1.12 that if we decrease the conductivity by reducing the impurity concentration, we increase the reverse bias leakage current by increasing the minority carriers. This has significance in the selection of material for intrinsic low bandgap detectors such as HPGe. The temptation is to choose very high purity (almost 'intrinsic') material, but if the impurity level is too low the reverse leakage current will increase even at low temperature. Another potential source of minority carriers and leakage current is from thermally generated carriers in the undepleted regions. If they are sufficiently close to the depleted region boundaries, they can diffuse across it and be collected by the field. This gives rise to 'diffusion current', which will be discussed in Section 1.11.2.

The growth of the depletion layer with the application of reverse bias applies to both the Schottky and p–n junction barriers. For example, consider a p-type metal contact or a B diffused abrupt p–n junction on n-type HPSi. In this case, virtually all the depletion is into the n-type silicon and we can neglect the depletion into the p-type contact. This is the ideal situation for an n-type radiation detector where the n-type material is the active volume and the diffusion or metal contact is merely the 'window'.

A rigorous treatment of the p–n junction (e.g., [4], p. 61) is given by solving Poisson's equation in the depletion layer. The result is that the depletion depth W into the n-type silicon is given by

$$W = \sqrt{[2\varepsilon\varepsilon_0(V_o + V_b)/eN_D]}. \tag{1.15}$$

There is an analogous expression for the depletion depth into the p-type silicon. It is usually more practical to express this in terms of resistivity ρ (the reciprocal of the conductivity) of the semiconductor that was given by Equation 1.7.

In n-type material, $n \gg p$ and

$$\sigma = e\mu(n)n \sim e\mu(n)N_D. \tag{1.16}$$

So, we can write

$$W = \sqrt{[2\varepsilon\varepsilon_0\mu(n)\rho(V_o + V_b)]}.$$

If $V_b \gg V_o$,

$$W \sim \sqrt{[2\varepsilon\varepsilon_0\mu(n)\rho V_b]}. \tag{1.17}$$

Equation 1.17 has many interesting properties. But first, let us estimate the extent of a typical depletion depth in silicon at room temperature. Using

$$\varepsilon_0 = 8.85 \text{ pF/m}$$

$$\varepsilon = 12$$

$$\mu(n) = 1350 \text{ cm}^2/\text{V/s}$$

$$\mu(p) = 450 \text{ cm}^2/\text{V/s}.$$

Note incidentally that the mobility of holes at room temperature is ~1/3 that of electrons. So, for the same bias and resistivity, the depletion depth is significantly greater for n-type silicon than for p-type silicon and it is generally preferred for radiation detectors.

If ρ is expressed in kΩ cm, V is in volts and W is in millimetres, for p-type silicon, Equation 1.17 can be rewritten as a simple practical approximation

$$W \sim \sqrt{(\rho V)}/100 \qquad\qquad (1.18)$$

and for n-type,

$$W \sim \sqrt{(3\rho V)}/100. \qquad\qquad (1.19)$$

Let us choose a relatively high resistivity, $\rho = 5$ kΩ cm, n-type Si and a bias of 60 V.

Using Equation 1.19, we find $W \sim 0.3$ mm. Thus, we see that we are getting depths that would already be sensitive to soft x-rays.

Considering the dynamic capacitance dQ/dV of the depleted junction layer, we note that it is free of mobile charge and thus the conducting p- and n-type silicon either side has capacitance/unit area given by

$$C = \varepsilon\varepsilon_0/W.$$

Using Equation 1.17

$$C = \sqrt{[\varepsilon\varepsilon_0 2\mu(n)\rho V_b]}.$$

For silicon, the capacitance/mm^2 is given by the practical approximation

$$C \sim 0.1 \text{ pF/W } (W \text{ in mm}). \qquad (1.20)$$

Let us consider an abrupt p-type contact (heavily doped p-type) on n-type bulk silicon. As noted, this 'asymmetric' type of junction is the type most commonly used for radiation detectors. When reverse biased, the field across the depletion layer can be calculated simply from the equation that equates the charge density to divD. In one dimension, we have

$$dE/dx = -eN_D/\varepsilon\varepsilon_0. \qquad (1.21)$$

If we integrate this, we get the field

$$E(x) = -eN_D x/\varepsilon\varepsilon_0 + c,$$

where c is a constant.

Because there is no field in the undepleted region, we have the boundary condition

$$E = 0 \text{ at } x = W$$

which gives

$$c = eN_D W/\varepsilon\varepsilon_0.$$

Therefore,

$$E(x) = eN_D[W - x]/\varepsilon\varepsilon_0 = [W - x]/\varepsilon\varepsilon_0\rho\mu(n). \qquad (1.22)$$

Note that the field at the junction ($x = 0$) is proportional to the square root of the bias (Equations 1.17 and 1.22) and the inverse square root of the resistivity of the silicon. This has implications for charge collection in a radiation detector where we want this field value to be maximised. Figure 1.23 shows the plots of field E and depletion depth W for various values of bias V_b using the values of our example.

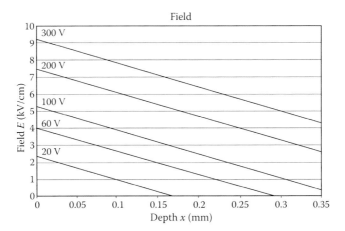

Figure 1.23 Variation of electric field with depletion width for various applied voltages.

Supposing the thickness D of the silicon wafer was only 0.35 mm. When the depletion depth reaches 0.35 mm, that is, to the back contact of the wafer, we get 'full-depletion' or 'reach-through'. Providing this contact, which is not a p–n junction and is termed an 'ohmic' contact, is properly manufactured, it will take up the field presented to it without breakdown as shown in Figure 1.23. For example, if this contact is made by heavily doping n-type silicon (as in a lithium diffusion) it will be rich in electrons, and hole injection into the field region of the semiconductor will be suppressed. Most x-ray detectors work above full depletion or in 'overdepletion' mode because it is beneficial for collecting the charge carriers generated by ionisation to have a high field throughout the crystal. Detector makers refer to the x-ray window contact, which is the p–n junction in this case, as the 'front', so we have to be a little careful in that in the planar processing of wafers this is usually regarded as the 'backside'. Figure 1.23 shows that the more we overdeplete the wafer, the more the field looks 'flat' and uniform and—as Equation 1.21 shows—the more it looks like very high resistivity or 'intrinsic' silicon (which has a resistivity of ~230 kΩ cm at 300 K). As the bias is increased and we have full depletion, the capacitance per unit area flattens to the value, we might anticipate

$$C = \varepsilon\varepsilon_0/D. \tag{1.23}$$

We have seen that n-type silicon is preferred to p-type because it depletes at lower bias and leakage currents can therefore be less. This is particularly important for room temperature or near–room temperature detectors. This n-type material can be produced by doping with phosphorus but is more usually produced from high resistivity p-type material by exposing it to neutrons from a reactor. One isotope of silicon captures a neutron to give a P atom in its place when it deexcites by β decay. This process, which is called neutron transmutation doping (NTD), is better controlled than doping during crystal growth and gives a more uniform doping. Figure 1.13 showed that the carrier mobilities increase rapidly as the silicon is cooled so that again less bias is required for the same depletion depth. In comparing silicon with germanium, the same figure shows that above 100 K (the temperature range usually achieved by LN cooling) the mobilities for silicon are lower than those for germanium. The respective dielectric constants of silicon and germanium are 12 and 16, so Equation 1.17 shows a 15% greater depletion depth for the same bias because of this effect. As a result of these effects, germanium has a significantly higher depletion depth for similar temperatures and resistivity. Although this is not a practical advantage for biasing, it will have consequences when we discuss surface charges.

1.7 Charge Generation by Radiation

The reverse bias semiconductor diode provides a medium approaching the ideal for converting x-ray photons to an electric charge signal. Let us look now at how the mobile charge is generated (or more precisely, separated as the net charge must still remain zero) and subsequently collected to provide a signal. We will turn our attention to the photoelectric effect as this is by far the most significant mode of charge generation for the x-ray energies which are of interest in spectroscopy, usually 100 eV to ~30 keV.

In the photoelectric effect, the x-ray as an entity is completely absorbed and its energy largely transferred to an atomic electron (the photoelectron) of the semiconductor. The energy will differ by the binding energy of the atomic electron. So, at this point, we have essentially a *charged particle detector*, albeit with the source *internal* to the detector. This, incidentally, is a great advantage that the x-ray detector has over conventional charged particle (α, β, etc.) detectors, as we are less concerned with the dead material of the window. A bigger concern is the efficient collection of all the charge generated in the depletion region of the semiconductor. The loss of this electron from the inner shell creates a vacancy that may be deep (a K-shell electron is the most probable). This can be regarded as a 'deep hole', although 'hole' is usually restricted to a missing electron in the covalent bond that can contribute to a flow of current. The inner shell vacancy is highly unstable and is filled as shown in Figure 1.24.

The semiconductor atom that was left in an excited state can relax back to the ground state by a number of mechanisms. An electron from a higher atomic level can fill the vacancy with the excess energy carried away by the emission of an x-ray (fluorescence) or the ejection of an electron. Overall, we see that there will be a cascade effect producing x-rays (which will generally be absorbed by secondary photoelectric interactions with other atoms) and Auger and 'shake-off' electrons (see Section 2.4.1) as well as the relatively energetic photoelectron. If sufficiently energetic, the resulting electrons and holes will cause further excitations and ionisations, leaving

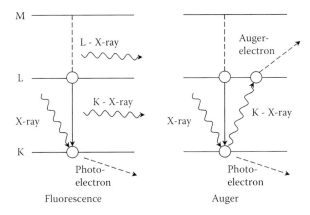

Figure 1.24 Production of characteristic x-rays.

secondary holes and electrons along their tracks until none of them (holes or electrons) have sufficient energy to produce any more. After ~1 ps, we are left with a charge cloud of 'thermal' electrons and holes, gaining as much energy from the lattice vibrations as they lose.

The maximum ranges of photoelectrons with energies of 1 and 30 keV are ~0.03 and 8.7 μm respectively in silicon and about half these ranges in germanium. The high concentration gradient of holes and electrons in the final thermal charge cloud causes rapid outward diffusion. The effect of this is far more important in terms of the charge distribution at low energies. The reason for this is that the photoelectron range is small, but the thermal (or near-thermal) carriers cannot lose further energy readily and continue to diffuse outward unchecked. The diffusion range against an applied field of 1.5 kV/cm is typically 0.2 μm in silicon. When moving with the applied field, the carriers continue this outward diffusion and by the time electrons reach the collection electrode after drifting a few millimetres, the cloud may have expanded to be many microns. Figure 1.25 shows the energy–band diagram in the crystal lattice of the semiconductor corresponding to this atomic description.

With all the energy E dissipated, how much charge is generated? After relaxation we expect all the electrons to sink to the bottom of the conduction band and all the holes to rise

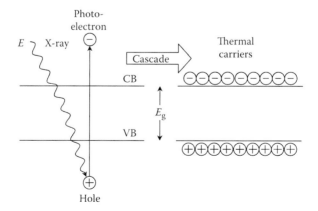

Figure 1.25 Illustrates how a photoelectron loses energy and populates both the conduction band (CB) and the valence band (VB) with charge.

to the top of the valence band. Naively, we might expect that the number of electrons and holes generated (which remember must be equal) would be just

$$n = E/E_g.$$

If this were true, we would always obtain the same number of carriers generated for a given energy x-ray, and the signal developed would always reflect this number. Unfortunately, this is not the case as not all the energy E goes into producing electron–hole pairs. A significant proportion is dissipated as direct heat, that is, lattice vibrations or 'phonon scattering' as mentioned in Section 1.4. Remember this is why the mobility of charge carriers 'μ' is finite. This has two effects: it reduces the number n to a value (Equation 1.3)

$$n = E/\omega.$$

This defines the constant ω that is related to E_g but 2 to 3 times larger. Second, because of the partition of energy, n is statistical in nature; the number n produced is not always the same for a given energy x-ray. As noted in Section 1.2, this gives rise to a statistical spread σ given by the Fano factor

$$\sigma = \sqrt{(Fn)}.$$

The charge signal distribution due to the photoelectric effect (the 'photopeak') is approximately Gaussian in shape even without any electronic noise (discussed below). Its width is significant and sometimes called the 'dispersion', 'Fano limit', or 'Fano noise'. Of course, eventually all the energy E is dissipated as heat, and this is the basis for the cryogenic 'bolometer' detectors. Because bolometers attempt to measure all the energy deposited, the statistical spread on the signal is in practice an order of magnitude less than for semiconductors. However, in semiconductor detectors, the goal is to move all the carriers generated by the x-ray to the detector electrodes and in so doing measure the signal in the Fano limit.

1.7.1 Carrier Diffusion

In the reverse bias diode we already have an electric field in the depletion layer creating the active volume for x-ray detection in the crystal. This field will compete with charge diffusion to eventually sweep the holes and electrons in opposite directions toward their respective electrodes unless those diffusing against the field impinge on these electrodes before the field reverses them (see Figure 1.26). As a consequence, the latter will be lost and will not contribute further to the signal.

Under the influence of field E, carriers will have moved a distance

$$x = \mu Et \tag{1.24}$$

toward the respective electrodes in a time t. In this time, the diffusion causes an outward flux such that carriers have expanded to a radius equal to the diffusion length

$$L = \sqrt{(Dt)}, \tag{1.25}$$

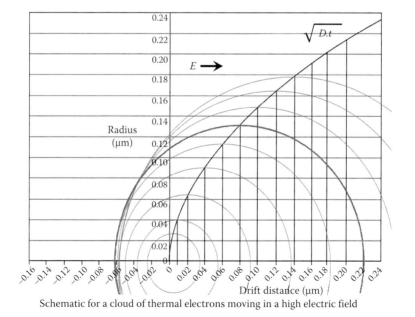

Schematic for a cloud of thermal electrons moving in a high electric field

Figure 1.26 Thermal electrons moving in a constant electric field.

where D is the diffusion constant and is the constant of proportionality between the diffusion rate and the concentration gradient (Fick's law). The diffusion radius of the charge cloud thus increases as $\sqrt{t}$. At some time t', the electrons will have drifted in a positive x-direction under the influence of the field by a distance equal to the distance covered by back diffusion in the negative x-direction. This time is given by

$$\mu Et' = \sqrt{(Dt')} \text{ or}$$

$$t' = D/(\mu E)^2.$$

Inserting this into Equation 1.25 gives the charge cloud radius at this point

$$R' = D/\mu E. \tag{1.26}$$

This schematic is for a cloud of thermal electrons moving in a high electric field with a drift velocity of 10^7 cm/s (close to the saturation velocity). The initial center of diffusion is $x = 0$, and the larger diffusion spheres are drawn at roughly 0.2-ps intervals. Time t' corresponds to the maximum back diffusion from the origin (and the bold sphere with $R' \sim 0.135$ μm). R' represents the maximum diffusion radius before the charge is reversed away from the negative electrode after about 1 ps. If the negative electrode was within the sphere radius R', that is, $x \geq -D/\mu E$, assuming it behaves as a sink for electrons, there would be charge loss as mentioned. This gives rise to a depth $D/\mu E$ below the electrode in which some of the charge generated is lost or leads to 'incomplete charge collection' (ICC). Note that (Equation 1.26) R' increases as the field decreases. This region is also erroneously referred to as a 'dead layer', but this expression should only be used for a region where the charge collection efficiency (CCE) is zero. However, as we shall see later, even the metal electrode and intermediate oxide can contribute. The same diffusion arguments hold for holes moving in the opposite direction that might be lost to the back (positive) electrode. However, as Beer's law (1.2) suggests, unless the x-ray energy is very high or the detector is thin, relatively few carriers are generated near the back electrode.

To see the temperature dependence of the region of ICC, we note the Einstein relation for diffusion at a temperature T,

$$D = kT\mu'/e. \qquad (1.27)$$

We have to keep in mind that μ in Equation 1.24 refers to the mobility of carriers at saturation velocity in the field and μ' in Equation 1.27 at thermal velocities at temperature T. However, because carrier mobilities decrease strongly with increasing temperature (Figure 1.13), Equation 1.27 indicates that D (and R') decrease with temperature and back diffusion is less of a problem at room temperature, for example, than at LNT. This will be discussed in Chapter 4.

1.7.2 Signal Development in Electric Field: Shockley–Ramo Theorem

Once the charge carriers start to move, a signal is induced onto one of the electrodes, usually arranged as the one connected to the FET in Figure 1.1 (the other is usually used to bias the crystal using a stabilised potential V as shown in Figure 1.27). This signal development can be demonstrated by a simple argument involving the energy stored on the reverse bias diode capacitor C_D and although the following is not a rigorous proof [4], it does give a flavour of the signal induction mechanism. For simplicity, let us consider the motion of one electron only drifting in the uniform field ε at a velocity v between the electrodes held with a constant bias potential V

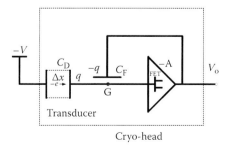

Figure 1.27 Charge flow in field ε in the detector.

between them. Consider the charge signal q induced in moving a distance Δx as depicted in Figure 1.27.

The energy given to the electron by the field in moving an increment dx is

$$E = -e\varepsilon dx$$

and the rate at which energy is given to the electron is

$$dE/dt = -e\varepsilon dx/dt.$$

This energy rate comes from the rate of change of energy stored on the capacitor C_{D}

$$dE/dt = d/dt[1/2Q^2/C_{\mathrm{D}}] = QI/C_{\mathrm{D}} = IV,$$

where Q is the total initial charge stored on C_{D} and I is the instantaneous current flowing from the electrode into G. Expressing the field ε as $-dV/dx$, the instantaneous current is

$$I(t) = -e(dV/dx)(dx/dt)/V = -e(dV/dt)/V.$$

If the electron, in moving through a distance Δx, moves through a potential difference ΔV, the charge signal q is the integral of $I(t)dt$

$$q = -e(\Delta V/V). \tag{1.28}$$

Equation 1.28 is, in essence, the Shockley–Ramo theorem (often shortened to Ramo's theorem). The induced charge signal at the electrode due to the motion of the charge is *the charge multiplied by the fraction of the potential moved through*. It can be shown that this is true even in the presence of the space charge in doped semiconductors. If the electron were to move through the total potential (from one electrode to the other), the integrated charge signal would be simply $q = -e$. The motion of the holes in the opposite direction adds to the signal by Ramo's theorem so that if both originate at *any* point between the electrodes, as they have equal but opposite charge, the contribution from both will again be just $q = -e$. This is depicted in Figure 1.28.

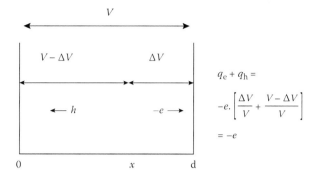

Figure 1.28 Movement of charge and signal development.

If the field is uniform, as it may be regarded in most Si(Li) detectors, we can replace the *fraction of potential* by the *fraction of thickness of the active volume d moved through*. It is important to realise that the electrons and holes do not have to reach the electrodes to induce a signal. However, in order for the signal to register the full charge ($-e$), both the electron and the hole from the same ionisation point *must reach their respective anode and cathode electrodes*, and the amplifier integration time must be long enough to allow this to happen.

In many cases, however, the situation is more complex. For example, multiple electrodes may be involved. Under such circumstances, the signal at each electrode has to be modified by a 'weighting potential'. Figure 1.29 shows the electron trajectories (broken lines) for some alternative electrode arrangements as follows:

(a) Guard ring detector (G-R)
(b) Coplanar strip detector
(c) Pixellated x-ray detector (PXD)
(d) Silicon drift detector

In all these cases, weighting is required and the signal is developed near the end on the electron trajectory, near the G-R, at the strips, close to the pixels (the smaller the pixels, the larger the weighting), and in the case of the SDD, close to the anode.

This weighting potential is calculated in practice by setting the potential of the measuring electrode to unity and

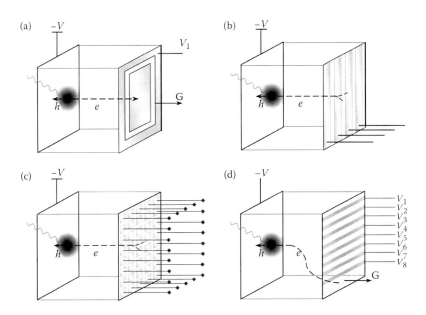

Figure 1.29 (a–d) Electron trajectories in various structures.

all the others to zero as the charge moves. Depending on the geometry, the signal is weighted toward different regions of the charge carrier's trajectory. The weighting potential calculated in this way is not the real potential existing between the electrodes but is a mathematical formalism for calculating the induced charge or current signal on any *particular* electrode.

Several remarks regarding Ramo's theorem are in order. It is seen that it is not essential that carriers actually reach the contacts in order to generate a signal. For example, if any carriers recombine, are lost from the active volume, or are trapped, we still obtain a signal. Whereas this might be desirable in some cases, it is a double-edged sword since the signal will not represent all the carriers that were generated and by Equation 1.3 will not therefore represent the x-ray energy absorbed. We will suffer the consequences of 'peak tailing'. Of course, we have no means of knowing which signals suffer from charge loss and which do not (but in some cases, efforts to discriminate on the basis of the pulse 'shapes' help with tailing). In good-quality silicon and germanium material recombination and trapping of charge carriers are not a problem. In x-ray detectors made from these materials, back diffusion to the contact can adversely affect the shape of the spectrum. In radiation-damaged silicon and germanium and other semiconductors where the material technology is not so advanced, recombination and trapping are important.

1.7.3 Recombination

It might be expected that the holes and electrons generated by the absorption of an x-ray would quickly recombine. In fact, this is found to be significant only in the case of short-range heavy ionising particles where the concentration of carriers is extremely high. There are a number of reasons for this. First, diffusion disperses them and second, the applied field separates them. Furthermore, in intrinsic and very pure silicon and germanium, the annihilation of electrons at the bottom of the conduction band and the holes at the top of the valence band (Figure 1.25) by a direct transition, is suppressed. This is because the bottom of the conduction band and the top of the valence band are offset in momentum space. This is shown in Figure 1.30a.

This type of band structure is known as an 'indirect gap'. For electron–hole recombination to take place, this momentum offset has to be found from another source and can involve momentum from lattice vibrations (phonons). Impurities giving levels in the forbidden gap where electrons and holes can be trapped (recombination centres) can also cause it to happen. In good-quality silicon and germanium, the 'lifetime' for recombination τ_r of the mobile charges can be very long (~ms) compared to charge collection times (~ns). Mid-band impurities, such as gold, can drastically reduce the charge lifetime by a 'stepping stone' effect. Compound semiconductors are mostly cases of 'direct bandgap' (Figure 1.30b), and recombination can take place directly with the emission of light as in a light-emitting diode (LED). Not only is recombination suppressed in indirect bandgap semiconductors such as Si and Ge, but so is electron–hole pair creation. This is why the direct bandgap materials have an efficiency advantage for optoelectronic devices such as solar cells as well as LEDs. The absorption of x-rays by the photoelectric effect also needs another source to conserve momentum. This is usually supplied by the recoil of the atomic nucleus and is true for all materials.

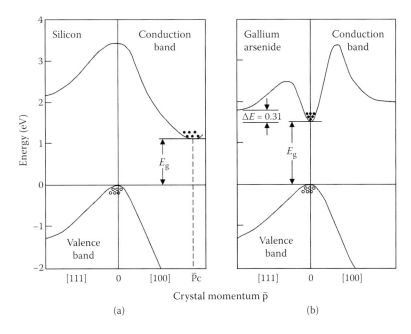

Figure 1.30 Offset shown between the bottom of CB and the top of VB in Si but not in GaAs. (a) Indirect band gap structure (e.g., silicon) and (b) direct band gap structure (e.g., gallium arsenide).

1.7.4 Trapping and the Hecht Equation

Besides loss of charge by recombination, charge carriers can also be trapped by imperfections, including impurities, in the crystal. As mentioned above, these can be represented by energy levels within the energy gap. Those close to the conduction band can transfer an electron to the conduction band, becoming positively charged and acting as an electron trap, and similarly those close to the valence band can become hole traps. These are termed 'shallow traps'. The most efficient traps are 'deep traps' located close to the center of the bandgap where they are almost equally accessible by both electrons and holes as mentioned above. The latter can also generate thermal electron–hole pairs, thus increasing the leakage current. In this respect, impurities such as Au in silicon and Cu in germanium are regarded as 'killers'.

We can define a trapping charge lifetime τ_t. If we deposit a charge q_0 into a semiconductor, by the definition of charge lifetime, it will decay as

$$q(t) = q_0\exp(-t/\tau_t).$$

Combining with the recombination lifetime, we have

$$q(t) = q_0\exp(-t/\tau_t)\exp(-t/\tau_r) = q_0\exp(-t/\tau)$$

where the total lifetime is τ given by

$$1/\tau = 1/\tau_r + 1/\tau_t.$$

Under the influence of a field E, the charge will drift a distance x in a time t given by Equation 1.5

$$t = x/(\mu E).$$

Therefore, as a function of drift distance,

$$q(x) = q_0\exp(-x/\mu\tau E). \tag{1.29}$$

The characteristic distance $\mu\tau E$ is known as the *drift length* L, and $\mu\tau$ can be used as a 'figure of merit' for semiconductor material. We can tabulate this figure of merit (values given in cm^2/V) for a number of semiconductors (Table 1.2).

The mobility is a fundamental property of the material, whereas the lifetime τ is influenced by using a direct bandgap and high-purity, defect-free material. It is immediately clear that on this basis silicon and germanium are far superior to all other semiconductors. It is not surprising that trapping is more likely in compound semiconductors of elements AB as there is more opportunity in the lattice for defects, such as A replacing the site usually occupied by B. We can rewrite Equation 1.29 as

$$q(x) = q_0\exp(-x/L). \tag{1.30}$$

TABLE 1.2
Material Properties Compared by Their μτ Product

Semiconductor	Electron Mobility, μ_e (cm²/Vs)	Hole Mobility, μ_h (cm²/Vs)	μτ (Product) Electrons cm²/V ×10⁶	μτ (Product) Holes cm²/V ×10⁶
Silicon	1350	450	>10⁶	>10⁶
Germanium	3900	1900	>10⁶	>10⁶
Cadmium zinc telluride	1000	120	4000	120
Cadmium telluride	1000	120	3000	200
Gallium arsenide	8000	400	80	4
Mercuric iodide	100	4	300	40
Silicon carbide (4H-SiC)	1000	115	400	80
Thallium bromide	30	4	500	2
Amorphous silicon	1 to 4	0.05	0.2	30

Note that if trapping occurs but the carriers are released in a time that is short compared to the total charge collection time, then they contribute to the signal. However, they do give rise to fluctuations in the signal ('trapping–detrapping noise').

Consider for simplicity the Shockley–Ramo theorem applied to the planar transducer depicted in Figure 1.28. The field is uniform and we can say that the signal induced by electron drift over an increment between x and $x + dx$ is

$$dq = q(x)dx/d.$$

If the electron charge cloud $-q_0$ starts at $x = x_0$, the total induced charge signal by Equation 1.30 is

$$q_e = (1/d) \int (x_0 \text{ to } d) \, q_0 \exp(-x/L)$$

$$q_e = q_0 L/d(1 - \exp[-(d - x_0)/L]).$$

There will be a similar contribution from the hole charge cloud q_0 starting from the same point and as the drift lengths will be different, the total signal from holes and electrons will give a charge collection efficiency (CCE)

$$\eta(x) = q/q_0 = L_e/d(1 - \exp[-(d - x_0)/L_e]) + L_h/d(1 - \exp[-x_0/L_h]). \tag{1.31}$$

This equation is commonly known as the Hecht equation. It is easily seen by expanding the exponentials in Equation 1.31 that, for negligible trapping (large L_e and L_h), the expression yields unity. However, for small L_e or L_h (as in compound semiconductors), the CCE will depend on the interaction depth x_0. The Hecht equation above neglects any detrapping that may occur within the charge collection time. In silicon and germanium, for most fields found in x-ray detectors the drift length for both holes and electrons is very long (approaching 100 m!). In the case of thick (>2.5 mm) silicon, the material is usually 'lithium drifted' to compensate for impurities and to achieve long drift lengths. In some compound semiconductors, the drift lengths can be relatively short, and bulk trapping effects over thicknesses of the order of d (~mm) are significant. Indeed, in some cases, the trapping of holes is so severe that the second term in Equation 1.31 can be neglected, and we have a 'single carrier detector'. Note that for low-energy x-rays the exponential range (distance over which number falls by factor $1/e$) is often much less than the active depth of the crystal, and we again effectively have a single carrier detector. For example, the exponential range of x-rays of energy below 6 keV is less than 30 μm in both silicon and germanium, and in a typical detector the signal is at least ~99% generated by electrons (for a negative window electrode as is usual). Any hole-trapping effects are therefore only apparent when high-energy x-rays are used, and detector testing usually includes the use of a radioactive source (or x-ray tube equivalent) such as Cd-109 that gives a line at 22 keV.

1.8 Pulse Height Analysis

Trapping and losses of carriers to electrodes or sidewalls are examples of ICC. Here, we briefly introduce pulse (or signal) height analysis (PHA) to form an x-ray 'spectrum' to indicate some of the effects of ICC and introduce the concepts and terms in common usage. This is a large topic that will be discussed in some detail in Chapter 2. The pulse signal $V_0 = -q/C_F$ (Equation 1.1) is usually amplified by a wide-band amplifier (preamplifier) that, apart from the FET, is usually external to the cryostat (if cooling is used). It is then 'integrated', 'shaped', or 'processed' with a time constant T (or 'peaking time') usually much longer than the charge collection time in the transducer (microseconds rather than nanoseconds). The main reason for doing this is to reduce noise (see Section 1.10). As we might expect, in general the longer the integration time τ, the more the random fluctuations of noise are smoothed out and the noise reduced. However, τ also puts a limit on the *rate* ('counting rate') at which signals can be processed without overlapping in time (known as 'pileup'). This 'shaper' also recognises the individual signals and rejects any pileup as well as acting as a discriminator against noise. The shaper may operate in analogue mode, but more usually the preamplifier output is digitised and the shaping is carried out digitally. The latter is termed a digital signal processor. The sequence is shown schematically in Figure 1.31.

The resulting voltage pulse height h (which is proportional to the signal q) is then sorted by amplitude in a pulse height analyzer (PHA; or multichannel analyzer, MCA) to produce a frequency spectrum or histogram. The channel h to $h + \Delta h$ is

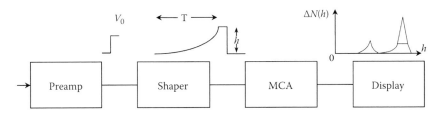

Figure 1.31 Pulse processing sequence.

interpreted as an energy by calibration with a known energy x-ray line, and the channel width Δh is usually 10, 20, or 40 eV. All this is usually carried out under computer control.

1.8.1 Trammell–Walter Equation

The Hecht equation (Equation 1.31) effectively gives the CCE $\eta(x) = q/q_0$ for collection from a generation depth x in the semiconductor. For simplicity, let us assume that we have only hole trapping and therefore Equation 1.31 reduces to

$$\eta(x) = 1 - x/d + L_h/d(1 - \exp[-x/L_h]).$$

This is plotted for $d = 3$ mm and $L_h = 5, 10, 20$ mm in Figure 1.32.

To construct a spectrum, we have to fold into the calculation the probability or fraction $g(x)dx$ of the charge packages being generated at a depth between x and $x + dx$, and in the simple case of x-ray absorption this is the same as the fraction of x-rays absorbed in the same element and is derived from Beer's law (1.2)

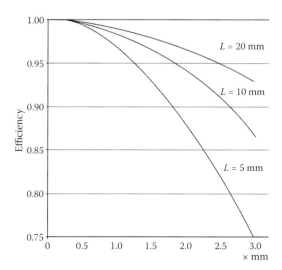

Figure 1.32 Charge collection efficiency in a 3-mm-deep detector for different hole-drift lengths.

$$g(x)\mathrm{d}x = (\mu_\mathrm{a}\rho)\exp(-\mu_\mathrm{a}\rho x). \tag{1.32}$$

If the pulse height h is measured on the PHA, the fraction of absorption events that result in heights between h and $h + \Delta h$ enter a channel in a pulse height spectrum. The fraction of events in the channel is given by $\Delta N(h)$. A series of 'slices' in x to $x + \mathrm{d}x$ will contribute to these, and we integrate over the appropriate limits of x

$$\Delta N(h) = \left[\int_{x_\mathrm{a}}^{x_\mathrm{b}} g(x)\,\mathrm{d}x \right] \Delta h$$

where   (1.33)

$$x_\mathrm{a} = \eta^{-1}(h) \text{ and } x_\mathrm{b} = \eta^{-1}(h + \Delta h).$$

The limits of the integration over x are the inverse functions of the efficiency function $\eta(x)$. The result can be visualised by the following graphic construction (Figure 1.33). In this example, it represents the response function of a 4-mm-deep silicon detector and a 22-keV x-ray (such as that given by a Cd-109 radioactive source) for a hole-trap drift length of 2 mm.

The shaded areas represent the fraction of the total events that enter the corresponding pulse height channel on the vertical axis. We see that the effect of hole trapping is to introduce a long quasi-exponential truncated tail if the x-rays penetrate the crystal appreciably as they do at 22 keV. Such spectra are common in compound semiconductors such as CdZnTe rather than in silicon or germanium. A form of this equation (Equation 1.33) was first written down by Trammell and Walter [5] while working at Ortec in 1969. They included a Gaussian statistical spread in h around the value given by $\eta(x)$ in the Hecht equation. This was to represent the Fano and noise fluctuations, which will be discussed below.

The Monte Carlo method [6] of calculating spectra has been used more widely than the 'analytical' one given by the Trammell–Walter equation. In these calculations, an initial photon vector is given random (hence 'Monte Carlo') coordinates to enter the detector. The photon energy is selected and

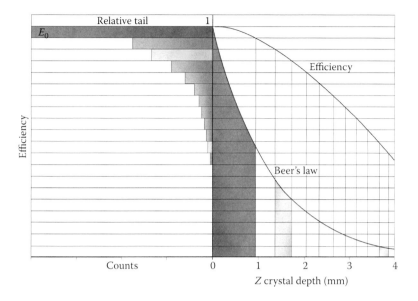

Figure 1.33 (See colour insert.) Generating a pulse height spectrum of a 22-keV x-ray incident on a 4-mm-deep detector in the presence of hole trapping.

all photon–metal and photon–silicon events are modelled on the probabilities of the various interactions which are stored in an array table along with charge collection efficiencies and any other data that need to be accessed for the generation and tracking of all emitted electrons and holes. Statistical broadening is added, and the charge is converted to energy to fill a spectrum channel. In this way, a spectrum is built up photon by photon until sufficiently good statistics are accumulated. This 'mixing' of probabilities makes it rather more difficult to identify the origin of features that occur in real spectra than the analytical approach.

1.9 X-Ray Spectroscopy

The great power of x-ray spectroscopy using semiconductor detectors is in its ability to identify the characteristic lines of the elements simultaneously. X-rays are emitted by atoms when bound electrons are displaced by any energetic impacting particles such as photons (e.g., x-rays) or charged particles such as electrons, protons and alpha-particles. The most energetic fluorescent x-rays are produced when the vacancy occurs in the most tightly bound atomic shell (the K-shell). The whole process can be thought of as being in three stages—excitation, relaxation, and x-ray emission—although of course the whole electromagnetic process occurs virtually instantaneously (within ~100 ps). This process is shown schematically in Figure 1.34.

To conserve energy, the energy difference between the two atomic shells is exactly balanced by the emission of the x-ray. The energy of such x-rays is therefore *characteristic* of the structure of the atom, the possible electron transitions in it and hence the element itself. There are *series* of transitions K, L, M, etc. (depending in which shell the vacancy occurred), and these fall naturally into energy groups. The energy of the x-ray lines emitted in these series is plotted in Figure 1.35 for a wide range of elements of interest.

We can see from this plot that although the K-shell lines run from very low energy up to around 100 keV, the heavy elements are also covered by L-shell lines up to around 10 keV. Thus, analysis of most elements can be accomplished in this most important x-ray range, 100 eV to 10 keV. The light elements provide particular problems. Their low-energy ('soft') characteristic x-rays are sensitive to physical parameters such as absorption in the detector entrance window, contamination of window or semiconductor, and semiconductor dead layers. Many general-purpose x-ray detectors use a beryllium

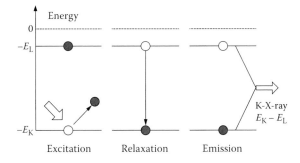

Figure 1.34 Generation of characteristic x-rays.

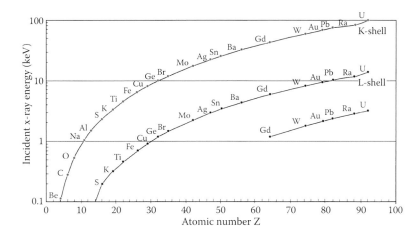

Figure 1.35 Characteristic x-ray line energies.

foil (8–12 μm) x-ray window, which limits the lower energy range to >1 keV. Sodium (Na) is usually the lower limit for element detection in such a detector. Many modern detectors, however, use very thin windows (TWs) that will also support atmospheric pressure (ATW). These allow the analysis of some light elements and avoid semiconductor contamination to an extent.

The x-ray lines also have fine details. Figure 1.36 shows this for the separation of the components in the K and L shells.

Figures 1.35 and 1.36 show that in general lines are less cluttered at higher energies so that it might be desirable, for resolution considerations, to extend the range of the detector beyond, say, 10 keV. By *resolution*, we mean the power to resolve x-rays of slightly different energies in the spectrum. The natural line width of the x-ray lines is of the order of 10 eV, so the *potential* for resolving them from one another even down to very low energies is enormous. Incidentally, the distinction between an x-ray and a gamma ray (which are both photons) is not their energy but their origin. Gamma rays are emitted from the atomic nucleus when nuclear energy transitions occur, which is why, in general, they have much higher energy. Gamma-ray spectroscopy can indicate the nuclear species as well as structure much as x-ray spectroscopy can for the chemical elements.

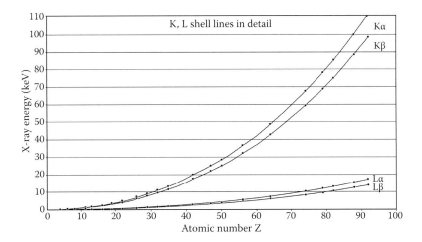

Figure 1.36 Splitting of K and L lines for various elements in the periodic table.

1.9.1 Line Broadening

We have already seen (Section 1.2) that the ionisation process results in a statistical broadening given by

$$\sigma = \sqrt{(Fn)}$$

We also have (Equation 1.3) that the number of carrier pairs n generated for an x-ray of energy E is E/ω, so we can calculate the equivalent energy broadening, known as the detector 'dispersion' (D) or sometimes 'Fano noise' (as it is the resolution limit when the electronic noise is zero).

$$\sigma(E)/E = \sigma(n)/n = \sqrt{(F/n)} = \sqrt{(F\omega/E)}$$

$$\sigma(E) = \sqrt{(F\omega E)}$$

It is usual to express energy peak broadening in terms of the full-width at half-maximum height (FWHM), ΔE, which is shown in Figure 1.37.

For a Gaussian distribution, this is related to $\sigma(E)$ by

$$\Delta E = 2\sqrt{(2\ln 2)}\sigma(E) = 2.355\sigma(E).$$

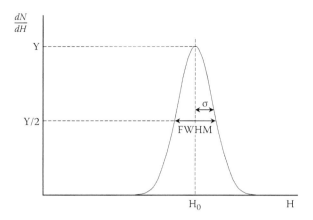

Figure 1.37 Definition of resolution.

We can therefore write the dispersion D (FWHM) as

$$D(E) = 2.355\sqrt{(F\omega E)} \qquad (1.34)$$

Values of F and ω are functions of the material, temperature T, and x-ray energy E. The measurements of ω and F are nontrivial in practice. The energy gap E_g can be measured by plotting the conductivity of a pure crystal as a function of $1/T$ and using the relation 1.10 and ω by measuring the photocurrent response [7] to soft synchrotron x-rays. The measurement of F requires the manufacture of a high-quality x-ray detector. In fact, it is the dispersion D that has to be measured from the peak broadening, and this only yields the product $F\omega$. Furthermore, for a measurement at a single x-ray energy, an accurate knowledge of the noise is required along with good counting statistics in the photopeak. Energy resolutions of x-ray detectors are normally specified at 5.9 keV (the Mn Kα x-ray from a radioactive Fe-55 source), and using the fairly well accepted values given in Table 1.1, we can tabulate D (see Table 1.3) at 5.9 keV for a number of materials and temperatures.

With these values of dispersion, we already have a resolution mismatch with the intrinsic x-ray line widths, but even after adding the electronic noise contributions (which are of

TABLE 1.3
Detector Dispersion, Fano Factor, and Activation Energy for Various Semiconductors

Material	Temperature (K)	E_g (eV)	ω (eV)	F	D (eV)
Si	300	1.12	3.64	0.118	118
Si	120	1.16	3.75	0.118	120
Ge	100	0.74	2.96	0.106	101
CdZnTe	300	1.57	4.64	0.11	129
GaAs	242	1.4	4.3	0.144	140
HgI_2	300	2.15	4.15	0.1	117

course independent of energy) the resolutions are orders of magnitude superior to gas proportional detectors. The importance of resolution to spectroscopy is twofold. Firstly, good resolution (high $1/R$) will help to resolve two closely spaced peaks, whereas poor resolution might only present them as one. Secondly, high sensitivity implies confident identification of peaks on a background level and hence a good peak-to-background factor (see Section 1.14.3). This is also improved by good resolution. Any 'figure-of-merit' for a detector is therefore likely to have a term $\sim 1/R^2$.

The noise N being random is added to the dispersion in quadrature.

$$R^2 = D^2 + N^2 \tag{1.35}$$

In Figure 1.38, we plot the relative contributions of dispersion and noise to resolution in typical Si(Li) and HPGe detectors using Equation 1.35.

A number of assumptions have been made in these plots. Firstly, it is assumed that the Fano factor F and the ionisation energy per electron–hole pair ω are independent of energy, and second, that there is no ICC. The latter will vary from detector to detector, but the dotted lines in Figure 1.38 indicate the type of deviations that might occur at low energies. Although the variation of F with energy can (and will later) be debated, measurements show that the variation of ω is not significant. As the number of carriers generated depends largely on the

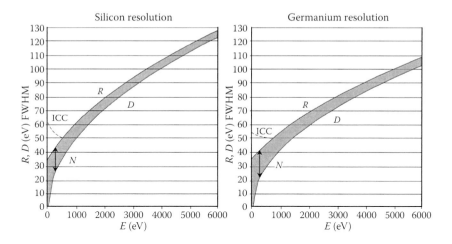

Figure 1.38 Resolution and detector dispersion as a function of energy for Si and Ge.

final stages of the ionisation cascade, neither would we expect it to be. Despite the relative low value of noise chosen for the plots (35 eV FWHM), Figure 1.38 illustrates the increasing importance of the noise contribution to resolution as the energy decreases. The lower energy limit is determined by the CCE at the entrance widow and signal recognition that requires low noise. We will now discuss the noise contribution.

1.10 Noise

This topic can be rather mathematical but, in the final analysis, is dealt with by the electronics in a rather pragmatic way. Various signal shaping techniques are used to optimise the signal/noise ratio depending on the noise conditions extant in the system. Here, we will present only the most significant results of the noise analysis of the x-ray detector circuit (Figure 1.27), emphasising the physics involved without the rigorous noise theory that is a significant topic in its own right. Good treatments of noise analysis can be found elsewhere [4,8]. We must consider all the possible sources of noise inherent in the circuit as a whole (including bias resistors and capacitors and the pulse shaper), and to clarify this, the circuit and its equivalent circuit for noise analysis is given in Figure 1.39.

We will assume that the bias resistor R_b is so large that it can be ignored as a noise source. We have replaced the thermal noise by a voltage noise generator e_n with a resistor R_s in series with it. It is important to realise that any such 'voltage noise' sources must be measured across the total input capacitance C_{TOT} to get the charge equivalent $e_n C_{TOT}$ for a charge-sensitive amplifier.

Before discussing this analysis, we will briefly review the types of noise that are likely to be important. The first thing to appreciate is that *electronics* is not 'psychic'. It treats noise like any other charge or voltage excursion. The difference between 'noise' and 'signal' is only in our perception of what we want to selectively study. If we have electronic responses to, say, incident x-rays, it is these particular responses that we wish to detect and analyze. Any other responses due to other physical effects such as temperature, vibration, fluctuation in electron arrival, charge trapping, and dielectric oscillations, merely get

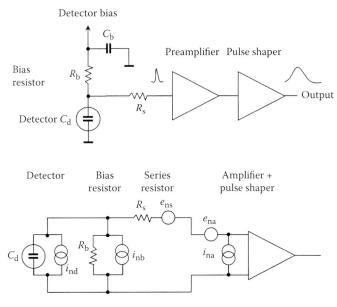

Equivalent circuit for noise analysis of the detector front–end

Figure 1.39 Amplifying section and its equivalent circuit.

in the way of the x-ray responses and as such are regarded as the unwanted 'noise'. Its importance lies in that it sets a lower bound on our ability to discriminate the 'signals' and measure them precisely.

The x-ray signal is generated by virtue of the Shockley–Ramo theorem, and we saw in Section 1.7.2 that this gives an induced charge q at G when an electron moves Δx between the electrodes separated by d. Assuming a constant field for n such electrons, we obtain a signal

$$Q = -ne\ \Delta x/d$$

The corresponding current I is given by

$$I(t) = \mathrm{d}Q/\mathrm{d}t = -ne\ \Delta x/\mathrm{d}t/d = -nev/d$$

where v is the velocity of the electrons. Fluctuations in $I(t)$ are due to fluctuations in n, for example, the variation of carrier numbers emitted over a potential barrier or junction, often referred to as 'shot noise', and in v (thermal noise, Johnson noise, or 'resistor' noise). 'Shot' is in analogy to the noise made by lead shot being dropped on a tin roof, the roof being the analogue of the diode's collection electrode. Shot noise applies when the carriers are injected *independently* of one another and does not apply where local fluctuation densities occur, for example, due to the presence of space charge or trapping. Number fluctuations also come from bulk carrier trapping–detrapping (as mentioned previously) and electron scattering in a conventional resistor, but these do not carry shot noise. The carrier concentrations can also fluctuate because of recombination (Section 1.7.3) normally associated with traps. This is known as generation–recombination noise and will be discussed further in connection with FETs (Section 1.13).

Signals and noise can be synthesised by a range of amplitudes at given frequencies (Fourier analysis) or can be analyzed as functions of time. Here, we will leave 'time domain' analysis to later on. To see the overall effect of these noise sources, an electronic system can be regarded as a 'black box' with a frequency response (or 'gain'), $A(f)$. We need to know the frequency spectrum (power spectrum $\mathrm{d}P/\mathrm{d}f$ or 'spectral density' per unit bandwidth df) of the various noise sources. Although the noise as a whole depends on its frequency spectrum ('bandwidth'), the noise density is independent of bandwidth. For example, thermal noise is constant for all frequencies (we say it is 'white' noise). Shot noise can also be 'white' (as for electrons being emitted over a potential barrier reaching the anode of a reverse bias diode). However, trapping–detrapping of charge results in a nonwhite '1/f' type of frequency spectrum. This is more prevalent in compound semiconductors.

Power is proportional to either the voltage squared (V^2) or the current squared (I^2). So, rather than a power density dP/df (measured in W/Hz), the noise is often given in terms of the corresponding voltage noise density (e_n) or current noise density (i_n). For example,

$$e_n^2 = dV^2/df$$

where e_n is measured in units of V/$\sqrt{Hz}$, or more commonly in low noise systems in nV/$\sqrt{Hz}$.

1.10.1 Thermal, Johnson, or Voltage Noise

When a current flows in a resistor, the electrons are scattered off the vibrating atoms in the resistor giving a fluctuation in the velocity of the electrons. This is also true in a semiconductor (e.g., in the bulk and surfaces of the semiconductor and in the FET channel), where electrons and holes are scattered giving rise to mobility and an effective resistance. As we have already noted, the degree of scattering depends on the energy kT of the vibrating atoms, hence the term 'thermal' noise. For a good treatment of noise, see Motchenbacher [9].

$$dP/df = 4kT$$

since

$$P = V^2/R$$

$$e_n^2 = dV^2/df = 4kTR \qquad (1.36)$$

Similarly, the spectral current noise density is

$$i_n^2 = \mathrm{d}i^2/\mathrm{d}f = 4kT/R \qquad (1.37)$$

Equations 1.36 and 1.37 are sometimes referred to as the Nyquist theorem. Note that they do not involve f and as such defines them in terms of white noise. Note also that, as in all high gain negative feedback amplifiers, the noise of the first stage dominates over the subsequent ones. The largest component of voltage noise in the detector circuit is usually the FET channel resistance which, because of the coupling to the gate by the gate-channel capacitance, is in series with the input. This resistance is proportional to the inverse of the FET transconductance, g_m. Hence, we can write the rms voltage noise density as

$$e_n^2 = K\,4kT/g_m \qquad (1.38)$$

The constant K is often taken as 0.7 but can deviate from this. The transconductance g_m depends on the mobility of electrons (in n-channel FETs) in the channel. As the mobility is a strong function of temperature and increases as T is lowered, the noise decreases correspondingly with cooling. However, there is a limit to how much the FET noise can be reduced by cooling. Below a certain temperature (usually $T \sim 120$ K), the numbers of electrons in the channel start decreasing because of the donor impurities having insufficient energy to be ionised ('carrier freeze-out') and the noise increases. Thus, there is an optimum temperature for the FET.

1.10.2 Dielectric Noise

This is a derivative of the thermal noise discussed above and is attributable to thermal fluctuations in the dielectric. Any fluctuations in electronic charge density due to polarisation in a dielectric will set up fields that draw current from the external circuit and add to system noise. For a low loss dielectric, a dimensionless 'dissipation factor' d is defined

$$d = G(\omega')/\omega'C$$

where $G(\omega')$ is the loss conductance at an angular frequency ω' and C is the capacitance associated with the *dielectric*. The best dielectrics have $d \sim 5 \times 10^{-5}$ at low temperature. $G(\omega')$ can be interpreted as the inverse of a frequency dependent resistor, so putting this into Equation 1.37,

$$i_{nd}^2 = 4kTG(\omega') = 4kTd\omega'C \qquad (1.39)$$

Because C depends on the dielectric constant ε of the dielectric, this noise depends directly on the product of ε and the dissipation factor, that is, on the 'loss factor' ($L = \varepsilon d$). In practice, the dielectric material (often the material of the FET 'package' or glues used to hold components in the package) is effectively inserted into a stray capacitance C_S, and the dielectric noise depends on the quantity and distribution of the dielectric in the vicinity of the FET gate and the transducer back contact. Table 1.4 gives values of εd (an 'inverse figure of merit') for some common dielectrics.

TABLE 1.4
Loss Factors for Different Materials Used in Cryo-head Construction

Material	ε	d	$\varepsilon d \times 1000$
Quartz	3.78	0.0001	0.38
Soda glass	4.84	0.0036	17.4
PTFE	2.1	0.0002	0.42
BN(HBT)	3.8	0.0002	0.76
BN(HP)	4.1	0.0001	4.1
BN (40% silica)	4	0.0076–0.023	30–90
Alumina	4.5	0.0002	0.9
Silicon	12	0.0002	2.4
Ice	3.17	0.004	12.7
Epoxy	3.6	0.005	18
PVC	4	0.02	80
Nylon	3.5	0.03	105
PCB	5.4	0.035	189

1.10.3 Resistive or 'Contact' Noise

R_S in Figure 1.39 is a resistance in series with the transducer and FET gate G. In practice, it is the sum of the resistance of the back layer in the transducer (diffusion and metallisation) and the physical contact to this electrode. As we have seen, this needs to be clear of lossy dielectrics and (as we will see later) of low stray capacitance. However, as also mentioned previously, the optimum temperature for the FET is usually higher than that for the transducer and the contact must support a temperature difference of the order of 10–20°C for a Si(Li) detector and usually more for a HPGe detector. In practice, this is an engineering challenge. If the resistance R_S is large, it will contribute to the voltage noise by (Equation 1.36)

$$e_n^2 = 4kTR_S$$

R_S is in series with the FET channel resistance discussed above.

1.10.4 Shot or Current Noise

The current that flows in diodes is not smooth and continuous but is the result of many individual moving charges. The spectral noise density of current or shot noise can be shown [4] as

$$i_n^2 = 2eI \tag{1.40}$$

where e is the electronic charge and I is the average current. Again, this does not involve f and so is white noise. For this to be considered valid, the charge carriers involved (holes and electrons) must be injected *independently* of each other into the medium. This is the case in a reversed bias semiconductor junction and thermionic and field emission from metal contacts but is not the case of current flowing through an ohmic resistance or along a surface where the charge motions are correlated. It is also not true for bulk trapping and release with one or a number of fixed time constants (trapping–detrapping noise).

1.10.5 1/f Noise

As noted above, if the charges are correlated as they are for trapping–detrapping and not injected independently, the noise is not white shot noise. If there are a few traps with time constants for release spread over a decade or two, the spectral density of the noise follows a 1/f distribution over the corresponding frequency range from f_1 to f_2. We can write

$$e_{nf}^2 = a_f / f, \tag{1.41}$$

where a_f is the 1/f noise coefficient.

For trapping–detrapping at time constant τ_t, the noise power density is constant (white noise) below a 'corner frequency' given by

$$\omega = 1/\tau_t.$$

Up to this frequency, it behaves like shot noise but falls off as 1/f^2 above it. Usually, there are many time constants τ and many corner frequencies. The envelope of these in the noise spectra falls off as 1/f. Because of the real complexity (e.g., different trap concentrations), the noise spectrum is more like

$$dP/df \sim 1/f^\alpha,$$

where α is in the range 0.5 to 2.

1.10.6 Circuit Noise Analysis and ENC

The total voltage noise v_n at the output is obtained by integrating the noise power e_n^2 with the dimensionless response function $A(f)$ for the pulse shaper over all frequencies

$$v_n^2 = \int_0^\infty A(f)^2 e_n^2 df \tag{1.42}$$

where v_n is the root mean square (rms) noise voltage or 'voltage noise'. In practice $A(f)$ only has nonzero values over a limited frequency range (the 'bandwidth' of the amplifier), and if this can be arranged to just overlap the signal bandwidth, we can optimise the signal/noise ratio in the pulse processing. Physically, the 'black box' (amplifier or 'processor') carries out the integration with the aim of optimising the signal/noise ratio. Naturally, the optimum shaper will depend on the frequency spectra of the signals and the various noise sources present. Any one type of shaper will have various characteristic shapes (Gaussian, triangular, cusp, etc.) and 'shaping times' τ, which for x-ray detectors is usually in the range 0.1 to 100 µs.

If a 'noiseless signal', such as a precision pulse, is presented to the input of the amplifier, the output will have on it the random, Gaussian distribution of the noise, and this can be measured on a PHA. The value of v_n can be expressed as the standard deviation σ of the Gaussian distribution or as is more often the case, the FWHM as shown in Figure 1.37. An equivalent method is to open the input of the black box when there is no signal present and measure both the positive and negative noise excursions. The resultant Gaussian peak centered at zero is known as the 'zero strobe' or, more loosely, the 'noise peak'.

In analyzing the contributions of each component to the noise at the amplifier output, it is useful to refer to the equivalent noise charge (ENC). This is defined as the charge that, if applied to the *input* of the system, would give rise to the rms output voltage due to the *noise alone*. Put another way, it is equal to the signal charge at the input that gives a signal/noise ratio of unity. If C_{TOT} is the total input capacitance, then ENC is just the *effective* charge at the input given by the amplifier rms voltage e_n across C_{TOT}. It is a *derived* rather than a primary quantity. It is appropriate for radiation detectors as the signal is usually an input charge on a capacitive sensor, and it can be measured by injecting a known precision charge to the input. The ENC is usually expressed as electrons rms (e-rms) or the equivalent deposited energy specifying the semiconductor, for example, eV (Si). The energy to generate a charge equivalent of one electron in a semiconductor is ω. For example, for silicon at room temperature noise can be expressed as

$$\Delta E \text{ (eV)} = \omega \times \text{ENC} = 3.64 \text{ ENC (e-rms)}.$$

Note incidentally that the lower value of ω for germanium implies that the noise ΔE (eV) (and therefore also the energy threshold) for germanium should theoretically be ~80% that for silicon. Let us now look at the ENC contributions of the various types of noise discussed above to the equivalent circuit shown in Figure 1.39. We derive ENC by calculating the *charge* at the input using the equivalent circuit that gives the corresponding rms noise component at the output. For example, in the equivalent circuit, we use e_n as a voltage generator in parallel with the total capacity across the input, C_{TOT}, to generate an equivalent charge ($e_n C_{TOT}$). We then calculate the ENC charge using Equation 1.42 and the result will depend on the pulse shaper and its peaking time τ. If the noise density is white, when we integrate over the bandwidth of the shaper the result will be proportional to τ. This is the case for current noise (Equation 1.40), where the resulting ENC is proportional to τ. For $1/f$ noise (Equation 1.41), as we saw it follows a $1/f$ form between frequencies f_1 and f_2. Integrating over these limits

$$v_{nf}^2 = \int_{f_1}^{f_2} \frac{a_f}{f}\, df = a_f \log\left(\frac{f_2}{f_1}\right). \tag{1.43}$$

Thus, for $1/f$ noise the ENC is independent of τ, and it depends on the ratio of the upper to lower cut-off frequencies rather than the absolute bandwidth. Like the thermal noise, e_{nf} is an rms voltage developed across the total input capacitance and so the ENC depends on the product ($e_{nf} C_{TOT}$). For the thermal noise given by the Nyquist theorem, ENC can be shown to depend on $1/\tau$.

The dielectric current noise given by Equation 1.39 is in series with the input capacitance C_{TOT}; it will give rise to voltage noise

$$e_{nd}^2 = i_{nd}^2/(\omega' C_{TOT})^2.$$

Note that this has a $1/f$ form so that the integration (Equation 1.42) for ENC will be independent of τ. Also, as the charge is ($e_{nd} C_{TOT}$), it will also be independent of C_{TOT}.

In general, we can write the four components of ENC (current, voltage, $1/f$, and dielectric noise) in the form

$$(\text{ENC})^2 = F_i i_n^2 \tau + F_v (e_n C_{\text{TOT}})^2 1/\tau + F_f a_f C_{\text{TOT}}^2 + F_d kTdC,$$
$$(1.44)$$

where F_i, F_v, F_f, and F_d are 'shape factors' depending on the particular pulse processor characteristics. Note also that C_{TOT} is the sum of the transducer, FET, C_F, and all stray capacitances at G. It is noteworthy that unlike the current, thermal, and $1/f$ noise, which may result physically from the bulk and surfaces of the FET and transducer, the pure capacitance C_{TOT} in itself is not a noise source although the dielectric in it may be. The noise, which involves the total input capacitance in a charge-sensitive amplifier with feedback, comes from the *amplifier* not the input capacitance itself. If the e_n of the amplifier was zero, there would be no noise in the system from C_{TOT}. Remember that e_n is the amplifier noise *as if* it originated from the input. Physically, it does *not*. It might be expected that a voltage-sensitive amplifier might therefore be preferable but, in fact, the signal/noise ratio is the same for both configurations. Note that the FET noise dominates at short processing times and that the current noise dominates at long processing times. The dielectric noise adds to the $1/f$ as 'excess noise' as a flat component. The term excess noise comes from the fact that it is usually the residual noise after the voltage and current noise have been accounted for. The form of Equation 1.44 also specifies an optimum shaping time τ for given noise conditions as Figure 1.40 shows.

1.10.7 Microphony

No discussion of noise sources would be complete without mentioning vibration-induced noise or 'microphony'. This is induced by vibrations external to the circuit and as such is not a fundamental noise source in the same sense that the above are. However, in a noisy (original sense of the word) environment, it can dominate the detector resolution. This is particularly true for the low acoustic frequencies. As an illustration as to how sensitive the circuit of Figure 1.39 is to vibration, we can note the following simple example. If we calculate the stray capacitance due to a 0.5-mm-diameter FET gate pin within 1 mm

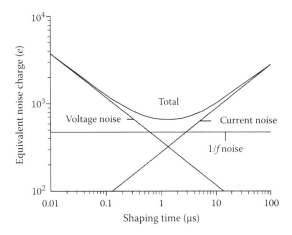

Figure 1.40 Equivalent noise charge vs. shaping time for a semiconductor detector connected to a shaping amplifier.

distance of a ground plane in a vacuum, the stray capacitance is ~0.045 pF. If the pin were to vibrate ±0.1 μm transversely, the capacitance would vary by ~5×10^{-6} pF. With the virtual ground G at =–1 V potential, this would inject a charge equivalent of ~30 electrons or >100 eV (Si) ENC of noise. With the addition of a dielectric or if the plane were carrying a 500-V bias, the effect would be amplified still further.

1.11 Leakage Current

Before looking at the relative contributions to the noise in semiconductor x-ray detectors, we need to discuss the origin of the so-called 'leakage current'. The transducer represented as a capacitor in Figure 1.27 in practice is always leaky. In the case of a reverse bias diode, it is an imperfect leaky valve for current. We have already implied (Section 1.6.1) that the minority carriers (thermally generated electrons in the p-type and holes in the n-type depleted regions) of a p–n junction will be a source of such leakage current because they see no barrier. The holes and electrons can also diffuse from the edges of the depletion region where they are swept away by the field giving rise to 'diffusion current'. Besides diffusion current, there are other sources (listed below), which we will discuss in turn

(a) Thermal generation current
(b) Diffusion current
(c) Surface leakage current
(d) Localised Injection current

1.11.1 Thermal Generation Current

We saw in Section 1.4 that n_i is dominated by the factor $\exp(-E_g/2kT)$ (Equation 1.11), and the intrinsic conductivity is proportional to n_i (Equation 1.9). This exponential term thus dominates the thermal generation leakage current and for Si it halves for every ~7°C of cooling. The ratio of the energy gap to absolute temperature in the exponential term has relevance to leakage current when we come to consider silicon in relation to germanium and the various WBGSs.

The depletion region acts as an intrinsic passive semiconductor and as such has thermally generated electron–hole pairs as described previously. Recombination in the depletion layer is suppressed by the electric field that sweeps the different carriers away from each other, giving rise to generation current before recombination can take place. As noted, this is a nonequilibrium situation and as generation dominates over recombination $np \ll n_i^2$. It is clear that the current so generated will depend on the intrinsic carrier concentration n_i and the *volume* of the depletion region. This component of leakage current can be minimised by choosing material of high intrinsic resistivity and long carrier lifetime and with suitable cooling can be made negligible even in the case of germanium. Note that if the semiconductor manufacturing involves high temperatures (such as the diffusion of boron to make a p-type material to form a p–n junction or the high temperature annealing after an ion implant), the lifetime may be compromised and high leakage currents as well as trapping may result. Obviously, WBGSs have reduced thermal leakage current and can often be used at or even above room temperature. These are discussed in Chapter 10.

1.11.2 Diffusion Current

Figure 1.41 depicts the depletion region W set up by the reverse bias applied to a symmetric p–n junction. The symbols n and p are used here to represent the *minority carrier* concentrations: electrons (in the p-type side) and holes (in the n-type side). The thermal equilibrium values deep in the field-free undepleted regions are represented by n_0 and p_0 where

$$pn_0 = n_{\mathrm{i}}^2 \tag{1.45}$$

$$np_0 = n_{\mathrm{i}}^2.$$

Equilibrium conditions do not apply in the depletion region where we assume n and p fall to zero as they are removed by the field.

Considering first only the diffusion current density (A/cm²) I_{e} of electrons injected into the depletion region, we note that this depends on the concentration gradient at the depletion region edge and this can be shown [10] as

$$I_{\mathrm{e}} = en_0 D_{\mathrm{n}}/L_{\mathrm{n}},$$

where D_{n} is the diffusion constant for electrons, L_{n} is the diffusion length $\left(= \sqrt{(D_n \tau_n)}\right)$ and τ_{n} is the recombination lifetime of electrons in the p-type material. L_{n} is a measure of the depth

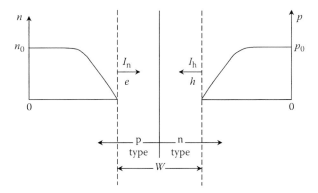

Figure 1.41 Currents in a symmetrical p–n junction.

from which electrons will diffuse into the depletion region, so that shallow junctions of small area give lowest diffusion currents. Also, as n_0 is inversely proportional to p (Equation 1.45 above) at a given temperature, this current is inversely proportional to the doping concentration N_A and $\sqrt{\tau_n}$, so both of these parameters should be high for low leakage. The total diffusion current density I_d will be given by

$$I_d = e[n_0 \times D_n/L_n + p_0 \times D_p/L_p]. \qquad (1.46)$$

If the p-type side has high doping and is shallow (as is usually the case if it is used as the x-ray 'window' in a radiation detector), the current is mainly due to holes injected from the n-type side. This will be the back contact in a fully depleted device. Low diffusion current is achieved with the use of long lifetime (long diffusion length) material. This is fortuitous as it is also a requirement for good charge collection in radiation detectors. Diffusion current also depends on the intrinsic carrier concentration as n_i^2 given by Equation 1.45. As n_i^2 is dominated by the exponential term $\exp(-E_g/kT)$, one consequence is that the diffusion current is smaller in WBGSs such as CZT and GaAs, and these can be operated at or near room temperature. The values of n_i at LNT and room temperature (RT) are given for silicon, germanium, and GaAs in Table 1.5.

In silicon, for ultralow noise, cooling below 300 K is usually required, and in germanium temperatures need to be of the order of 100 K or lower. Table 1.6 and Figure 1.14 show that n_i can be made very low by cooling. Hence, by suitable cooling diffusion, current should not technically be a serious problem.

TABLE 1.5
n_i at LNT and RT Temperature

Semiconductor	n_i at Liquid Nitrogen Temperature	n_i at Room Temperature
Silicon	2.3×10^{-20}	1.45×10^{10}
Germanium	3.17×10^{-7}	2.4×10^{13}
Gallium arsenide	7.74×10^{-32}	1.79×10^6

1.11.3 Surface Leakage Current

Currents can flow on, or rather just below, the surface of the walls of the depleted volume of a p–n junction. For selected material and appropriate cooling, surface currents can be by far the most significant of the four current sources discussed here. This is abundantly clear to detector manufacturers as surface etching and surface passivation have enormous influences on the leakage currents encountered. To see how this can happen, let us first look at the possible charge states that can occur at the surface of a semiconductor. We should prelude this with the remark that the semiconductor surface is an extensive and complex but important topic and will be discussed in more detail in Section 5.1.9. Here, we will give a brief outline.

The most interesting situations often occur at surfaces and interfaces. This is as true of the surface of semiconductors as it is with the surface of the earth or the event horizon of a black hole. The well-behaved regular crystal lattice terminates in a rather unstable broken 'free-for-all'. We will see in Chapter 11 that many years of effort went into taming the vagaries of such surfaces. Let us consider the surface of a depleted n-type silicon or germanium volume. The first thing to note is that, by cleaving a crystal in a perfect vacuum, the covalent bands are broken, leaving a deficiency of electrons at the surface. The electrons that are left form what are often referred to as 'dangling bonds'. In practice, there is no 'perfect vacuum', and these dangling bonds will endeavour to be satisfied by combining with any atoms they encounter, thereby forming 'adatoms' bound to the surface. The most likely candidates in the ambient will be water, which gives rise to a fixed positive charge at the surface, or oxygen, which gives rise to a fixed negative charge at the surface. The former results in an n-type surface and the latter in a p-type surface. This is shown schematically in Figure 1.42.

We see that this results in three basic possibilities.

(a) In the case of fixed positive charge, electrons are drawn toward the surface from the bulk forming an 'accumulation layer' or 'n-type surface channel'. We refer to this as an 'n-type surface'.

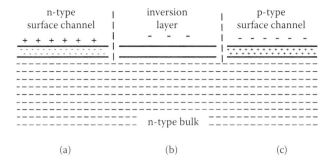

Figure 1.42 Surface channels and inversion layers. (a) n-type surface channel, (b) near intrinsic or inverted surface, and (c) p-type surface channel.

(b) In the case of a moderate fixed negative charge, a layer below the surface has been inverted from n-type to 'intrinsic' (or slightly p-type) by an influx of holes from the bulk. These holes neutralise the majority electrons that were originally in excess. We refer to the surface being 'inverted'.

(c) In the case of further fixed negative charge being added, holes will gather in the subsurface channel. This is a 'p-type channel', and we refer to the surface as being a 'p-type surface'.

Just as in the case of surface barrier contacts (see Section 1.5.2), the fixed surface charges give rise to band bending. Where the surface is near neutral as in case (b) above, there is no bending and we say it is 'flat band'. These possibilities have relevance to the surface conductivity, for when a reverse bias is applied to the junction, it gives rise to a tangential field at the bare walls of the device that will cause current flow in the surface channels that have been formed. It is clear that for a low surface current a light surface inversion or flat-band condition might be desirable, and it is the achievement and maintenance of such a surface that is termed 'passivation'. Silicon is particularly blessed with a natural 'native' oxide that is always present. In a dry atmosphere, an etched surface will be driven p-type. However, thermally grown oxides (typically ~1000°C in steam) and glasses will drive them n-type. In planar detectors manufactured by photolithography techniques, a thermally grown high-quality oxide is very effective and is the usual method of passivation. However, where the high temperatures for this process are not possible, the minimum currents are often achieved by the 'pinching off' of any surface channels that exist, for example, by increasing the component of the field *normal* to the surface in a localised region. This can be accomplished by giving the surface concave curvature or 'contouring' (see Section 1.16.2). In planar detectors, 'guard rings' or 'guard anodes' can also be used to suppress surface leakage current.

1.11.4 Localised Injection Current

So far, we have treated the p–n junction as being an ideal barrier. We have seen that, in practice, one side of the junction is very shallow and is often used as the 'window' for x-rays to enter the depletion region. Such very shallow junctions—whether diffused, ion implanted, or surface barrier—may in practice have imperfections that cause the barrier to inject carriers into the depletion region. This will give rise to significant shot noise. The sources of injection tend to be localised and in the extreme, such as a physical scratch or a pinhole, the barrier may break down completely and the rectifying contact becomes an 'ohmic' contact. The problem of injection current is more prevalent in the thin surface barrier contact, and we will discuss later the technical problems in reducing the barrier thickness while avoiding injection current.

1.11.5 Summary

The bulk leakage components of diffusion and generation current can be controlled technically. Because of the n_i^2 dependency, the fractional increase in diffusion current with temperature coefficient varies as E_g/KT^2, whereas for the generation current the fractional increase varies as $E_g/2KT^2$ [4]. Thus, the diffusion current usually dominates at high temperatures. However, even at room temperature diffusion currents in silicon can be controlled to 1 nA/cm^2 or lower. Useful devices can be operated at or near room temperature either by reducing the element area (as in CCDs or PXDs) or by reducing the input capacitance (as in the

SDD) and operating away from current noise at short processing times as indicated in Figure 1.40. As a rule of thumb, the leakage current per unit area in silicon devices reduces by a factor of 2 for every 7°C decrease in temperature from room temperature, so that moderate cooling, for example, by an electric Peltier cooler (see Section 11.25.2), can often reduce the noise and significantly improve performance. In Si(Li) detectors, the much more difficult technology lies in controlling the side-wall surface currents and surface barrier injection currents. Even at 200 K (still in the range of Peltier coolers), surface leakage current dominates.

1.12 Typical Noise Values

As Figure 1.40 shows, for low leakage current devices (e.g., those cooled by LN), the noise is dominated by the voltage ($1/\tau$ or 'series') noise and the excess ($1/f$) noise which, being independent of τ, really sets the ultimate minimum at long τ (>20 µs). The voltage noise is dominated by the FET capacitance C_{FET} and the diode capacitance C_{D}. Table 1.6 lists the parameters of various FETs that have been used as the first stage of the preamplifier.

The last entry is 'The Cube' [11], which was designed by a group from Italy. This is a recent innovative development and uses a p-type MOSFET for the 'front–end' first stage of the preamplifier. The developers have also integrated this FET with the rest of the charge-sensitive preamplifier including a restore mechanism and feedback capacitor. This whole assembly can now be mounted close to or on a bond pad on the detector.

Yet, again, bold development has dispelled one of the myths that have held back the use of MOSFETs in low-noise charge-sensitive circuits because of the known excess of $1/f$ noise. A precedent for the use of such FETs in CCDs has been available for many years. Here, the MOSFET is formed directly on the CCD and is connected to the low capacity pixel at the end of the row feeding the output of this structure.

In The Cube, the developers have cleverly made use of the low capacity matching requirements needed to optimise the signal/noise ratio when interfacing to an FET. This low capacity coupling makes amplification possible at short time constants and,

TABLE 1.6
Transconductance (S) and Input Capacity (pf) for Various
FETs measured at their optimum temperature

FET Type	Transconductance g_m (Ω^{-1})	Capacity C_{FET} (pf)	g_m/C_{FET} (Ω^{-1})/(pf)
2N4416	6.0	3.5	1.7
Harwell 8553[a]	5.0	3.5	1.4
Harwell 8552[b]	6.0	2.5	2.4
De Witt[c]	4.0	4.0	1.0
Pentafet[d]	6.0	1.0	6.0
Moxtek[e]	4.5	0.4	11.3
Interfet NJ 14AL[f]	5.5	2.3	2.4
SDD Integrated[g]	0.3	0.1	3.0
Semifab SR51[h]	10	0.4	25
The Cube[i]	2.3	0.4	5.75

[a] Similar in performance to the 2N4416 but available in chip form and with a high yield.
[b] New, circular concentric structure, with lower input capacity and an independent second gate, the substrate, which could be adjusted independently from the main or top gate.
[c] First recorded on-chip electronic restore (see Chapter 11, ref. [73]).
[d] Oxford Instruments trademark for a device with on-chip electronic restore and feedback capacitor.
[e] U.S. supplier of FETs with on-chip electronic restore.
[f] U.S. supplier of specialist FETs.
[g] See the products from PN Sensors in Germany.
[h] UK supplier of low capacity FETs with on-chip electronic restore.
[i] Italian supplier of MOSFETS integrated with the preamplifier and supplied as a component for mounting close to the sensor chip.

as often stated in papers on SDDs, moves the optimum amplifier time constant to lower values with much reduced sensitivity to leakage current. It also considerably reduces the sensitivity to $1/f$ noise. As yet, details of how the restore mechanism works have not been published, but the results in recent papers indicate a clean restore in less than 4 us.

As mentioned previously, g_m/C_{FET} is a constant 'figure of merit' for a particular FET geometry. The choice of which FET is governed by C_D as the noise is a minimum when it is matched to C_{FET}. This is seen by the following.

Let the 'figure of merit' be M

$$M = g_{\mathrm{m}}/C_{\mathrm{FET}}.$$

The second term in Equation 1.44 is the voltage noise due to the FET, and using Equation 1.38 and ignoring C_{F} and any stray capacitances

$$(\mathrm{ENC})_{\mathrm{F}}^2 = F_{\mathrm{v}}(4K\tau/g_{\mathrm{m}}C_{\mathrm{TOT}}) = F_{\mathrm{v}}(K4K\tau/M)^2(C_{\mathrm{FET}} + C_{\mathrm{D}})^2/C_{\mathrm{FET}}. \tag{1.47}$$

It can easily be shown that the term involving the capacitances is a minimum when

$$C_{\mathrm{FET}} = C_{\mathrm{D}}.$$

Equation 1.47 can be written in simplified versions depending on the characteristics of the pulse shaper. Spieler [4] offers the following for a quick estimate (differentiatior followed by an integrator) using a resistive-capacitance shaper

$$(\mathrm{ENC})_{\mathrm{F}} \sim 6e_{\mathrm{n}}C_{\mathrm{TOT}}/\sqrt{\tau} \quad \text{e-rms}, \tag{1.48}$$

where e_{n} refers to the FET noise in nV/$\sqrt{\mathrm{Hz}}$. The appropriate units for C_{TOT} and τ are pF and μs, respectively.

The first term of Equation 1.43 gives the current noise

$$(\mathrm{ENC})_I \sim 3.46 \sqrt{(I\tau)} \quad \text{e-rms}. \tag{1.49}$$

The appropriate units for I and τ are pA and μs, respectively.

The $\sqrt{(I\tau)}$ dependence is easily explained as if we integrate the current I over a time τ, we have a number of electrons N_{e} proportional to $I\tau$, and the variance of this number is just $\sqrt{N_{\mathrm{e}}}$.

Let us now look at some examples.

Example 1.1: Typical 10 mm² Si(Li) Detector

$$e_{\mathrm{n}} = 1.5 \text{ nV}/\sqrt{\mathrm{Hz}}$$

In units of pF,

FET = 1, det = 0.6, FBC = 0.05, stray = 0.3 CTOT ~ 2

If τ = 10 μs, using Equation 1.48,

$$(\text{ENC})_F = 6 \times 1.5 \times 2/\sqrt{10} \text{ e-rms} = 5.7 \text{ e-rms}$$

In Si, 1 e-rms = 3.75 × 2.355 eV FWHM

Voltage noise ~50 eV at τ = 10 μs

For the current noise, we can use Equation 1.49. The current in a typical Si(Li) detector is commonly in the range of 1 fA (10^{-15} A) or less, and the FET gate current is usually less than this. However, if we take a worst possible case, where I = 100 fA and τ = 100 μs, the current noise is ~94 eV. This would present a problem for high-resolution work.

Example 1.2: A 5-mm² Photodiode at Room Temperature

It is possible to achieve a leakage current of ~1 nA/cm².

I = 1 nA/cm² (giving 0.05 nA for the diode) and τ = 10 μs

We can use Equation 1.49

$$(\text{ENC})_I \sim 3.46 \times \sqrt{50}/\sqrt{10} = 77 \text{ e-rms}$$

Current noise ~660 eV

Such a diode could only be useful if cooled below room temperature.

Example 1.3: A 10-mm² SDD with Discrete FET as in Example 1.1

Peltier cooled to ~230 K.

The anode capacitance in an SDD is extremely low, and we can neglect its capacitance compared to the FET. We will assume $C_{\mathrm{TOT}} \sim 1$ pF (for an integrated FET, this would be much smaller). Comparing with the 2 pF in Example 1.1, we can say that the voltage noise at $\tau = 10$ μs is ~ 50 eV/$\sqrt{2} \sim 35.4$ eV.

Assume a current of ~ 0.1 pA when Peltier cooled ($T \sim 230$ K).

Using Equation 1.49, at $\tau = 100$ μs

$$(\mathrm{ENC})_I \sim 3.46 \times \sqrt{0.1 \times 10} = 11 \text{ e-rms} = 94 \text{ eV.}$$

In modern detectors, the dielectric noise $\sqrt{(F_{\mathrm{d}} \cdot kT \cdot dC)}$ (Equation 1.44) has been reduced to almost negligible levels by the careful use of materials. The major breakthrough was taken with the removal of the FET chip from its lossy glass encapsulation in the late 1960s. By the use of high-grade sintered boron nitride, minimum use of epoxies, and reduction in stray capacitance, this component has been reduced to the level of a few eV FWHM. We are therefore left with the other component in Equation 1.44 that is independent of τ, referred to as $1/f$ noise or, because its origins are so ubiquitous, the catch-all expression 'excess' noise. It is of no recompense that it has been suggested that $1/f$ noise is a universal characteristic of existence and that music and language [12] have their roots in it! As mentioned earlier, at high tangential fields surface currents give rise to a $1/f$ component of noise, which can be important in cooled Si(Li) and HPGe detectors. This is usually suppressed by contouring, and we are left with $1/f$ noise from the FET. It has been estimated that for a triangular shaper, this is given by

$$(\mathrm{ENC}_{\mathrm{f}})^2 \sim 2e_{\mathrm{n}}^2 f_{\mathrm{c}},$$

where f_{c} is the 'corner frequency' (in kHz). In silicon, this leads to

$$\mathrm{FWHM}_{\mathrm{f}} \sim 2.4 e_{\mathrm{n}} C_{\mathrm{TOT}} \sqrt{f_{\mathrm{c}}}.$$

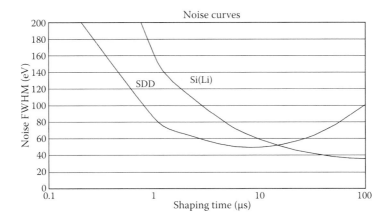

Figure 1.43 Noise vs. shaping time for an SDD and a Si(Li).

Using f_c = 30 kHz, e_n = 1.5 nV/√Hz and C_{TOT} = 2 pF for a 10-mm^2 Si(Li) detector

$$\text{FWHM}_f \sim 40 \text{ eV.}$$

The $1/f$ component of noise in silicon does seem to be stubbornly pinned at about 30–40 eV.

In Figure 1.43, we have combined noise components to arrive at noise curves for a Si(Li) detector and an SDD of roughly 10 mm^2.

The capacitances have been chosen as 2 and 1 pF and the leakage currents as 1 fA and 0.1 pA for the Si(Li) and SDD respectively, with 31 eV $1/f$. We see that the optimum shaping time τ is reduced for the SDD, and 50 eV or more of noise can be achieved at shorter time constants (and therefore at higher input rates) in the SDD. This plot is for illustrative purposes only. A more detailed comparison of SDD and Si(Li) performance is given in Chapter 9.

1.13 FET

The FET is intimately connected to the x-ray transducer. For example, any noise analysis must consider them as a combined unit. We have already seen the need to match their capacitances (Equation 1.47). Physically, they also need to be very closely linked both to reduce stray capacitance and microphony and avoid lossy dielectrics surrounding them. They also usually share a common cooling circuit. There are principally two types of FETs in common use, the MOSFET and the JFET. We will now look briefly at the physics involved.

1.13.1 MOSFET

Looking at Figure 1.42, we can see how the surface channel conductivity can be influenced by the presence of surface charge. It was realised in the 1920s (see Section 11.5.1 for details of FET development) that a biased metal contact on such a surface could be used to control the surface leakage current and therefore act as a sensitive charge amplifier. However, the control of the surface states to make this achievable came much later with the application of high-quality oxide layers to give the metal–oxide–semiconductor (MOS) structure shown in Figure 1.44.

The p-type wafer is coated with a thin oxide layer (other insulators can be used) and two n⁺ regions, the source and drain, diffused through to make p–n junctions. When these are reverse-biased appropriately, depletion occurs into the bulk, isolating them except for a surface current set up between the two. The conductivity of this shallow region, known as the 'channel', can be controlled by charge on the metallised region on top of the oxide, the gate, as explained in Figure 1.42. The

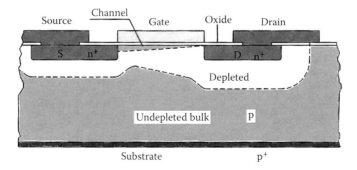

Figure 1.44 Metal oxide semiconductor field effect transistor.

channel is tapered to a 'pinch-off region' because of the biasing of the source relative to the drain. The gate (G in Figure 1.1) accepts the charge from the transducer and the modulated drain–source current amplifies it. The device depicted in Figure 1.44 is an example of an n-channel MOSFET.

1.13.2 JFET

The p–n junction FET or JFET was developed before the MOSFET because of the problems caused by surface charge states. It avoided these by burying the conducting channel as shown in Figure 1.45.

We see that the metal–oxide gate has been replaced by a p⁺ diffusion (or implant) pushing the tapered conducting channel below the surface. The substrate is tied to the gate so that the current through the channel is modulated by charge on the gate through the depletion regions from the top and the substrate below. This is an example of a three-terminal n-channel JFET.

1.13.3 FET: General Discussion

Note that in the n-channel FETs described here, the current is attributable to the majority carrier. This is true also of a p-channel FET, and they are termed *unipolar* devices. If the signal on the gate is a negative charge (negative biased transducer as in Figure 1.27), then the depletion layers are expanded and the electron current in the channel is reduced, causing a positive step at the output. Eventually, such signals

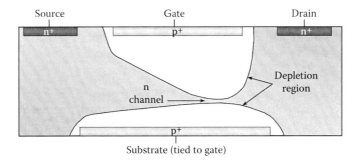

Figure 1.45 Structure of junction field effect transistor.

will cause the channel to 'pinch off' and the output will saturate or go 'high'. The FET gate is very high impedance, and if there is no leakage resistor connected to it (which would add noise), there must be a mechanism to restore the charge on the gate to zero in order to accept further signals. This is effectively achieved by shorting out the gate-drain by supplying carriers to the bulk in a pulse mode. One option is to use a pulse of light or infrared known as 'pulsed optical restore' and another is to inject carriers using transistor action known as 'electronic restore'. The implementation of these options is discussed further in Section 1.16.4 and their development in Sections 11.18 and 11.24, where references are supplied. The leakage current of the transducer is usually much greater than that of the FET, and the quiescent current is therefore in the same charge sense as the signal.

The output is thus a 'ramp' with signal steps superimposed (see Section 1.16.5). If the leakage current does swing in the opposite direction, the FET can be reactivated by putting x-rays (or light) into the *transducer*. We have considered here negatively biased transducers, and it should be clear that if the bias is positive p-channel FETs would be needed in order to effect a restore. In Figure 1.45 it is assumed that the silicon substrate was tied to the gate. They can, however, be isolated and separate contacts are made in a 'tetrode'. The substrate bias can therefore be varied independently, whereas the 'top gate' is connected to the transducer. This not only reduces the capacitance, but the position of the channel in the bulk can be adjusted to minimise noise (see below).

1.13.4 FET Noise

We have already discussed the voltage noise of the FET given by Equation 1.38. In practice, there is also an 'excess' $1/f$ noise component due to what is called Shockley–Reed–Hall generation–recombination (SRH noise). Being $1/f$, it can be dominant at low frequencies as Figure 1.40 shows. It is caused by impurities and defects in the silicon. These can act as charge generation–recombination centres fluctuating between a charged and a neutral state. Their effect is shown [13] in Figure 1.46.

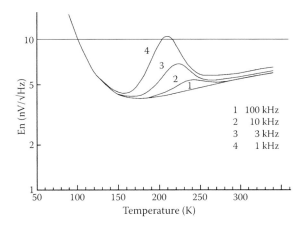

Figure 1.46 Noise measured at different frequencies over a range of temperatures.

At short τ ($f \sim$ 100 kHz), the noise is essentially thermal noise. The increase below ~150 K is due to carrier freeze-out (see Section 1.10.1). The noise peak shown for longer τ is due to SRH noise. When the density of defects is low, individual sites close to the depletion region can be recognized in the noise as 'telegraph signals' and can be avoided in the tetrode by moving the depletion boundary by adjusting the substrate bias. In MOSFETs, the source of $1/f$ noise can be due to oxide traps and generally JFETs have lower $1/f$ noise than MOSFETs, but recent improvements in the quality of oxides and epitaxial silicon have lowered the noise in both types of FET.

1.13.5 FET Structures

It can be shown [4] that the transconductance g_m is given by g_m = constant × (W/L), where W is the width of the channel (referring to Figure 1.45, W is the depth of the channel into the plane of paper) and L is its length. Thus, the geometry is chosen to optimise g_m with large W and small L. Two examples that do this are the 'interdigital electrodes' and the 'concentric ring electrodes', examples of which are shown in Figures 1.47 and 1.48. Microscope views are given in Figure 11.63 (Section 11.19.3).

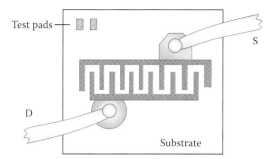

Figure 1.47 Interdigital or interleaved structure of an early field effect transistor (FET) design.

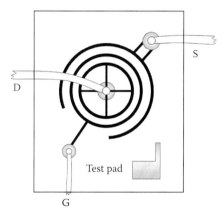

Figure 1.48 FET structure with a concentric geometry.

The reason the top gate and substrate are connected in this interdigital architecture is because the gate diffusion runs linearly (vertically between the digits in Figure 1.47) into the substrate, and the gate is accessed through a metal pin onto which the chip is mounted.

In the concentric geometry FET, the top gate is separate from the substrate usually mounted onto a metal pin. Because there are now four connections, it is termed a tetrode. Later advances (see Section 11.24) integrated an 'emitter' for electronic restore and an on-chip FBC.

MOSFETs have been used extensively where the transducer itself is manufactured lithographically such as CCDs where they can be integrated onto the same epitaxial silicon wafer during the processing. JFETs have been used exclusively with Si(Li) and HPGe detectors as discrete chips. This is because of the technical difficulties of integration and because of the different temperature requirements. SDDs have until recently also used discrete JFETs, but now integration on the same wafer is fairly common [14]. Very recently, discrete MOSFETs have become available [11]. It should be remarked that because of the low total capacitance of FETs integrated on SDDs and CCDs, the actual quality of the FET is less important than in the case of discrete FETs.

1.14 Detector Response Function $F(E)$

The semiconductor response to x-rays or 'response function', $F(E)$, is the frequency distribution in equal width channels dE of charge signals induced by single discrete energy x-rays of energy E. This is discussed later in Chapter 2 in some detail, and here we will just introduce the concept. The x-ray carries no charge and the semiconductor is electrically neutral, so the response function represents all the electrons and holes generated in the solid but not any neutral entities that also carry energy such as phonons or escaping fluorescence or scattered x-rays. In this respect, the phonons and escaping photons are 'lost'. However, electrons and holes can also be lost (e.g., to the front contact or walls of the semiconductor), and $F(E)$ must account for these. It is likewise possible that the x-ray can *add* charge by means other than the ionisation of the semiconductor by the photoelectric effect. For example, the contact, collimator, and window material can fluoresce or emit photo- and Auger electrons, and these can be 'detected' in the active volume of the semiconductor. Thus, $F(E)$ depends on all the possible physical processes (not just ionising processes) that can take place both in the bulk of the semiconductor and at its boundaries. Ideally, $F(E)$ would be a delta function $\delta(E)$ at a charge $Q(E)$ corresponding linearly to E, the x-ray energy. In practice, in its simplest approximation, it is spread statistically to higher and lower energies by the broadening of the 'photo-peak' as discussed earlier. Associated with the peak is what may be termed a 'low energy tail' and a more-or-less level 'background' (or 'shelf') running from zero to E as shown in Figure 1.49.

No actual x-ray spectrum is exactly like this, but the above terms are nonetheless used to describe the main features. The FWHM of the photopeak is obtained by fitting a Gaussian distribution usually to just the high-energy side of the photopeak in the x-ray energy spectrum. This is termed the 'resolution' (usually referring to the 5.9 keV Mn Kα line; see below).

The tail represents small fractions of the total charge lost at the end of the cascade (usually due to bulk trapping during the drift of carriers to the electrodes or the diffusion of charge from the active volume before the field drifts it to the electrodes). The background represents the loss of relatively large amounts of energy due to escaping high-energy electrons at the start of the cascade, but can also be attributable to the trapping of charge at the sidewalls of the crystal preventing it from reaching the electrodes. The latter is expected to have a radial dependence and is often reduced by the collimation of the incoming x-rays. At high energies, Compton scattering will also contribute to the background.

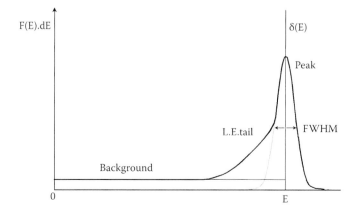

Figure 1.49 Detector response function.

1.14.1 Escape Peaks

The primary x-rays will give rise to fluorescence x-rays (XRF) in the detector itself. Although the fluorescence of materials used in the construction (window material, electrode material, collimators, etc.) can be made negligible either by making them very thin or of very low atomic number Z e.g., polytetrafluoroethylene (PTFE), boron nitride, graphite-covered collimators, the fluorescence of the semiconductor material itself cannot be avoided. Deep in the crystal, fluorescence x-rays from the semiconductor are highly likely to be reabsorbed by the photoelectric effect resulting in no reduction in the ultimate charge count, and the event will go into the photopeak as if fluorescence had not occurred. Even in pixel-array detectors (PXDs) and CCDs, the fluorescence x-ray may be caught by an adjacent pixel and can be summed back into the peak. But near the front or rear surface, the escape of some of these x-rays is unavoidable. However, collimation can usually reduce any loss to the sidewalls, and the majority escape to the front surface. Thus, we can expect 'escape peaks' displaced in energy by the characteristic x-ray energy E_x of the material. For example, each photopeak at energy E in a Si(Li) detector will have a smaller peak displaced to the energy ($E - E_x$), where E_x is 1.74 keV (Si Kα). Other such 'satellite' peaks in silicon are normally negligible. In germanium and other higher Z semiconductors, the situation is more problematical because there are more significant escape peaks. For example, in Ge, E_x = 9.87 keV (GeKα), 8.1 keV (GeKβ), and 1.2 keV (GeLα). In addition, they are more intense as the fluorescence yields are higher, and the higher E_x results in a greater range of the x-ray. Fluorescence yields are defined as the probability that a photoabsorption and subsequent atomic decay will result in a fluorescence x-ray (rather than an Auger electron). Table 1.7 gives the K-shell fluorescence yields for a number of relevant elements.

Escape peaks in silicon detectors can be stripped from spectra by software, but there is an inevitable loss in sensitivity in the spectrum in their vicinity. The problem increases in the higher Z elements such as used in compound semiconductor detectors. The argon in air is often fluoresced by low-energy x-ray sources.

TABLE 1.7
Fluorescence Yields for Selected Elements

Z	Element	Percentage Yield
4	Beryllium	0.04
6	Carbon	0.2
13	Aluminium	3.6
14	Silicon	4.7
18	Argon	11.5
31	Gallium	51
32	Germanium	54
33	Arsenic	56.7
48	Cadmium	84
52	Tellurium	87.5
53	Iodine	88.2
80	Mercury	96.6

1.14.2 Low Energy Efficiency

The response function at very low energies (<1 keV) is determined by two parameters, one physical and the other electronic. Firstly, the efficiency with which the 'soft' x-rays enter the active volume of the detector, and secondly the electronic efficiency of the amplifier in discriminating the signal induced by the x-rays from noise. The first parameter is governed largely by the type of x-ray window separating the detector from the source. In cooled detectors such as Si(Li) s, HPGe, and SDDs, this separates the cryostat vacuum from the source and for all practical purposes (transportation, testing, etc.) must be able to support atmospheric pressure. Such windows are termed 'atmosphere thin windows' (ATWs). In the past, completely 'windowless' detectors were widely used, mechanically removing the ATW (e.g., by a gate valve) once vacuum was established in either side (such as in SEMs). These have lost favour because of the development of good ATWs with the advantage that no cross contamination is possible between the two vacuum regions. There is another type of window used that does not support atmospheric pressure but will isolate the vacuum regions once they have been established. They can be thinner and allow for better x-ray transmittance and are referred to as 'ultrathin windows' (UTWs). These options and their history will be discussed in detail in Section 11.21. The attenuation or fluorescence in the 'window electrodes' or 'dead layers' of the detector itself is not normally an issue.

It is quite clear that it is futile to attempt the analysis of ~100 eV x-rays by reducing the electronic threshold of the amplifier if the Gaussian noise tail extends into this energy region: the noise will simply swamp any signals. The obvious answer is to reduce the noise; this, however, leads to another problem. We have seen from the noise curves that we can only reduce the noise by increasing the shaping time τ, and as a result the throughput rate for analysis is reduced as a significant number of x-rays arriving within 2τ of each other must be rejected as 'pileup', where one event contaminates the previous one. The way around this problem is to split the amplifier into two (or more) inputs. Let us say, one channel (the 'processing channel') with long τ_L for spectral analysis and an added one with short τ_S (the 'fast channel'), for recognising the signal above noise and guarding against pileup. This is shown schematically in Figure 1.50.

The result of the pileup rejection is that the throughput curves shown in Figure 1.51 peak, because increasing the input rate further results in *decreasing* the output rate.

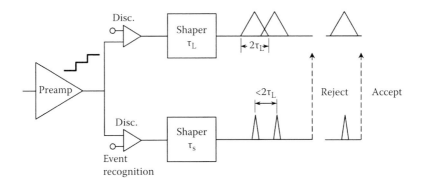

Figure 1.50 Pileup rejection.

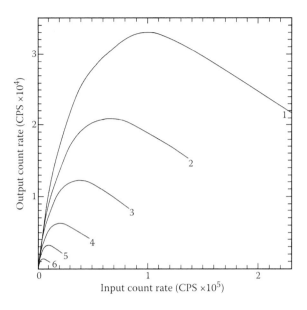

Figure 1.51 Throughput response.

This shows that the maximum output rate decreases as τ increases from 2.5 µs (curve 1) through to 80 µs (curve 6).

The arrival of x-rays randomly in time follows Poisson statistics and a more thorough treatment of pileup [15] shows that the observed count rate m is related to the true count rate n by $m = n \exp(-n\tau)$.

From a practical point of view, one must realise that a low observed count rate may not imply a low true input rate but can also result from a very high true input rate that partially paralyzes the pileup rejector. The other consequence of pileup rejection is that the noise at the shortest peaking time ($\tau = \tau_S$) puts a low energy threshold for event recognition in the processing channel, as depicted in Figure 1.52. This low energy threshold not only reduces the efficiency for the detection of low-energy x-rays but can also distort the photopeaks as shown.

Thus, the fast channel controls what the processing channel feeds to the PHA, and as the low energies are not swamped with noise, photopeaks of low Z materials (such as Be, B, C) can be recognized—albeit often with reduced sensitivity and some distortion.

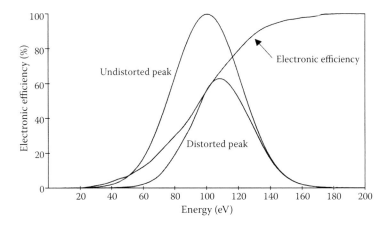

Figure 1.52 Peak distortion at low energies.

1.14.3 Standards

Over the years, the radioisotope Fe-55 (an electron capture source that gives Mn x-rays) has been adopted as the standard for testing x-ray detectors. The Mn Kα line at 5.9 keV is relatively pure (although it is actually two lines separated by 11 eV), and the background is relatively free from other x-rays (although there are weak 'radiative Auger' bands between the Kβ escape peak and the Kα peak). Furthermore, the source can be used with an air path between it and the detector window. It was adopted by the IEEE as a standard for resolution (FWHM on Mn Kα) and peak-to-background (P/B) (approved in 1981) [16]. Figure 1.53 shows a log plot schematic of the Fe-55 spectrum obtained with a Si(Li) detector. The P/B was defined as the Mn Kα peak channel count divided by the average background count around 1 keV as shown. As we shall see later, this is far from being an ideal measure, and additional parameters have been introduced.

1.14.4 Summary

To summarise, the detector response function will obviously be different for different types of semiconductor detectors. What is perhaps not so obvious is that in detail, it is different

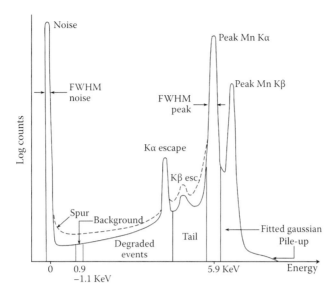

Figure 1.53 Iron-55 spectrum shown in log format to reveal tail and background features.

for the *same* type of detectors. Fluoresce of materials surrounding the transducer, differences in geometry, in surface processing, in passivation, etc., make no two detectors alike. This has consequences where the deconvolution of spectra by software depends on the knowledge of the response function profile.

A graphic illustration of $F(E)$ can be made by plotting the intensity of processed events in an input/output energy space. Ideally, they would lie on a diagonal line of unity gradient. Because of the low trapping probability and low background levels in the case of Si(Li) and HPGe detectors, such plots do look rather 'clean'. As an illustration where this is not the case, Figure 1.54 shows a simulation of the response function for a 2-mm-thick CdZnTe detector [17], where features such as escape peaks, hole trapping, and Compton scattering are enhanced over those in Si(Li) and HPGe detectors.

In this plot, the yellow region marks high output counts, whereas the red region indicate the low output counts. The tails for low energies and the increased hole trapping for more

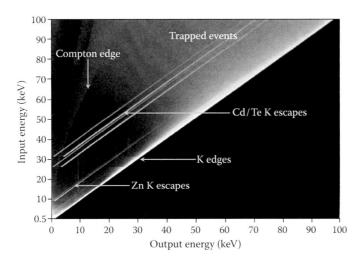

Figure 1.54 (**See colour insert.**) Simulation of the response function for a 2-mm CdZnTe detector.

penetrating x-rays, show up clearly as do the prominent escape peaks.

1.15 Categorisation of Semiconductor Detectors

The SSD and various other types of detectors reduce the noise by reducing the input capacitance C_{TOT}. They achieve this by making the charge collection electrode, or 'node', more or less independent of the field by which the charge is transported. Ways in which this can be done were illustrated in Figure 1.29. This leads to a natural categorisation of detectors depending on the way the signal is generated. We have seen that in the simple planar detector (and its guard ring version) the charge carriers move essentially in the field between the two plates of a capacitor, and the signal is developed during this transit time on one of the plates. When all the holes and electrons have reached their respective electrodes, the *integral of induced* charge from both holes and electrons exactly equals the electron charge initially liberated by the x-ray. The exact form of the signal in time reflects the difference in the mobility of the holes and electrons. Thus, in the transit time it takes for this to happen, which is typically <50 ns, a signal is integrated onto one of the electrodes (the other electrode is usually held at constant bias potential), and it is only then that the detector is clear to measure the next x-ray signal without any 'pileup'. We will refer to this type of detector as *charge induction detector.*

The alternative is to transport and deposit *one type* of carrier generated (usually the electrons) by means of 'transport electrodes' in a packet of charge to a specialised low-capacitance measuring node (usually the anode) effectively screened from the bias electrodes. This can be termed single carrier detection. During this transport process, any charge induced on the other electrodes is not used as a signal and, because of the geometry, virtually no signal is generated on the measuring electrode: it is only at the very last part of its travel when the charge is physically *deposited* onto the measuring electrode that a signal is generated. In other word, the charge packets can be regarded as being *stored* in the transducer while in transit until they reach the measuring node and more than one charge packet can be stored at any one time. We will refer to this type of detector as a *charge delivery detector.* It is sometimes also called a 'memory' detector and the first type as a 'classical' detector [18].

An extreme example of a charge delivery detector is the CCD. This is shown schematically in Figure 1.55.

The CCD is essentially an array of semiconductor pixels in which any charge generated (e.g., by x-rays) is stored until a command is received to transfer it to the next pixel. The charge is held temporarily by a holding field applied to each pixel in a row and by 'channel stops' between the rows. The charge transfer is achieved by switching the holding field to the next pixel a process known as 'charge transfer' or 'clocking'. The efficiency of the charge transfer between pixels is called the charge transfer efficiency (CTE) and is typically 0.999999 (or $1-10^{-6}$) in good-quality silicon. In this way, the charge is transferred along the rows to a serial line readout register where it is transferred pixel by pixel to an FET readout node.

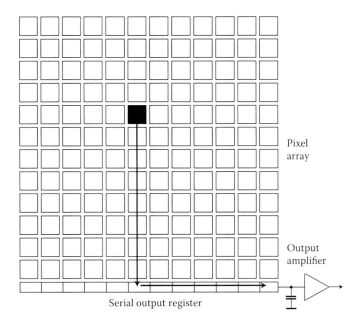

*Pixel
array*

*Output
amplifier*

Serial output register

Figure 1.55 Pixel array in a typical charge-coupled device.

It is clear that in the CCD more than one charge packet can
be stored until it is read out, and the charge measurement
node is completely independent of the mode of transport of the
charge delivered to it.

Note that the rate at which x-rays can be analyzed (throughput rate) depends on the shaping time τ of the analyzer in
both types of detector. In the case of CCDs, the readout noise
increases with the clocking rate. The fact that the charge
delivery type of detectors can store more than one signal at
a time is not an intrinsic advantage in practice. The advantage they have is that the voltage noise is lower, and a shorter
value of τ can be chosen for the *same resolution*. It is this factor that leads to a greater throughput rate. Examples of the
charge induction type of detector include the planar, coaxial,
hemispherical semiconductor detectors, Si(Li) detectors, p–i–n
diodes, photodiode arrays, and conventional avalanche diodes.
Examples of the charge delivery type are SDDs and CCDs
(conventional and p–n junction). Coplanar grid and small pixel
(dimensions of pixel are much less than the detector thickness) also act effectively as charge delivery detectors, because

the signal is developed only when a single carrier is very close to the electrode. This is termed the 'small pixel effect' (see Section 7.3.3). Note that no signal strength is lost in the charge delivery detectors as a result of them being only *single carrier*. Physically, only electrons (or holes) are presented to the node electrode, whereas in the charge integration detector both electron and hole signals must be integrated to get just *one* unit of charge so the resulting signal is the same. The difference is that we *choose to ignore* the contribution from one carrier in the charge delivery detector. For example, if the hole is much more likely to be trapped, we have by default a single carrier detector that is position (depth) sensitive. This may not be a problem if the x-ray range is much less than the detector thickness, but if it is not, it is better to deliver electrons only to the node electrode.

Yet another alternative is to make the charge carriers pass through a very high field region in which they gain sufficient energy to cause further ionisation in the semiconductor. This *avalanche* process would give the detector internal *charge gain*. The noise does not undergo this gain process, so in principle the threshold levels of detection are much improved over a conventional device. These *avalanche photodiodes* (APDs), can be referred to as *charge multiplication* detectors. In silicon, the field in the entrance window is adjusted to be high enough to produce mainly the multiplication of *electrons*. The positive bias is such that electrons produced in the bulk move through this high field n–p region, avalanche, and with the holes moving in the opposite direction induce a signal on the back contact. APDs are discussed in detail in Section 7.2.

Moreover, in principle, we could use this avalanche region to count each charge carrier as it passes through as long as the carriers have been separated out sufficiently. This technique has, in fact, been realised for a gas detector [19] but not yet for a semiconductor detector. Table 1.8 summarises some of the features of this categorisation.

We will complete this introduction to semiconductor x-ray detectors by discussing in more detail the two detectors that are presently the most popular for x-ray analysis. The first, the Si(Li) detector, is of the charge induction type, and the second, the SSD is of the charge delivery type.

TABLE 1.8
Classification of Semiconductor Detectors

Examples	Signal Developed by Charge Induction	Signal Developed by Charge Delivery
	Conventional planar Si(Li)	CCD (conventional and pn depleted)
	Conventional planar HPGe	SDD (conventional and controlled readout)
	Conventional compound detectors	Very 'small pixel' and coplanar grid detectors
	p–i–n diodes, photodiodes	
	Pixel array detectors (PXDs)	
	Coaxial and hemispherical detectors	
	Microstrip detectors	
	Avalanche diodes (APDs)	
Characteristics	Charge induction	Charge delivery
Electrode structure	Simple (planar)	Complex (pixel, strips, grids)
Carrier sensitivity	Holes and electrons	Single carrier (usually electrons)
Capacitance	Relatively high	Relatively low
Noise	Dominated by capacitance	Dominated by current
Processing times	Relatively long	Relatively short
Events in device	One at a time (possibility of pile up)	Multiple (storage during charge transport)
Rate limitation	Processing time	Charge diffusion (SDD) readout noise (CCDs)

1.16 Lithium-Drifted Silicon Detector (Si(Li))

Since the 1970s, the Si(Li) x-ray detector has been the 'work-horse' semiconductor detector for energy dispersive x-ray spectrometry (EDXRS). Indeed, over the years this type of detector has also set the gold standard by which other such detectors are compared. This has been achieved by a radical transformation of the silicon material. We have already mentioned that in order to achieve carrier drift lengths of greater than ~2.5 mm with the high-purity silicon that is readily available, the silicon must be lithium drifted. In the case of germanium, there was good incentive to produce material of sufficient purity (germanium dedrifts at room temperature) and the lower melting point made it easier, but even so most of the commercial germanium detector manufacturers have been forced to grow their own material. In the case of silicon, this situation has been avoided by lithium drifting, but there only a few suppliers of 'detector-grade' material. The mainstream semiconductor industry has not demanded in any quantity the very high resistivity silicon that would be required for x-ray detectors. This is because they have striven toward microminiaturisation (submicron dimensions), whereas detector technology has striven toward thicker detectors (many mm). Furthermore, large silicon foundries are not particularly interested in producing the small quantities (on a world scale) that go into detector production. Before discussing lithium-drifting technology, let us look briefly at how HPSi is manufactured.

Crystalline silicon and germanium is produced by slowly pulling a small 'seed' crystal vertically from the melt in an atmosphere of hydrogen. Hydrogen is used to exclude oxygen from the crystal. This is called the *Czochralski process*. Because the melting point of silicon is much higher than that of germanium (1410°C and 937°C, respectively) and is generally more reactive, there is a greater chance of contaminants entering the ingot, and germanium ingots at this stage are already much purer than silicon ones. The silicon at this stage contains impurities and defects at a level unsuitable for detector-grade material, although it can be used for standard silicon IC manufacture. For detector-grade material, the ingot is further purified by a process called *float-zone* (FZ) refining. Here, the ingot (or 'boule') is held vertically at both ends as a molten layer, held together by surface tension, is drawn down through the crystal by an inductive heating coil. The contaminants tend to accumulate in the molten zone, thus purifying the material above. The process can take place in a vacuum or in an inert atmosphere, and because of the vertical arrangement there is no contact with the walls of the furnace. Only one floating zone is possible at a time, and many 'passes' are usually necessary. The efficiency of the process depends on the degree of segregation of the impurity into the molten zone. Because the segregation coefficient of boron is close to unity, it is usual to end with high-purity silicon that is naturally doped and slightly p-type. Such material of 1–2 kΩ cm resistivity is commercially available. Higher resistivity wafers of 0.3–0.5 mm thickness are also possible for fully depleted silicon p–i–n diodes and SDDs. For depleting 2–3 mm, very high purity silicon made by the decomposition of silane is required. This is expensive and not readily available. The Japanese company Komatsu and the x-ray detector company Horiba (subsidiary of Hitachi) have pioneered such materials. Most x-ray detector companies successfully lithium drift the 1–2 kΩ cm resistivity FZ material.

1.16.1 Lithium Drifting

This is where nature comes to the rescue. The small atoms of H, He, and Li readily diffuse interstitially into silicon at reasonably low temperatures (~500°C). Whereas the gases are easily removed by heating in a vacuum, the interstitial Li is relatively stable at room temperature (as stated this is not the case for germanium). Thus, a deep (up to 0.5 mm) diffused p–n junction is easily formed on a p-type silicon slice as shown in Figure 1.56a.

The donor level of Li is a mere 30 meV below the conduction band, and even at room temperature (average thermal energy 25 meV) most of these levels are ionised leaving copious Li⁺

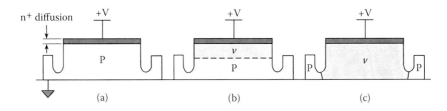

Figure 1.56 Cross section of a Si(Li) at various stages of drift. (a) Deep diffused p–n junction, (b) partially drifted, and (c) totally drifted.

ions. By raising the temperature to between 100°C and 150°C and applying a moderate reverse bias (500–1000 V across a 4-mm-thick slice), the Li^+ ions can be persuaded to 'drift' into the silicon under the influence of the field. In doing so, there is a natural tendency (LeChatalier's principle) for the Li^+ ions to migrate in such a way that the space charges neutralise and the field tends to become smooth. This is achieved by the Li^+ ion forming a complex with the B^- ion impurity acceptors. This process is called *lithium compensation* and is illustrated in Figure 1.57.

The 'lithium-drifted' or Si(Li) material, has many of the qualities of the truly intrinsic semiconductor (and indeed was historically referred to as 'intrinsic'), particularly when cooled. Note that n-type material cannot be lithium compensated. As the drifting process progresses, usually over a period of several days, the junction moves through the slice, leaving behind an expanding compensated layer. Figure 1.56b shows partial drift and Figure 1.56c illustrates what occurs after 'punch-through' to the opposite side. With an appropriate contact, it would finish at the other side of the slice where lithium would accumulate. In planar semiconductor processing, this would conventionally be referred to as the 'back side', but because radiation detector manufacturers usually use this side as the 'window' for the radiation we will refer to it from now on as

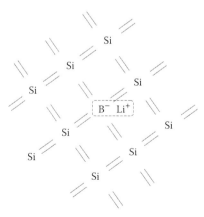

Figure 1.57 Lithium ions seeking out boron ions and electrically neutralising these charged sites.

the 'front side'. If we assume a constant bias, the field profile during drift resembles that of Figure 1.58.

Although larger depths can be drifted, the thickness of the drifted material is usually 2.5–4 mm, which is adequate for EDXRS up to ~30 keV. However, the story is not quite complete as at the temperatures mentioned for the drift the concentration of intrinsic carriers in the silicon is very high and the thermally generated electrons and holes in transit across the junction cause high currents. These currents increase with the drifted depth and reach a few milliamperes, for example, in the case of a 30-mm² crystal. Because of the negative 'space charge' effect of the electrons, there is a tendency for overcompensation by lithium. An added drifting process is applied at a lower reverse current (temperature reduced to 80–100°C) so that compensation is more complete and the field becomes more uniform. This process is called 'cleanup' or 'levelling' (see Figure 1.58). The currents during this time are a few 100 µA, and the time for cleanup can be 10 days or more depending on the material.

The drifted volume ends effectively intrinsic with resistivities in excess of ~200 kΩ cm and fully depleting at ~20 V bias. Being very close to intrinsic but slightly n-type, it is referred to as ν-type. Perfect compensation would only be achieved by

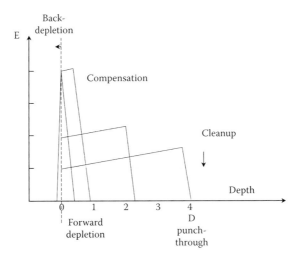

Figure 1.58 Schematic of field profile during drift.

a cleanup at very low temperature, but it would take many years to achieve this. The initial lithium diffusion for the junction is normally carried out on a silicon slice up to say ~10 cm diameter depending on the original ingot size, but the drifting process takes place on individually diced crystals as depicted in Figure 1.56. This has the advantage that an undrifted 'periphery' of the original p-type silicon is left with surfaces that are insensitive to handling and allow a continuous metal contact to be made on the front face, which avoids high fields at the junction edge. Indeed, the cleanup process is usually carried out with such a p^+ surface barrier contact (e.g., Au) in place. The 'groove' depicted in Figure 1.56 was historically [20] introduced by Miller at Bell Laboratories to confine the 'punch-through' of the lithium drift to a well-defined area, but as we will see later, this had other consequences. The flat lithium–diffused region defined by the groove is referred to as the 'mesa' (Spanish or 'table' or 'flat-topped table-land', a carryover from early transistor technology). The lithium diffusion is not normally removed and the mesa acts as a robust n^+ contact. A generous evaporation of Al completes a low resistance contact for the FET to the compensated volume. Being thick (~0.3 mm), the lithium diffusion also acts as a shield for low-energy XRF from materials at the back of the Si(Li)—an advantage not shared by p–i–n diodes, SDDs, and CCDs. Heavy metals close behind the crystal in detectors (such as In, Pb, Sn often found in solders) are to be avoided because they contaminate x-ray spectra. The Li^+B^- ion pair complex has a much smaller electric field associated with it than either the Li^+ ion or B^- ion in isolation. Therefore, the B^- acceptor site is effectively removed as a scatterer of carriers. Furthermore, they are in effect removed as traps from the bandgap. Note that the Si(Li) crystal temperature cannot be raised to such high temperatures again without 'decompensating'. Room temperature without bias is acceptable, but thermally grown oxide passivations are ruled out. In the case of a Ge(Li) crystal, this applies also to room temperature, and this explains why HPGe has been used almost exclusively since the early 1970s.

Various attempts were made to lithium drift other materials without success. Today, lithium drifting is unique to the production of thick silicon radiation detectors and is still, to some extent, an art. Many techniques are in use, for example, drifting at constant bias, constant power, computer-controlled current, in air or in liquid. The process is further examined in detail in Chapter 5.

1.16.2 Contouring

After lithium drifting, the front face of the silicon crystal is etched to form a recess defined by the lithium drift profile (Figure 1.59). This removes excess lithium and any diffusants such as gold that may have entered during the high temperatures of the drifting process. After a final etching of the sidewalls to remove excess lithium and a suitable surface barrier front contact has been applied, the crystal is essentially a finished x-ray p–n diode transducer. As pointed out earlier, surface leakage currents are a major problem in introducing noise in the finished system. Groove and 'top-hat' (or 'shelf') contouring as shown in Figure 1.59 are used almost exclusively today to control these.

The p^+ surface barrier (SB) junction is terminated by the undrifted p-type periphery, which takes the p–n junction to the surface of the groove as shown. This junction will act as a barrier to majority carrier (hole) emission but may be a source of diffusion current (electrons). As indicated, the field lines are brought

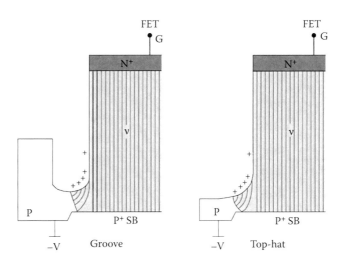

Figure 1.59 Cross sections of both a grooved and a top-hat structure.

close to normal to the surface by the groove or top-hat contouring and if strong enough, the n-type surface channel (electron current) can be 'pinched off'. Figure 1.59 is schematic and exaggerates this effect (the field lines will only be absolutely normal if the density of the surface charge is very high as in a metal).

The p-type periphery will back-deplete into the groove or shelf from the drifted profile at the surface by very little because the field *normal* to the junction there is weakened by the surface charge. Although an n-type surface is usual after an etch, the desired surface charge state is most often achieved by using a *passivation* technique to set the surface particularly near to the junction. Such passivations are usually proprietary, and many different materials and techniques are used. Passivation in general is discussed in Section 5.1.9. We will also see in Chapter 11 that some alternative geometries have been tried. The pinching off effect of the surface channel by contouring is also effective for both p-type and n-type HPSi and HPGe crystals.

1.16.3 Guard Rings

Under circumstances where surface leakage is a significant noise source (such as at higher temperatures in a Peltier cooled detector), a guard ring can be used as shown in Figure 1.60.

In the Si(Li) detector, the guard ring groove must be sufficiently deep to penetrate through the Li diffusion and to electrically isolate the guard ring from the mesa. In this configuration, the sidewall surface currents are shorted to ground and do not enter the equivalent circuit. The surface currents in the guard ring groove, that has only a few volts at most across it, replace them. It is also worth noting that the surface states in the guard ring groove are less critical because the bulk field is largely normal to the surface of the groove. The depth of the guard ring groove is a compromise between shallow (minimising the intrinsic surface area that generates the currents) and deep (reducing the detector capacitance). The disadvantages of the guard ring are the increased capacitance when it is added, the reduction of active to physical area, and the increased complexity in mounting and connecting. Today, guard ring structures are more commonly used for room

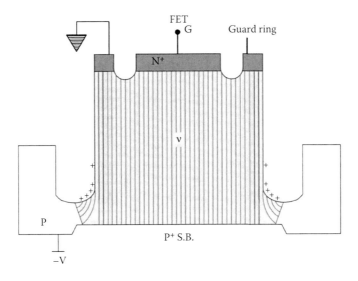

Figure 1.60 Cross section of a guard ring detector.

temperature or near–room temperature applications such as p–i–n diodes and SDDs, where rather than grooves they are made by metallising or implanting regions around the active area using photolithographic techniques. Such guard rings may be grounded or 'floating'. This is discussed in Section 7.1.

1.16.4 Mechanical Construction: Cryo-Head

The Si(Li) crystal must now be mounted in close proximity to the FET and any other components required to be close (such as heater resistors, reset LEDs or diodes and FBC) ready for cooling in a cryostat. Cooling is achieved by fixing the package or cryo-head to a cold-finger connected to an LN reservoir or some alternative such as a cooling engine. These alternative cooling techniques are discussed in Section 11.25.2. In mounting, the following criteria must be met:

(1) The components must not be able to work loose on cycling between operational temperature and ambient or baking temperature (to minimise microphony).
(2) The stray capacities at the FET contact must be minimal (minimise voltage noise).

(3) The thermal contact to the cold-finger or alternative heat sink must be efficient and emissivities (cryo-head and cold-finger) high (to minimise heat absorption).

(4) The FET temperature must be controlled with a heater to minimise the FET voltage noise and must be sufficiently thermally isolated from the Si(Li) crystal so as not to raise its temperature significantly.

(5) The cryo-head must be isolated vibrationally from outside influences (including boiling LN).

(6) Materials used in the FET gate and crystal mesa area must be minimal and not dielectrically lossy (PTFE, BN, alumina are acceptable, whereas epoxies and glasses are to be avoided).

(7) The Si(Li) crystal mount must act as an adequate collimator to prevent events from occurring in the peripheral area where the field is distorted (Figure 1.60). In transmission electron microscopy (TEM) and XRF applications, this usually involves backing an aluminium cup with a layer of tantalum or tungsten.

(8) For TEM applications, there must be no magnetic materials or insulating materials that will charge up and cause astigmatism.

In considering these criteria, it is useful to keep in mind that on cooling Al shrinks onto Cu, and PTFE shrinks onto all metals and becomes rigid. Silicon and BN have very good thermal conductivities. Al is lighter than Cu for a cold-finger (reduced microphony), and at low temperatures (100–120 K) the thermal conductivity of Al is almost as good as that of Cu (the factor is ~1.6). The FET header pin is often kovar (relatively high thermal conductivity and magnetic), and the bond wires are usually 25 μm Au and carry very little stray capacitance or heat flow. Figure 1.61a–d provides some flavour of how these criteria can be—and have been—met to a greater or lesser degree.

There are of course a myriad of engineering solutions, and their evolution is discussed in Section 11.19.3. The schematic in Figure 1.61a, in fact, uses a three-terminal FET (where the gate is physically the kovar pin on which the chip is mounted) and an LED restore, which have been almost

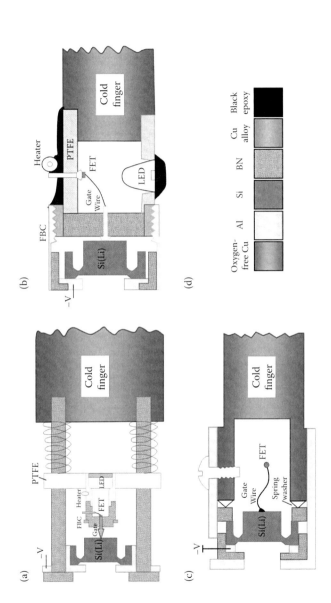

Figure 1.61 Various 'front-end' arrangements. (a) Sprung loaded FET package, (b) wire connection between FET and back of detector, (c) electronic restore, and (d) materials used in front-end construction.

exclusively replaced today by the tetrode FET (where the chip substrate is connected to the pin) and electronic restore, both of which are used in the schematic shown in Figure 1.61c. If an LED is used, the FET package must be light-tight because the Si(Li) crystal is a very sensitive photodiode! The FBC and heater resistor are also often now integrated onto the FET chip.

The cryo-head is mounted on a cold-finger that, in the simplest case, is attached via a copper braid to an LN reservoir. The copper braid is flexible and takes up movement during vacuum pumping, baking, and cooling. It also acts to decouple any external vibrations. The simplest arrangement is shown in Figure 1.62a and b.

Considering microphony (Section 1.10.7) and Figure 1.62b, one would not ideally choose to put the sensitive transducer on the end of a long, heavy metal cantilever (the cold-finger), intimately connected thermally to a large resonant cavity (the LN dewar) full of a boiling liquid (LN), and then to place it in a long narrow metal tube (detector vacuum tube) isolated mechanically and thermally from it, so that it is free to vibrate! One fairly effective solution is to fix the cryo-head relative to the detector tube (e.g., by an intermediate stainless steel tube in the vacuum) and decouple vibrations from the cold-finger by a short length of copper braid or bellows.

1.16.5 Output Waveforms

For x-ray analysis, the charge restore is usually applied at a preset output voltage level ('limit restore'). Hence, the restore frequency depends on the quiescent current flowing into G (Figure 1.1) when there are no x-rays or on the energy-rate product when x-rays are present. For large signals, as for high-energy x-rays or gamma rays, it is often more convenient to restore after each event, but because there is a 'dead time' during restore for lower energy x-rays a limit restore is preferred. When the bias is first applied, because of the changing potential, displacement current flows through the Si(Li) crystal and this is reflected in the output voltage gradient dV_o/dt or 'ramp' at the preamplifier output. The restore frequency is at first

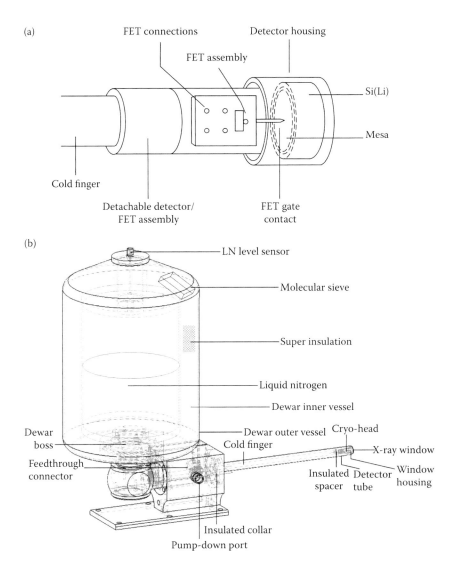

Figure 1.62 (a) Cryo-head attached to a cold-finger. (b) Full cryostat assembly.

rapid, slowing as the bias settles as shown in Figure 1.63a. When the ramp length has settled (usually ~few seconds), the leakage current can be estimated as (after allowing for the gain of the preamplifier) it simply given by

$$I = C_F \times dV_o/dt$$

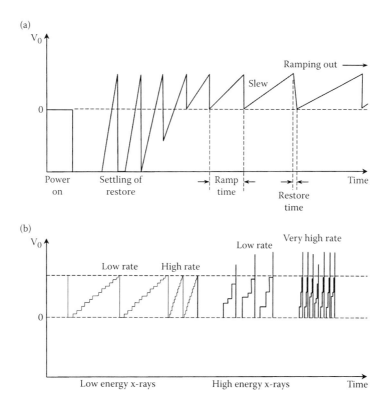

Figure 1.63 (a) Preamp output after initial switch on. No x-rays present. (b) Preamp output once settled. X-rays present.

Figure 1.63b shows the steplike signals for each x-ray event, the height of the step being proportional to the charge collected and the x-ray energy. By observing such steps for a given x-ray (such as the 5.9 keV Mn Kα x-ray emitted by Fe-55), it is possible to obtain an estimate of the signal/noise ratio before shaping.

The manufacturing technology of the Si(Li) detector is similar to that of the HPSi detector (see Chapter 6) and does not lend itself to large-scale automatic production. Many of the processes involved, such as individual masking and wet chemistry, are labour-intensive, and photolithography is not an option on such thick substrates. The SDD, to be introduced next and in detail in Chapter 9, moves partway toward standard industrial techniques by using lithography on silicon wafers up to ~500 μm thick. This thickness is adequate for EDXRS up to ~10 keV. However, unlike standard silicon IC foundry techniques, it does use double-sided wafer processing and specialised ion implantation on detector-grade FZ HPSi.

1.17 Silicon Drift Detectors

Consider the structure depicted in Figures 1.64 and 1.65.

This represents concentric p⁺ implant rings on an n-type Si wafer. In the center, there is a small n⁺ implant (the anode). On the other side of the wafer there is a shallow uniform p⁺ implant across the wafer. This completes a p–n diode structure and will constitute the front 'entrance window' electrode for x-rays when the device is fully depleted. In this particular arrangement, the bias is applied across the n⁺ anode and the outermost p⁺ ring as shown in Figure 1.65.

In this diagram, we also show how depletion develops when the bias is increased. The back face, across which the bias is applied, depletes from the outermost ring toward the center, and the rings in between will 'float' at potentials stepping up toward the anode. Figure 1.65b shows the point where the

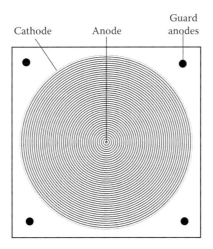

Cathode   Anode   Guard anodes

Figure 1.64 Drift detector with radial symmetry and without an integrated FET. Rings are biased to achieve drift field.

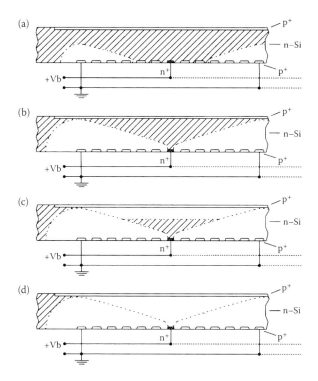

Figure 1.65 Application of bias to ring structure depletes volume of drift detectors. (a) Partial depletion, (b) depletion from rings reaches front face, (c) increasing the voltage depletes from front face towards anode and (d) full depletion (with permission from Springer, *Semiconductor Radiation Detectors*, Lutz, p. 135).

depletion meets the top p⁺ surface. This causes this implant to 'float' at some potential that is negative with respect to the anode. Further increasing the bias initiates depletion from the top toward the anode at bottom center (Figure 1.65c) until the whole device is fully depleted (Figure 1.65d). The guard anodes shown in Figure 1.64 are placed external to the ring structure and are biased to collect any electrons generated outside the ring structure. The fully depleted device has a potential minimum along the dotted line shown in Figure 1.65d and forms a 'channel' or 'tilted trough' as depicted in Figure 1.66, along which any electrons generated in the volume will flow toward the anode.

This ingenious device therefore will act to collect all majority carriers (electrons) generated by an x-ray in a similar way

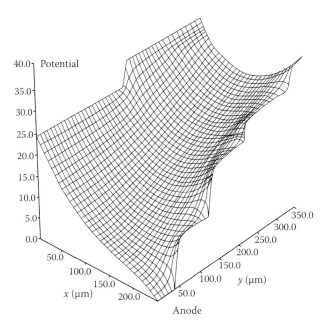

Figure 1.66 Equipotentials in a biased SDD showing 'tilted trough' that feeds electrons to anode.

to a CCD but without the complexity of 'clocking'. As in a CCD, the holes will be lost so it is a single carrier detector. The effect of the p⁺ rings is to screen the electrons from the anode during their trajectory, and the signal is only developed when they enter the vicinity of the anode. It should be noted that in the small central region, the device approximates to a simple planar p–i–n diode where the signal is induced on the anode as described by the Shockley–Ramo theorem. In this way, we can consider a large area SDD as an example of a charge delivery detector such as a CCD. There are other biasing arrangements that can achieve the same results.

1.17.1 Mechanical Construction

Much of the pioneering work after the initial concept [21] has been carried out in Germany, most noticeably at the Technical University of Munich [22]. The spectral quality has been improved by reducing noise (by integrating the JFET and improving the passivation) and improving charge collection (by moving the JFET to the edge of the active area and with innovative implanting of the window contact). Details and history of these developments are given in Chapters 9 and 11.

The leakage currents in SDDs are low enough for them to operate at temperatures well within the limits of a few stages of Peltier cooling (180–240 K). The Peltier effect is the inverse of the thermoelectric (Seebeck) effect whereby a potential difference can be obtained across a p–n junction by applying a temperature gradient across it. The diffusion of carriers produces the potential difference. The converse (Peltier) effect is that if a current is made to flow across the junction, a temperature gradient can be supported across the junction, and we can absorb heat from one side to the other. This makes an ideal cooling engine for x-ray detectors because it is vibration free and needs only a current supply. They are widely used in the electronics industry and are commercially available. Technically, the temperature differences are limited by the number of junction stages and the rate at which the heat can be removed to a heat sink. The mounting of the chip on the cooler is shown schematically in Figure 1.67 (see also Figure 1.68).

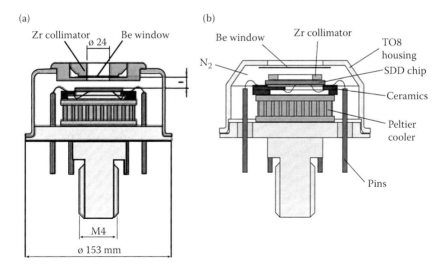

Figure 1.67 Schematic representation of two commercial detectors. (a) Example of a drift detector in its package and (b) another example of a drift detector with a Zr collimator mounted very close to the silicon detector. (Courtesy of Ketek, GmbH.)

Figure 1.68 A commercial detector arrangement. (Courtesy of Ketek, GmbH.)

The Peltier cooling stages and copper cold-finger are clearly seen. Much progress has also been made in the encapsulation of p–i–n diodes and SDDs. The use of getters and the choice of materials with minimum outgassing have made reliable packages even with UTWs. Usually, the capsules are back-filled with a low-pressure inert gas such as nitrogen to further inhibit outgassing.

1.17.2 Summary

The advantages of the SDD over the Si(Li) are as follows:

(1) The very low capacitance achievable for the same active area. As we have seen, this leads to high count rate capability.
(2) The low physical volume results in lower thermal leakage current and noise. They can be operated at much higher temperatures well within the reach of Peltier coolers. This lack of reliance on LN is greatly valued both in the laboratory environment and for field and space operation.
(3) Foundry manufacturing leads to mass production and reduced cost.
(4) The application of standard lithographic oxide passivation and guard rings leads to low leakage currents and again higher temperature operation.

There are, however, some downsides:

(1) They are more or less restricted to silicon technology at the moment.
(2) The thicknesses are limited (<1 mm), and sensitivity falls off above ~10 keV.
(3) The active areas are limited as the SDD requires the same full depletion bias as a conventional p–i–n diode but this limits the practical drift lengths possible. This is overcome by incorporating four or more SDDs in an array at the expense of increased complexity.
(4) The lack of a thick back contact results in spectral contamination from materials behind the SDD.
(5) Manufacture is restricted to a suitable silicon foundry, whereas Si(Li) detectors can be manufactured under reasonably clean conditions (class-100 conditions an advantage but not an essential).

Appendix 1A: Published Books on Semiconductor Detectors in Order of Publication Date

Dabbs JWT and Walter FJ (Eds.), *Semiconductor Particle Detectors, NAS-871*; *Nuclear Science Series*, Report Number 32 (Asheville Conference, Washington), 1961.

Taylor JM, *Semiconductor Particle Detectors*, Butterworths & Co. Publishers Ltd., 1963. Suggested U.D.C. Number 539.12.074: 537.311.33.

Price WJ, *Nuclear Radiation Detection* (2nd ed.), McGraw-Hill, New York, 1964, ISBN 0070508707.

Sharpe J, *Nuclear Radiation Detectors*, Methuen & Co. Ltd., London, 1964.

Dearnaley G and Northrop DC, *Semiconductor Counters for Nuclear Radiations*, E&FN Spon Ltd, London, 1966. ISBN 0419034501.

Bertolini G and Coche A (Eds.), *Semiconductor Detectors*, American Elsevier (John Wiley), New York, 1968, ISBN 044410142X.

Brown WL and Higginbotham W (Eds.), *Semiconductor Nuclear-Particle Detectors and Circuits*, NAS-1593, National Academy of Sciences, Washington, DC, 1969.

Eichholz GG and Poston JW, *Principles of Radiation Detection*, Ann Arbor Science Publishers, Ann Arbor, 1979, ISBN 0873710622

Tait WH, *Radiation Detection*, Butterworth, London, 1980. ISBN 9780408106450.

Cooper PN, *Introduction to Nuclear Radiation Detectors*, Cambridge Univ. Press, Cambridge, UK, 1986, ISBN 0-521-28132-6.

Debertin K and Helmer RG, *Gamma and X-ray Spectrometry with Semiconductor Detectors*, North Holland, Amsterdam, 1988, ISBN 0 444 871071.

Mann WB, *Radioactivity Measurements, Principle and Practice*, Pergamon Press, Oxford, 1988, ISBN 0-08-037037-3.

Fraser GW, *X-ray Detectors in Astronomy*, Cambridge University Press, Cambridge, 1989, ISBN 0 521 32663 X.

Delaney CFG and Finch EC, *Radiation Detectors: Physical Principles and Applications*, Oxford University Press, New York, 1992, ISBN 0 19 853923 1.

James RB, Schlesinger TE, Siffert P and Franks L (Eds.), *Semiconductors for Room-Temperature Radiation Detector Applications*, Materials Research Society MRS 302, San Francisco, 1993, ISBN 1-55899-198-0.

Lutz G, *Semiconductor Radiation Detectors*, Springer, Berlin, 1999, ISBN 3-540-64859-3.

Knoll G, *Radiation Detection & Measurement*, 3rd ed., Wiley, New York, 2000, ISBN 0-471-07338-5.

Spieler H, *Semiconductor Detector Systems*, Oxford University Press, Oxford, 2005, ISBN 0-19-852784-5.

Tsoulfanidis N and Landsberger S, *Measurement and Detection of Radiation*, 3rd ed., Taylor & Francis, CRC Press, Boca Raton, 2010, ISBN 10 1420091859.

Owens A, *Compound Semiconductor Radiation Detectors*, Taylor & Francis, CRC Press, Boca Raton, 2013, ISBN 978-1-4398-0.

Appendix 1B: Books Focusing on X-ray Applications

Frankel RS and Aitken DW, *Applications of X-rays and Gamma Rays*, Gordon & Breach, 1971.

Woldseth R, *X-ray Energy Spectrometry (All You Ever Wanted to Know about XES)*, Kevex Corporation, Burlingame, CA, USA, 1973.

Reed SJB, *Electron Microprobe Analysis*, Cambridge University Press, Cambridge, 1975, ISBN 0 521 20466 6.

Bertin EP, *Introduction to Spectrometric Analysis*, 2nd ed., Plenum Press, New York, ISBN 0-306-3109-0, 1980.

Jenkins R, Gould RW and Gedcke D, *Quantitative X-ray Spectrometry*, Marcel Dekker, New York, 1980, ISBN 0824712668.

Heinrich KF, Newbury DE, Myklebust R and Fiori CE (Eds.), *Energy Dispersive X-ray Spectrometry*, NBS (604), 1981, U.S. Department of Commerce, National Bureau of Standards 1981.

Williams DB, Goldstein J and Newbury DE, *X-ray Spectrometry in Electron Beam Instruments*, Plenum Press, New York, 1995, ISBN 0-306-44858-0.

Van Griecken R, Tsuji K and Injuk J (Eds.), *X-ray Spectrometry: Recent Technical Advances*, Wiley, Chichester, 2004, ISBN 0-471-48640-X.

References

1. Korff S.A., *Electron and Nuclear Counters—Theory and Use*, D. Van Nostrand Company Inc., 1946, ISBN 117849571X, 9781178495713.
2. Goulding F.S., *IEEE* NS 17, 1, 218, 1970.
3. Fano U., *Phys Rev*, 70, 1, 1946.
4. Spieler H., *Semiconductor Detector Systems*, Oxford University Press, 2005, ISBN 0-19-852784-5.
5. Trammell R.C. and Walter F.J., *NIM* 76, 317, 1969.
6. Joy D.C., *Monte-Carlo Modelling for Electron Microscopy and Microanalysis*, Oxford University Press, 1995.
7. Scholze F., *NIM* A439, 208, 2000.
8. Nicholson P.W., *Nuclear Electronics*, Wiley, 1974.
9. Motchenbacher C.D. and Fitchen F.C., Low Noise Electronic Design, Wiley, 1973, ISBN 10 0471619507.
10. Lutz G., *Semiconductor Radiation Detectors*, Springer, 1999.
11. Bombelli L., *NSS Conference Record*, IEEE, ISBN 978-1-4673-0119-0/11, 2011.
12. Voss R.F., *Science and Uncertainty*, Nash S. (Ed)., Science Reviews Ltd., 1985.
13. Kandiah K., *NIM* 326, 49, 1993.
14. Eggert T., *NIM* A512, 257, 2003.
15. Knoll G., *Radiation Detection and Measurement*, 3rd ed, Wiley, 2000.

16. IEEE Standard 759, 1984 (approved 1981), *Standard Test Procedures for X-ray Detectors*, IEEE.
17. Bale G., PhD thesis, University of Leicester, 2001.
18. Rehak P., *NIM* 289, 410, 1990.
19. Panski A., *NIM A*392, 465, 1997.
20. Coleman J.A., *IEEE* NS 11, 3, 213, 1964 (private communication from G.L. Miller, Bell Labs).
21. Gatti E., *NIM* 225, 608, 1984.
22. Kemmer J., *NIM* A253, 378, 1987.

2

Detector Response Function

We introduced the concept of the detector response function, $F(E)$, in Section 1.14. The formulation of a response function might be seen by university workers as an academic challenge, by detector manufacturers as a means of measuring and improving detector performance, and by users as a means of providing profiles for speedy quantitative analysis. The deconvolution of x-ray spectra in analytical applications has gone far beyond 'hunt and identify the peak' and needs to be purely computational in real time. There are essentially three methods of generating the response $F(E)$ for a monochromatic x-ray line of energy E. First, one can attempt to fit the spectrum with a series of empirical functions, one for each x-ray line. The broadening of the line is the simplest to deal with because of the random nature of the ionisation (Fano factor) and the electronic noise. As explained in Chapter 1, these follow well-known dependencies on energy E and signal processing time τ. Often, the noise peak can be monitored during spectrum accumulation by processing and storage into a separate computer memory partition during the absence of any x-ray. In its simplest form, the result from noise or an x-ray is a Gaussian peak with one parameter [variance $\sigma(E)$ or full width at half-maximum (FWHM)] that is then determined experimentally. It can, of course, also be measured using monochromatic synchrotron radiation or more easily from a quasi-monochromatic radioactive source, usually Fe-55. However, because of the complex physics involved, the photopeak deviates considerably from a pure Gaussian. One empirical function that has been developed over the years is the Hypermet function (discussed below). Second, one can use a Monte-Carlo approach where all the processes involved in the absorption, ionisation, and charge collection are simulated in a computer program, event by event, inserting the probabilities at each stage using a random number generator. Third, one can attempt an analytic approach to produce a physical model. This method results from a better understanding of the underlying physics involved as improvements in detector performance have revealed more structure in the spectral shape. The random Gaussian dependence is often folded into the analytic expressions at the end of the calculation.

2.1 Hypermet Function

A modified error function was introduced as an empirical model function by Phillips and Marlowe [1] in 1976 and has widely been used in gamma-ray spectroscopy. As discussed in Chapter 11, this was refined over the years for x-ray spectroscopy by workers such as Ian Campbell [2] at Guelph University, Canada, and Yosuke Inagaki [3] at Kyoto University, Japan, and is still used extensively. Besides the Gaussian peak, a very short exponential low energy tail and a much longer one are added. Below the photopeak, a flat 'shelf' extends the response function to very low energies. The functions involve up to seven free parameters for each excitation energy E. However, it is difficult to relate these parameters to physical processes, and this makes it difficult to extrapolate the models. Such functions do have an advantage for the deconvolution of spectra by not being statistical in nature and not requiring lengthy computations, as in the case of Monte-Carlo calculations (discussed in Section 2.2).

2.2 Monte-Carlo Calculations

In Section 1.8.1, we introduced the concept of the Monte-Carlo calculation as applied to the generation of the detector response function. The approach has been very successful with the main proponents being David Joy [4] at the University of Tennessee, USA, Manfred Geretschlager [5] of Johannes Kepler University, Austria, and Ian Campbell [6]. The idea is to simulate the actual statistical process event by event, incorporating all the processes that are considered likely to influence the final spectral distribution. The results should always converge on a given form as the number of events simulated increases, hopefully indicating the origin of any features. It, of course, relies on some analytical expressions, probability distributions, and look-up tables for experimental values. Often, Monte-Carlo calculations are used as part of an analytical approach as, for example, done by Shunji Goto [7] of Fujitsu Ltd. in Japan, in describing the energy loss process of the photo and Auger electrons. Both the empirical approach and the Monte-Carlo approach rely on a good understanding and proper application of the all physical processes involved. Before discussing the results of such models, we will now look at these processes.

2.3 Physical Processes and the Analytical Approach

Besides Goto, the main other proponents of the analytic approach go back to Fred Goulding's group [8] at the University of California, Berkeley, USA, and developed by Inagaki and by Kalinka [9] at the Institute of Nuclear Research, Debrecen, Hungary. More recently Campbell [10], Lowe [11] (Oxford Instruments, UK), Eggert et al. [12] (Ketek, Germany), and Scholze and Procop [13] (Helmholtz Zentrum, Berlin, Germany) have developed the analytical approach further.

For an ideal detector for an x-ray of discrete energy E_0, all counts would fall into a single energy channel dE forming a 'delta function photopeak' in the spectrum. The photopeak channel would correspond to a charge $Q = ne$, where

$$n = E_0/\omega \qquad (2.1)$$

and e is the charge of the electron. We pointed out that this is a non-physical situation as there are many processes that occur that distort the 'photopeak', giving rise to what we loosely termed a 'peak broadening', 'low energy tail', and 'background or shelf'. We will look closer at the physical processes that occur during the photoelectric effect and charge collection in a semiconductor and their effect on the response function.

2.4 Peak Broadening—The Fano Factor

We briefly discussed the Fano factor in Section 1.7, where it was largely presented as a 'fudge factor' to obtain the experimentally measured resolution of a detector. This is demonstrated in the equation for the resolution R at energy E_0 in a system with electronic noise N

$$R^2 = 2.355^2 \, \omega F E_0 + N^2. \tag{2.2}$$

No other topic in x-ray detectors has generated such controversy and passion as the Fano factor, and the reasons are easy to see. Equation 2.2 shows F (along with the easier-to-measure ω) to be the basic limiting factor for the resolution R of the x-ray detector. In fact, we see from this equation that what is actually measured from the resolution R of a detector and its noise N is the product ωF. It does not help the controversy that the measurements are very difficult to make, being calculated from the difference in the widths of two peaks *in quadrature* which, as a result, need to have very good stability and statistics. This leads to large errors particularly at low energies when we are subtracting two relatively large similar numbers. From a practical point of view, monochromatic sources are necessary such as high-power x-ray tubes with monochromators or synchrotron radiation, although the radioactive isotope Fe-55 can be used for measurements at 5.9 keV providing corrections are made for the contribution of the Mn Kα2 line, which is 11 eV above that of the MnKα1 but half its amplitude. The noise *during* the spectrum acquisition must be constant and accurately determined. Some researchers, for example, Lepy et al. [14] working at CEA, Orsay, France, have avoided the noise issue by making measurements at different energies. They assume that the noise does not change over the measurement times and ωF is constant within the energy range chosen (1–10 keV). Furthermore, any broadening due to incomplete charge collection (ICC) must be either negligible or accounted for. ICC leads to an 'effective Fano factor' particularly at very low energies or in areas of the crystal where the field is weak. The fall-off in the electronic efficiency of the processing system at low energies set by the noise in the discriminators can also influence the peak width (see Figure 1.52). This causes, for example, the width of the BKα peak at 185 eV to appear *narrower* than that of the CKα peak at 277 eV [11,15]. Another reason for controversy is that a recognized consensus for both the theoretical and experimental value of F for any kind of medium and ionising particle is very difficult to achieve. Figure 2.1 shows some of the theoretical and experimental values reported for x-rays in silicon over the years.

Arguments about the limiting values of F started with early measurements [16] and continue to this day [17,18]. Detector manufacturers [19] were not averse to claim finding lower values of F for their products than their competitors. The theoretical difficulties lie in the necessity to include all significant mechanisms and products involved in the complex ionising process and after doing this, to insert the 'correct'

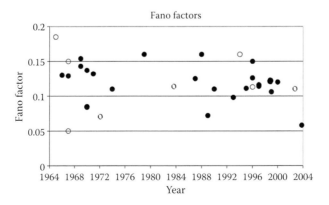

Figure 2.1 Experimental and theoretical values of the Fano factor *F*. Open points are calculations and filled points are experimental.

values (theoretically or experimentally derived) of the parameters involved. We will see later that ω and F are intertwined theoretically, and usually both are calculated together for a particular model. In stating an experimental value for F, ω must be assumed, and most commonly the value measured by Pehl et al. [20] (3.81 eV at 77 K) at Berkeley way back in 1968 is regarded as the standard although more recent values [21,22] may be more reliable. For Peltier cooled detectors such as charge-coupled devices (CCDs) and silicon drift detectors (SDDs), the room temperature measurement (3.66 eV) of Scholze et al. [22] using a silicon pin diode can be used. Most measurements of ω have been made by comparing the charge signal generated in detectors by gamma-rays or x-rays with the charge signal from injected precision pulses through a calibrated capacitor to the field effect transistor (FET) gate of the detector. However, Scholze et al. measured it from the spectral responsivity of photodiodes to synchrotron radiation.

Some of the experimental results, for example, those of Bilger [23] and Palms et al. [24], were extrapolations to an infinite electric field. It was hoped that this would eliminate the effects of ICC. Another approach is to select the very best detectors from a large sample [25], particularly those that show good linearity down to very low energies (C Kα 277 eV). Taking the advantage that the escape peaks in germanium are relatively intense and much lower in energy than the parent peak (by 9.886 and 10.982 keV for the GeKα_1 and Kβ, respectively), another approach has recently been taken by Papp et al. [26]. For parent peaks just above the Kα excitation energy, the escape peak originates from a photoelectric absorption relatively deep in the germanium. Thus, it is assumed that any effects of ICC at the surface can be ignored even for the very small quantities of charge deposited. The parent photoelectric peak can be tuned on a synchrotron to obtain escape peaks over a range of very low energies. They used a parent peak of 11.136 keV with the Kβ escape peak at 154 eV and assumed that this is close enough to zero to represent the electronic noise. Thus, the value of ωF can be calculated for the 11.136-keV and the Kα_1 escape peak. They then shifted the parent peak up by 1.1 keV so that the Kβ escape lines up with the previous Kα_1 escape peak. The new noise can now be calculated because they have already calculated ωF at this energy. They then measured ωF for the Kα_1 escape peak and so on in an iterative process. They obtained results for ωF of about 0.3 at 11.1 keV, which is similar to other measurements at 5.9 keV [25], but results of ~0.2 at 1.25 keV, which is much lower than those obtained with other measurements (e.g., ~0.35 at 1.48 keV). Using the 'standard' value of ω for germanium [27] of 2.96 eV gives $F \sim 0.075$, whereas the more conventional methods give similar values to silicon (0.12) at 1.48 eV. The method is ingenious and casts some doubt as to the role of ICC even in state-of-the-art Si(Li) and high-purity germanium (HPGe) detectors. However, there are several problems with their measurements. As mentioned previously, the noise needs to be known very accurately and the statistics need to be very good. The escape peaks are, in fact, complex and likely to have insufficient statistics and sit on a relatively high background.

2.4.1 Theoretical Calculations of ω and F

To illustrate how such calculations are approached, we can consider for simplicity the generation of electrons only in the ionisation process (as Fano originally did for a gas). Consider an electron of energy E_p moving through a silicon crystal rapidly and losing energy in each interaction, producing secondary electrons in a cascade-like process. Each secondary electron requires a fixed energy (the ionisation energy I) for its production and the energy lost to the crystal after a *fixed* number of such interactions (or a fixed track length) would obviously vary statistically in a random manner about a mean. The interactions can, in this case, be regarded as independent of each other, and Poisson statistics would give the yield of electrons and its variance. The Fano problem is the *reverse* of this. We let the process finish, and the number of interactions (and the track lengths) varies but the total energy lost E_p to the crystal is *fixed* by the conservation of energy. The problem then is to calculate the yield of electrons and its variance with this constraint. This is the same as saying that the interactions are *correlated*. As we come to the end of the process approaching the ionisation energy I and thermalisation of the electrons in the lattice, it becomes increasingly more obvious that the interactions cannot be independent as the total energy constraint gets tighter and tighter. Thus, the Fano factor is predominately determined by the final interactions in the cascade and is more or less independent of the detailed physics of the ionisation (type of ionising particle and the initial energy E_p). However, when E_p approaches I, at very low energies, we might expect the variance in the number of electrons produced to increase, along with the Fano factor.

One of the first to apply Fano's ideas to a semiconductor was van Roosbroeck [28] at Bell Laboratories, using a geometrical analogy that he called 'crazy carpentry'. He used both an analytical and a Monte-Carlo approach. He first considered a simple case where there is only one type of secondary carrier produced (electrons, say). This is similar to Fano's model, where the energy of the positive ion of the gas was neglected. Representing the available incident energy E_p by a wooden board of length L (in units of the ionisation energy I) and the impact ionisation by a random division of the board at a point x along it, the two remaining boards represent the energies of the two secondary electrons available for further ionisation. To cause further ionisation, the electron must have energy $E > I$. At first, $L \gg 1$ normally, but of course if any random cut gives rise to a board of length $L < 1$ it cannot go to the next stage and is discarded. This discarded energy will be dissipated by other unspecified mechanisms. Because the ionisation energy (I) was required in the ionisation process, a unit length of board must be cut off (but not discarded) from one of the remaining boards. This cut is not discarded, because it represents an electron of the threshold ionisation energy (and hence produces no more secondary electrons) but adds 1 to the final yield, n_0, of electrons. Thus, after the first ionisation, we have one unit length of board (or one unit of electron yield) and two boards of lengths x and $L - x - 1$. We apply the same rule with random cuts to each of these boards with unit length of board being cut off as one unit is added to the yield. It is clear that for large L the 'waste' will initially be negligible at each random cut, but as the process is repeated and the lengths get shorter, the waste will then become appreciable. Toward the end of the process, when there are very many boards remaining having lengths of just a few units, the waste will be a significant part of the total lengths. This is, as already stated, because after each cut all lengths less than unity will have to be discarded. We now compute the total yield, n_0 of electrons, after the whole ionisation cascade by counting the final number of boards of unit length (but not those less than unity). Obviously, the yield or number of boards each of unit length will be proportional to the starting length L but will be significantly less than L itself because of the waste involved.

Following van Roosbroeck, we can write

$$n_0 < E_0/I$$

and define a relative yield y by

$$n_0 = yE_0/I.$$

In fact, the average energy required to produce one secondary electron, ω, is given by Equation 2.1 above and

$$n_0 = E_0/\omega, \text{ where } \omega = I/y.$$

This crazy carpentry view is consistent with the idea that ω and F come about because of the partition of energy into ionisation (which is 'detected') and other unspecified processes that are not. The relative yield y gives the partition fraction not going into ionisation. For example, for silicon where $\omega \sim 3.6$ eV and $I \sim 1.1$ eV, this fraction not going into ionisation is $y \sim 1/3$. Of course, with the random nature of the board cutting, there are many possible outcomes for n_0. There will be a scatter of values with a variance that defines the Fano factor.

$$F = \langle n_0 \rangle^2/n_0 \qquad (2.3)$$

The significance is that ω and F can be calculated *mechanically* from the model independent of the details of the physics. In fact, van Roosbroeck carried out this crazy carpentry calculation in 1965 and obtained a yield of 0.56. For silicon, he obtained a value of ~3 eV for ω and ~0.12 for F from this simple model. The fact that these results are fairly close to the experimental values for Si at room temperature accepted today is remarkable and somewhat fortuitous. The model goes a long way to explain why the figures are so similar in terms of the ionisation energy for different semiconductors and indeed even the gases in proportional counters. The model is almost *independent of the physics*. The values of ω and F are seen to be determined largely in the final stages of the cascade when most of the electrons are generated. The final stages are relatively insensitive to the details of the high energy process and are largely independent of the type and energy of the incident radiation.

The above simple model was only used to illustrate the general nature of the problem. Van Roosbroeck went on to refine it to more closely describe the situation in real semiconductors. This allowed for the facts that (1) holes as well as electrons are produced during ionisation, (2) plasmons (coherent oscillations of the valence electron plasma of the semiconductor induced by the hot electrons) are also excited and have enough energy (16.9 eV in Si) to cause ionisation when they decay (but can produce no further ionisation), and (3) phonons are produced not only when the electrons and holes thermalise but also during the cascade process itself. In terms of the crazy carpentry, the board lengths are divided again to represent the energy taken by holes as well as electrons (he assumed an equal division that will not be correct for an indirect gap semiconductor such as silicon). Besides cutting off a unit length after a random division, we have a probability r that an extra length is cut off to account for loss to phonons. This refined model applied to silicon gave $y = 0.75$, that is, ¾ of the energy goes into phonon production and $F \sim 0.05$. Some researchers believe that the true value may in fact still be close to this value. All the energy is eventually dissipated as phonons (heat) and cryogenic bolometers working at liquid helium temperatures actually attempt to measure this quantity of heat. This results in energy resolutions of a few eV.

Publications of calculations of ω and F are not very frequent, but as we might expect they have become more refined over the years, taking into account more effects and details of the processes involved in the ionisation. Monte-Carlo calculations making some simplifying assumptions have taken advantage of the greater computational powers available and more accurate experimental values for the parameters involved in the calculation. Alig [29] used the same basic assumptions of 'crazy carpentry' but put more detail into the interactions. In a Monte-Carlo calculation, he assumed that in any scattering, all possible product particles are equally probable. For simplicity, he assumed that the ratio of the scattering cross sections for electron–phonon and electron–electron interactions were the same for all energies, for electrons and holes, for all materials, and for all ambient conditions. He calculated that ω is constant (3.64 eV) for silicon above 100 eV and $\omega = 2.99$ eV for Ge. The agreement with experiment is generally also good for other materials. He obtained an F value of 0.113 ± 0.05 for silicon and 0.126 ± 0.05 for germanium.

Fraser et al. [17] of the University of Leicester, UK, took a more fundamental approach starting from basic principles. They followed the various relaxation pathways in silicon in detail resulting from an electron vacancy produced by the photoelectric effect or charged particles in the various atomic shells of silicon. The atoms relax to the ground state by the migration of the initial vacancy to the outer shells through the emission of characteristic x-rays and/or Auger or Coster–Kronig electrons (see below). For x-rays of an energy E that is above the K absorption edge at 1.84 keV, ~92% of these occur in the K-shell. Besides a photoelectron of energy $E = 1.84$ keV, there will be numerous other 'primary' energetic electrons to be considered. Most numerous will be the Auger electrons ejected from the less tightly bound shells. These have energies between 1.5 and 1.8 keV with varying emission probabilities. There will also be a small (~4%) probability of a K-shell fluorescence x-ray of energy 1.74 keV and an even smaller probability of L-shell fluorescence. As Fraser was mainly concerned with spectra from thin CCDs, he assumed these fluorescence x-rays escaped but in general they may be absorbed and contribute to the photopeak. The relaxation process is followed until all vacancies are transferred to the outermost M-shell. Coster–Kronig transitions modify the primary vacancy distribution by filling vacancies with electrons from the *same* shell. The electrons ejected in this relaxation are called Coster–Kronig electrons. Another effect to be taken into account is attributable to the sudden charge loss of the atom on the creation of the electron vacancy. This causes an abrupt change in the Coulomb potential well experienced by all the other electrons and finally results in the probability of 'shake off' electrons from the outer shell. The consequence of all these sources of 'primary electrons' and their respective probabilities is taken into account in the calculation.

The energy loss mechanisms of these primary electrons include phonon generation, K- and L-shell ionisation, valence band ionisation (hole creation), and plasmon excitation. Plasmons can be considered quasiparticles that couple the fast electrons with electrons in the valence band and the resonance decays creating electron–hole pairs. However, these have insufficient energy to create further ionisation. It is found that despite being ignored in most previous calculations, plasmon excitation is the most probable energy loss mechanism for electron energies above about 80 eV. The actual energy loss in the plasmon excitation varies from 25.6 eV at its threshold and falls asymptotically to the optical plasmon energy of 16.6 eV for high energies. The average electron–hole yield for plasmon decay is estimated as 5, but these have insufficient energy to be further cascading. Thus, although it is most probable, as Alig also showed, it does not dominate the energy loss along the track.

The Monte-Carlo calculation uses the cross sections for the various primary electron energy loss processes listed above to calculate the number of 'thermal' electron–hole pairs (and hence ω) in the residual charge cloud. The spread in values found in the Monte-Carlo calculations gives a measure of the Fano factor F by Equation 2.3. They found that low energy peaks ($E < 500$ eV) were not symmetric having excess events in the high side of the peak. If this is indeed true, the measured values of F at these energies are underestimated by fitting a Gaussian shape to the photopeaks [15].

The agreement between the results given by Fraser and Alig is not good. Fraser et al. gave $F = 0.16$ for silicon at 5.9 keV at liquid nitrogen (LN) temperatures, whereas Alig gave a value of 0.113. Further theoretical and experimental values of F for silicon are summarised in Figure 2.1.

2.4.2 Behaviour of ω and F with Energy

Because most electron–hole pairs are generated toward the end of the cascade, we do not expect a strong dependence of ω

or F on the initial energy of the x-ray, or indeed a dependence on the mode of creation of the initial shell vacancy. We might expect that at very low energies where the cascade is relatively short there may be some effects, and Fraser et al. calculated an increase in ω from 3.65 eV (at 8 keV) to 3.80 eV (at 280 eV). Alig calculated a more modest increase with 3.70 eV above 1 keV, increasing to 3.73 eV at 200 eV. For further discussions, see Lechner and Struder [30]. Recently, a Monte-Carlo calculation [31] showed fluctuations below the silicon L-shell energy (110 eV) but a constant value of ~3.7 eV (at T = 150 K) above this and indeed above the K-shell energy. The spectral deviation from linearity in Si(Li) detectors is not observed experimentally across the 1.84 keV K-edge and probably not across the Al L-line at 66 eV with data from a CCD [32]. The deviations commonly observed at low energy, for example, in C Kα peak at 277 eV, are due to ICC as higher energy peaks do not show them.

Turning to the Fano factor, Fraser et al. calculated an increase of ~14% in F from 300 eV going down to 150 eV. An increase in F has also been reported experimentally for Si(Li) detectors at low energies [25], where the increase was ~15% at 677 eV and 35% at 277 eV from that at 5.9 keV. However, the results from CCD measurements [32], albeit with poor statistics, and recent resolution results from some SDDs (Chapter 9) do not generally support such large increases at low energy. The values at 5.9 keV were found to be 0.116 for silicon. For germanium, it was found to be 0.106 at 5.9 keV but followed closely that of silicon below 1 keV. Studies of escape peak widths might be useful in confirming or otherwise the reality of these increases, but such studies are made difficult by rising background at low energy and insufficient counts in the escape peaks.

2.4.3 Behaviour of ω and F with Temperature

The value of ω is related theoretically to the value of the indirect bandgap E_g. A linear relationship

$$\omega = aE_g + b \tag{2.4}$$

was suggested by Alig et al. [33]. The general linear relationship is shown in Figure 10.2 for a number of semiconductors. Values of a have been measured [34] using a γ-ray source and silicon detectors and range from 2.77 to 2.87. It has been known for many years [35] that E_g decreases with temperature, as would be expected because of the increase in lattice spacing (see Figure 1.9). Varshni [36] proposed that, for silicon,

$$E_g(T) = E_g(0) - \beta T^2/(T + 1108),$$

where $\beta = 7 \times 10^{-4}$ eV/K.

The theoretical and experimental dependence of ω with T have been reviewed by Groom et al. [37] and agree reasonably well with the above relations, that is, a linear dependence on E_g rather than T. Most of the measurements have been made for high-energy γ-rays or charged particles. Within the range achievable with LN (80 K to room temperature), some researchers [20,21] have not detected any deviation from linearity with T. Within this temperature range the slope is measured to be -0.013%/K.

Fraser et al. [17] point out that, apart from E_g, the phonon emission and absorption coefficients also influence ω, and they calculated a slope of -0.01%/K. As a consequence of the decrease in ω, we might expect improved energy resolution because $\sqrt{\omega}$ occurs in the detector dispersion (see Equation 2.2). Also, as we have seen in Equation 2.1, the inverse of ω can be regarded as part of the intrinsic detector gain, so that as the temperature increases the photopeak shifts up in the channel spectrum in proportion to the change in ω. Recalibration then results in a decrease in the electronic noise by the same factor. This applies to noise both in the processing channel and in the threshold discriminators. These remarks, of course, also apply to materials having a lower bandgap (and ω) than silicon such as germanium. Thus, we might expect that, all other things being equal, the resolution and low energy sensitivity should improve as the temperature *rises*. Of course, this improvement is generally not realised in practice as in silicon and germanium and many other semiconductors, the thermally generated leakage currents dominate the noise at higher temperatures. However, SDDs have now been developed (see Chapter 9) that take advantage of the lower ω near room temperature and exhibit state-of-the-art resolutions at process times that avoid current noise.

As the gain changes with temperature, temperature stability is important. In the case of detectors cooled by LN open to the atmosphere in Dewar flasks, this is achieved in practice by the constant boiling point of LN; however, other cases, such as Peltier cooling, require special consideration. One must also keep in mind that outgassing in a cryostat over time can increase the temperature.

No dependence of F with temperature has been predicted [17,31] or indeed measured, and it is usually assumed to be constant with temperature.

2.5 Charge Collection Efficiency Function

The charge collection efficiency (CCE) function $\eta(z,r)$ can be defined as the charge induced onto the detector FET gate, that is, 'collected', divided by that generated by the x-ray ionisation event (q) at depth z from the crystal front face and radial distance r from its axis of symmetry. This is represented in Figure 2.2 where the thickness of the detector is D.

Neglecting bulk trapping (as described by the Hecht equation) for the moment, the function will vary, in general, from zero—where the charge lifetime is negligible—to unity in the bulk crystal lattice of the semiconductor. Examples of zero lifetimes are in the front metal contact (which is as thin as possible

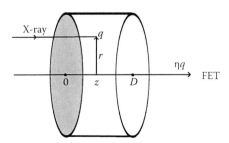

Figure 2.2 Charge deposited in the detector.

without adding resistive noise) and the back metal contact, relatively thick lithium diffusion or implant. Between these extremes there will be regions of ICC. For example, we expect a gradual fall off in the magnitude of $\eta(z,r)$ as z approaches zero (charge loss by charge diffusion to the metal contact or because of diffusion of the metal into the semiconductor) and as r approaches the crystal diameter (edge effects where the field weakens).

The CCE function $\eta(z,r)$ associated with each photo-event will, in general, cause the count to fall into a channel below that corresponding to E_0. For example, if $\eta = 0.5$ for a particular photo-event, then this will be counted in the channel corresponding to $E_0/2$. Taking all photo-events into account, with all values of η represented in the bulk, we will not have all counts in a single channel at E_0 but a spread of channels occupied from E_0, in principle, down to zero energy. The Gaussian spread discussed in Section 2.4 will, of course, be superimposed on this distribution.

2.6 Generation of Spectral Response Function

In Section 1.8.1, we demonstrated a geometrical interpretation of the Trammell–Walter equation. There, we took an example of hole-trapping to generate the efficiency function $\eta(z,r)$ from the Hecht equation and generated the response function. In general, the actual form of $\eta(z,r)$ will depend on the physical model of the detector used. The most appropriate models will depend on the geometry, physics of the charge generation and loss, and also the details of the manufacturing processes used. These models mostly included a parameter d describing the depth from the front surface in which there is some ICC. This depth can easily be estimated if the tail component can be separated from the peak and background components in a first approximation spectrum as shown in Figure 2.3.

Calculations are simplified by assuming a homogeneous medium. Then, the fraction of incident x-rays at a depth z after removal by the photoelectric effect is predictable and given, as we saw in Section 1.1, by Beer's law:

$$I = I_0 \exp(-\mu_a \rho z) \tag{2.5}$$

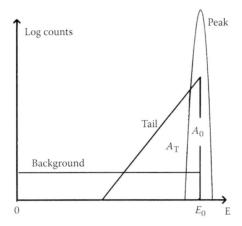

Figure 2.3 Spectral components.

This is often written in terms of a linear absorption coefficient $\mu = \mu_a\rho$, and the 'exponential range' (given by the depth by which the intensity has fallen to a value $1/e$ of its original incident magnitude) is thus $1/\mu$ with units of the depth z. The number of counts falling into the 'tail', A_T in Figure 2.3, will be found by integrating Equation 2.5 between $z = 0$ and $z = d$. Assuming $1/\mu \gg d$, as is usually the case, and the number in the photopeak, A_0, is large compared to both the number in the tail and the background

$$A_T/A_0 = 1 - \exp(-\mu d) \sim \mu d.$$

Here, μ must be selected for the material of the detector and energy of the x-ray. Such crude approximations generally lead to $d \sim 50$–300 nm in Si(Li) detectors and is more or less independent of energy. There is usually also some radial dependence in d, even with collimation.

Whatever form $\eta(z,r)$ takes, the response function (or energy spectrum), resulting from a monochromatic x-ray line energy E_0, can be generated from it, and the energy axis of the spectrum is a linear mapping of the efficiency function $\eta(z,r)$ for each event with the statistical broadening, already discussed in Section 2.4, folded in. We will discuss bulk carrier trapping and the radial effect in detail later and begin by looking at the depth z relation only.

By differentiating Equation 2.5, the fraction of x-rays absorbed between depth z and $z + dz$ is given by

$$g(z)dz = \mu \exp(-\mu z)dz. \qquad (2.6)$$

Here $g(z)$ can be regarded as a 'charge generation' function as each x-ray absorbed gives rise to a charge cloud q centered at depth z. This assumes that the charge cloud is centered, on average, at a point close to the interaction in comparison with the dimension D of the semiconductor crystal. Because of the short range of the Auger and photoelectrons in most x-ray applications (as distinct from γ-rays where scattering is more dominant), this assumption is usually valid. We will assume that the charge cloud at any time extends from the origin of the photoelectric event at z by a range R. This will be dominated by charge diffusion and the electric field (see Section 1.7.1), rather than the photoelectrons' range.

Let us consider as an example an x-ray of energy 1 keV. If we build a histogram using 10 eV/channel and we consider just one channel, say from 500 to 510 eV, the corresponding value of $\eta(z)$ changes from $\eta' = 0.50$ to $\eta'' = 0.51$. To count the fraction of events that fall into this range of $\eta(z)$, we need to integrate the charge generation function 2.6 over the range of z between the limits z' and z'', where z' and z'' are the *inverse functions* of the two boundary limits η' and η''.

$$z' = \eta^{-1}(0.50)$$

$$z'' = \eta^{-1}(0.51)$$

The fraction of events falling into this channel is thus given by

$$f' = \mu \int_{z'}^{z''} \exp(-\mu z)\,dz. \qquad (2.7)$$

This process is repeated for all values of z up to the active depth of the crystal D. In general, numerical methods of integration are necessary to evaluate 2.7. The alternative technique is a Monte-Carlo calculation, which will be discussed below. The analytical approach has the advantage of making all the processes rather more transparent than in the Monte-Carlo method.

The process of integration to generate the frequency spectrum can be seen graphically in Figure 2.4.

This particular arbitrary form of $\eta(z)$ was first used by Bloomfield et al. [38] to explain the low energy 'spur' in the spectrum seen in Figure 2.4 and in some x-ray spectra (see Section 3.2). The areas under the exponential term on the right represent the integral Equation 2.7, and these values are entered into the channels of equal efficiency on the vertical axis on the left. This builds up the frequency spectrum with the 'E_0 full energy' channel (containing the fraction of events with 100% CCE) at the top and the 'zero' channel (very low charge collection) at the bottom. Once this distribution has been generated from an analytic form of $\eta(z)$, the statistical variance due to the dispersion at the energy and noise (Equation 2.2) can be added.

Kalinka [39] proposed a more elaborate model using the Hecht equation to include both electron and hole trapping in Si(Li) detectors and a variance σ^2_{coll} taking the effect of trapping into consideration. The variance σ^2_{coll} therefore depends

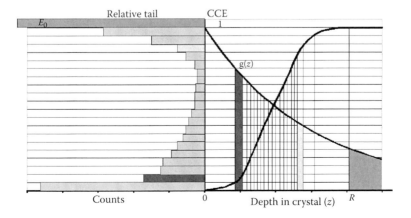

Figure 2.4 (See colour insert.) Spectrum generated from arbitrary charge collection efficiency (CCE) function $g(z)$ on right-hand side of figure.

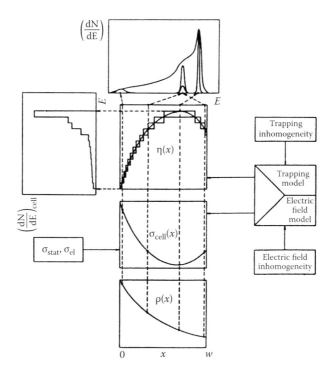

Figure 2.5 Line shape formation. (After Kalinka, G., Atomki Annual Report, 91 [Debrecen, Hungary], 1991.)

on depth x (in his notation) of the interaction shown in his schematic for the line-shape formation in Figure 2.5.

2.7 Charge Loss Mechanisms

A fortuitous advantage of energy dispersive x-ray spectrometry (EDXRS) over charged particle spectrometry is often not fully appreciated: the x-ray does not, in general, start the ionising process immediately on entering the solid and thus is less influenced by the entrance window surface. Scattering and straggling is not the same issue in EDXRS. By Beer's law, most penetrate into the bulk completely unaffected by the medium until the photoelectric absorption takes place. Except for the very lowest energy x-rays, the majority of the ionisation takes place well away from the influences of the window. Semiconductor detectors are therefore less sensitive to absorption in the materials used in the engineering of the window contact and indeed of any vacuum window material present. Having said that, Beer's law also states that $z = 0$ is the *most probable depth* for an ionising event, that is, *at the entrance window*, but the probability is low. This makes the form of $\eta(z)$ critical as z approaches zero and the entrance window. To avoid tailing at any x-ray energy, $\eta(z)$ needs to approximate to a step function and requires special consideration and will be discussed in Section 2.7.3.

2.7.1 Escape Peaks

The topic of escape peaks was introduced in Section 1.14.1. Here, the penetrating nature of x-rays works against us, and the escape of the fluorescence x-rays, generated in the detector material at some depth, to the surface or boundary is reasonably probable. In most cases, however, the peak is predictable and can be easily accounted for. For example, at 5.895 keV (Fe-55 Mn Kα) in silicon, the escape peak is at 5.895–1.740 keV below the parent x-ray line and ~2% of its intensity. This is because the silicon Kα x-ray has an energy of 1.740 keV and the fluorescence yield in silicon (see Table 1.7, Section 1.14.1) is ~4.7%, but only about half of these will radiate toward the entrance window. The stripping of the escape peak from the spectrum is also made easy by the fact that the escape peak FWHM will be approximately the same as that of the parent line being dominated by noise at low energy and close to the parent at high energies. For heavier semiconductors such as germanium and CdZnTe, this is not the case and the relative intensities are much higher. Furthermore, there are usually more of them to take into consideration (as noted in Section 1.14.1). There has been some discussion over the years on apparent anomalies regarding the position and intensity of escape peaks in silicon [40,41].

2.7.2 Loss of High-Energy Electrons

The low fluorescence yield in silicon implies that the probability of the production of only high-energy (or 'hot') electrons (photo- and Auger electrons) is the dominant process in photo-absorption. Although these electrons do not have the penetrating power of the fluorescence x-ray, they too may escape from the active volume of the semiconductor. For collimated crystals, the most probable loss will be to the front contact. If the event occurs within the range R of the metal–semiconductor interface, not only is the charge of the hot electron lost but also its potential to produce multiple ionisations by virtue of its energy. In order to estimate the charge that will be lost, we need to estimate the energy lost by an electron of energy E_e along such a track length t, and in this respect the rate of energy loss with distance along the track is relevant. This is given by the 'stopping power' $S(E_e)$ of the material defined as

$$S(E_e) = dE_e/dt \qquad (2.8)$$

The stopping power of electrons in the semiconductor can be taken from numerous sources, both experimental and theoretical, the most commonly cited being Fitting's formula [42]. There are many others [43–46], all giving reasonable agreement, and the form is relatively flat increasing only as the electron approaches the loss of all its energy as it slows, as shown in Figure 2.6.

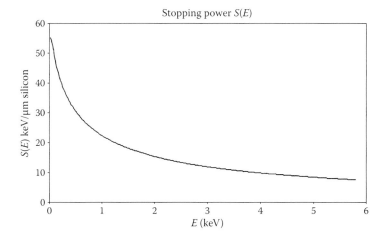

Figure 2.6 Stopping power $S(E)$ (given by Iskef, H. et al., *Phys. Med. Biol.*, 28, 535, 1983) for Si.

This relatively constant loss of energy of the hot electron for most of its range gives the background due to primary electron escape a characteristically (and fortuitously) 'flat' form. Indeed, as we have remarked, it is often referred to as a 'shelf'.

2.7.3 Straight Electron Track Model

Figure 2.7 [11] depicts the possible trajectories to the metal front face of the detector by one of these hot electrons.

This model assumes, for simplicity, straight-line electron tracks of range R_{ex} in silicon, but in practice both scattering and straggling will affect these tracks. R_{ex} therefore represents the maximum (or 'extrapolated') electron range. It is instructive to first consider cases where the range of the electron R_{ex} is much shorter than the exponential range of the incoming x-ray, which as we have seen is $1/\mu$. This occurs for higher energy x-rays as shown for silicon and germanium in Table 2.1.

Here, we use the values of μ taken from Henke [47] and R_{ex} the 'extrapolated electron range' given by Iskef et al. [43]. We see that for all but the lowest energy lines the approximation is good. In such cases, the charge generation function $g(z)$ given by Equation 2.6 will approximate to μ.

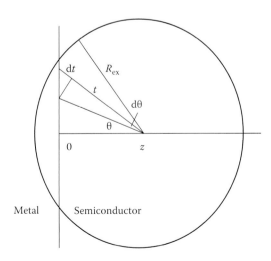

Figure 2.7 Representation of hot electron trajectory to front face.

TABLE 2.1
μR Product for Different Energy Photons in Both Si and Ge

X-ray Line	Energy (eV)	μR_{ex} (Silicon)	μR_{ex} (Germanium)
B	183	5	4
C	277	2	2
O	525	0.5	0.7
Al	1487	0.035	0.5
Si	1740	0.005	0.3
Ca	3691	0.01	0.04
Mn	5895	0.009	0.014

If the electrons are emitted isotropically, the probability $P(t)$ of an electron track length between t and $t + dt$ before escape is independent of the value of t. This can be shown analytically from Figure 2.7 [11] and statistically [10] in Monte-Carlo calculations.

In general, it can be shown by simple geometry that, for an isotropic distribution at any depth $0 < z < t$, the probability of any track length between t and $t + dt$ is given by $z/2t^2 \cdot dt$ [11,13]. Integrating over the whole range of z, from 0 to t (the maximum value), gives a value of 1/2. This is true of all track lengths but will not be true for nonisotropic distributions. There will be an additional factor 1/2, because only half the electrons travel back toward the metal surface contact. In the approximation described above, the probability of x-ray absorption (and charge generation $g(z)$) has the value μ over this range of z; thus, the probability $P(t)$ of an x-ray giving electron track length between t and $t + dt$ is given by the product of the probability of x-ray absorption and the probability of finding an electron track of length t.

$$P(t)dt = \mu/4 \ dt \qquad (2.9)$$

If the electron has energy E_e on escaping and the initial energy was E_p, then the energy deposited in the semiconductor is

$$E = E_p - E_e. \qquad (2.10)$$

The energy lost by an electron of energy E_e over an element of track dt is given by Equation 2.8, and from Equation 2.7 the probability of electrons depositing energy between E and $E +$ dE in the crystal on escaping is given by

$$P(E)dE = \mu/4S(E)dE, \qquad (2.11)$$

where E is given by Equation 2.10.

Thus, we see that the magnitude of the background generated in an x-ray spectrum by the escape of primary electrons is dominated by μ (and is therefore larger for low energy x-rays), and the shape is largely dependent on the form of the inverse of the stopping power of the electrons in the media.

For example, for 5.9 keV x-rays in silicon $\mu = 0.035$ μm^{-1} and at 3 keV, $S(E_p - E) \sim 15$ keV μm^{-1}.

Thus, for $dE = 10$ eV (10 eV/channel),

$$P(E)dE \sim 0.035/4/15,000 \times 10 = 5.8 \times 10^{-6}.$$

Figure 2.8 gives the result of calculations for the contributions to the background for primary electron escapes (L- and M-shell photoelectrons, Auger electrons, and photo-Auger electron coincident escapes) in germanium at 5.9 keV using the above simple model.

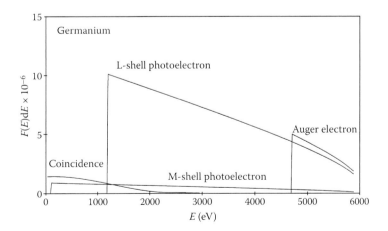

Figure 2.8 Relative contribution of different electron energy losses in germanium.

Figure 2.9 shows the measured (7 million counts in the Mn Kα peak) and calculated background with the predicted step for a HPGe detector at ~1.3 keV (L-shell absorption edges).

The peak at about 4.7 keV in Figure 2.9 is just above the germanium L-escape peak and is due to electrons from the metal contact. It will appear in all spectra using this contact (10 nm Ni) and is explained in Section 2.8.1.

The situation for silicon detectors is improved upon this as the photoelectrons from the K-shell are emitted preferentially at right angles to the original x-ray direction [48]. This has the effect of reducing the probability of escape and therefore the background magnitude by a factor of 2. Figure 2.10 shows the measured (4 million counts in the Mn Kα peak) and calculated background for a Si(Li) detector with the predicted step at ~1.84 keV (silicon K-shell absorption edge).

Figure 2.11 shows the Monte-Carlo simulation [10] for a Si(Li) detector for 5-keV x-rays again using the electron range formula of Iskef et al. [43]. This again clearly shows the background step due to the escape of K-shell photoelectrons, and the magnitude is in good agreement with experiment.

Naturally, in many applications where x-rays are excited by electrons [scanning electron microscopy (SEM), transmission electron microscopy (TEM)], this is largely academic because the background is dominated by Bremsstrahlung radiation in the stopping of the electrons. It is also rather academic where

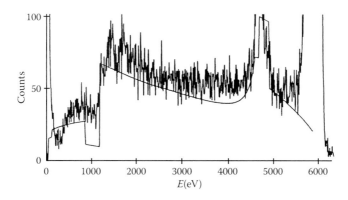

Figure 2.9 Comparison between observed and predicted backgrounds in an HPGe detector using the simple model.

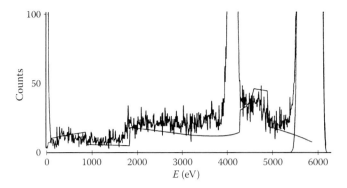

Figure 2.10 Comparison between observed and predicted backgrounds in a silicon detector using the simple model. (From Lowe, B.G., *NIM A*, 439, 247, 2000. With permission.)

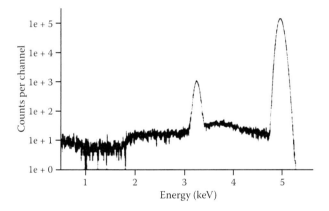

Figure 2.11 Monte-Carlo simulation for a Si(Li) detector. (From Campbell, J.L. et al., *NIM A*, 418, 394, 1998. With permission.)

compound semiconductor detectors are concerned because other factors such as trapping usually dominate.

2.7.4 Simple Charge Diffusion Model

Most researchers agree that the last process in the ionisation regime to influence charge collection is the simple diffusion of the charge carriers, electrons, and holes, in all directions before they are separated by the electric field. This was first proposed as a significant mechanism in 1970 [49] but put on a sound basis in 1977 in a classic paper by the Berkeley group [8]. The concept was discussed in Section 1.7.1. As an example of how spectra can be generated from the efficiency function $\eta(z)$, we will now consider a simple charge diffusion model. The photoelectric effect produces energetic or 'hot' electrons (photo- and Auger electrons) as discussed above. These rapidly lose energy by ionisation until the electrons generated can lose no further energy by ionisation and their range increases only by their diffusion. These 'thermal' electrons and their corre- sponding holes are eventually swept apart by the electric field. If any hot electrons escape the active volume of the detector, they produce 'background' counts in the spectrum as we have seen. Here, we will restrict ourselves to those events in which no *hot* electrons escape, and all the energy E_0 goes into gener- ating n quasi-free thermal electrons and holes (Equation 2.1). We will have a significant carrier concentration gradient in the crystal as it is depleted of free charge in the quiescent state, and the overall tendency is for the cloud of carriers to diffuse outward very rapidly. The effect of the field, electro- static repulsion, and recombination of electrons and holes are secondary to this diffusion effect at this stage. We will assume that they diffuse to a sphere of radius R before the field sepa- rates the carriers. In Section 1.7.1 (Equation 1.26), we derived the expression $D/\mu_{mob}E$ for this radius, where μ_{mob} is the carrier mobility in field E and D is the diffusion constant given by the Einstein equation. In the energy ranges we are considering (0–20 keV), the range of the hot electrons is small compared to R, which will not be a strong function of E_0 but will depend on the applied field E. We expect that if the radius of the sphere R impinges the metal contact or detector walls there is a poten- tial for ICC, and it is therefore closely related, if not identified with, the depth d of ICC (discussed in Section 2.6). For fields of ~100 kV/m, it is of the order of 200 nm in silicon. Because the field usually falls off in the detector radially, R will also increase radially, giving rise to edge effects that are discussed in Section 3.1. Holes generated near the front contact will not contribute significantly to the signal (Shockley–Ramo theo- rem), so we are really concerned with the diffusing electrons as depicted in Figure 2.12.

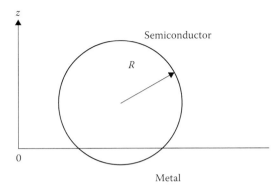

Figure 2.12 Loss of electrons to metal contact due to diffusion sphere of radius R.

The charge lifetime of thermal electrons in the metal contact is negligible, and therefore we will assume that all charge in the region $z < 0$ is lost and does not contribute to the signal. This is not true, however, for hot electrons generated *in the metal*, as we will see later. It can be shown geometrically that the fraction of volume of the sphere outside the metal in Figure 2.12 is given by

$$f(z) = 1/4(Z^3 - 3Z + 2), \qquad (2.12)$$

where $Z = -z/R$.

If we assume that we have a *uniform* distribution of electrons in the sphere (a better approximation would be a Gaussian distribution), this is also the fraction of charge collected by the field and, neglecting bulk trapping, Equation 2.12 represents the CCE $\eta(z)$.

The maximum electron loss occurs when $z = 0$, that is, when half the electrons (those diffusing into the metal) have been lost. We have seen that under normal operating conditions, R is less than ~1 μm. For all but the lowest energies, we can assume that within this range the charge generation function 2.6 is approximately constant (~μ). This situation is illustrated graphically in Figure 2.13.

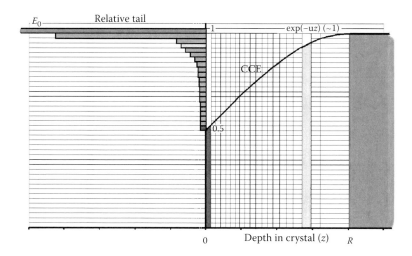

Figure 2.13 (**See colour insert.**) Populating bins or channels in a simulated spectrum.

The 'count frequencies', as a function of CCE (or x-ray energy), are represented by the areas of the rectangles as shown. There are no counts below the channel corresponding to energy $E_0/2$.

For $z \ll R$ to a first approximation, $\eta(z) \sim 1/2 + 3z/4R$.

Thus, $\eta(z)$ increases linearly from 0.5, the rectangular areas are equal and the truncated tail is flat. As z approaches R, $\eta(z)$ curls over with zero gradient at a value of unity making the inverse function increase rapidly into the peak. The response function for this simple model is thus a low-energy tail rapidly falling from the photopeak at E_0 to a 'shelf-like' tail truncated at energy $E_0/2$. When the Gaussian, due to Fano and electronic noise is added, the variation with R can be calculated at a given energy and is shown in Figure 2.14 for a 5.9-keV x-ray in silicon. An R value of ~0.2 μm fits for most Si(Li) detectors.

This is, in fact, very similar to the spectra obtained with a very poor or contaminated detector, or one run at very low bias, as shown in Figure 2.15.

At low bias, R (and the tail) increases and spurious peaks ('ghost peaks') appear because of the presence of undepleted material. The latter artefacts are discussed in Section 3.4.

The large white rectangle on the right of the vertical axis up to the orange region in Figure 2.13 has an area representing

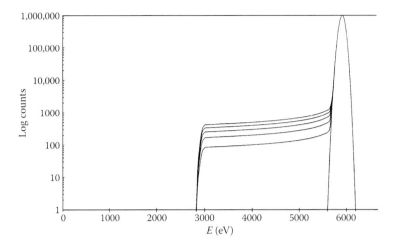

Figure 2.14 Calculated spectrum for different values of R. (From Lowe, B.G., *NIM A*, 439, 247, 2000. With permission.)

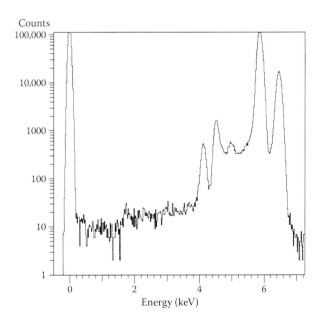

Figure 2.15 Example of an Fe-55 spectrum obtained from a Si(Li) detector operated at very low bias.

the counts falling into the tail ($\eta < 1$). However, the *total* area (including $\eta = 1$) is the integral

$$\int_0^\infty \exp(-\mu z)\,\mathrm{d}z = 1/\mu.$$

Thus, the fraction of events in the tail is given simply by μR, confirming our result given in Section 2.6. This will be true of any form of $\eta(z)$ or tail shape because—irrespective of the form of η—regions where $\eta < 1$ will generate counts in the tail. It is clear that the form of η near the front contact has a large influence on the photopeak tail and the response function. The engineering of this contact is of critical importance and will be discussed in Chapter 4.

2.7.5 Dead Layers

The ICC described above is the result of the geometry of the charge distribution near the semiconductor boundary interface and does not involve any changes in the general physical properties of the semiconductor, the metal, or the field in the region. In this sense, it is not a genuine 'dead layer'. Care should be taken because in many publications, this effect is lumped together with any region of ICC, including ones giving a low field or low charge lifetime because of physical changes. For the sake of argument, consider a true dead layer of depth z_0 in the surface of the silicon where the CCE after the rapid charge diffusion is negligible for $0 < z < z_0$ (see Figure 2.16). This could be, for example, due to very low charge lifetime in the layer even though it supports some field. Such a region could be, for example, a charge-trapping region because of a poor-quality thick oxide, an implanted or uncompensated layer, or a diffused contaminant in the surface. Kalinka [50] considered the effects of a dead layer caused by the out diffusion of Li in a Si(Li) detector using an arbitrary form of $\eta(z)$. For x-ray interactions in this region, similar charge clouds, as discussed, can diffuse back into regions of good charge collection, and some fraction of the charge can be collected. We

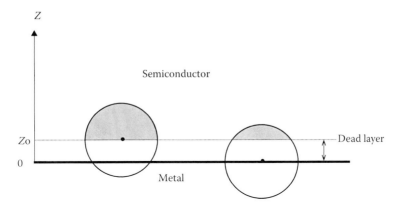

Figure 2.16 Dead layer below metal contact.

can see from Figure 2.16 that the maximum charge loss fraction can exceed 0.5 in this case, and the cut-off energy of the truncated tail will be depressed below $E_0/2$. This is commonly observed in practice.

We will see in Section 2.8.1 that photoelectric interactions in the metal can result in fast electrons entering the detector and can contribute significantly to the background in a spectrum. We will see in Chapter 4 that a good-quality thin oxide layer will not produce a field-free region and will not act as a 'dead layer' in this sense.

2.7.6 Very Low Energies

We define very low energies as those where the product μR is comparable to or greater than unity (see Table 2.1). If the x-ray energy is <850 eV in silicon or <3 keV in germanium, $1/\mu$ will be of the same order as R, and the exponential term in Equation 2.5 cannot be assumed to approximate to unity. Such a case is depicted graphically in Figure 2.17.

Again, unless we introduce a dead layer, there are no counts below $E_0/2$. We see that the effect of the 'linear' rise of $\eta(z)$ is now a tail peaked at the $E_0/2$ truncation and falling exponentially as $\exp(-\mu(E - E_0)/2)$. The tail then increases rapidly toward E_0. In the extreme case of low energy where very few x-rays penetrate to a depth greater than R, the whole

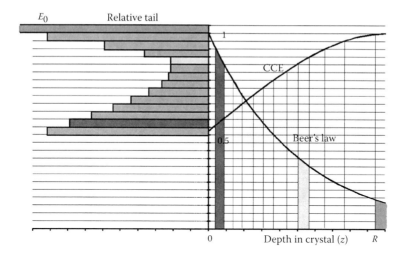

Figure 2.17 **(See colour insert.)** Spectrum resulting from low energy x-rays.

photopeak can essentially become its own tail (see Figure 2.18). Folding in the Gaussian distributions due to Fano statistics and noise, we effectively get double peaking or a downward peak shift. This will occur when the field is very low or at these very low energies. In practice, if we do not collimate

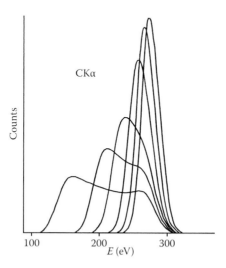

Figure 2.18 Result obtained from reducing detector bias on a spectrum of carbon x-rays. (The peak starts to move down and eventually is dominated by its tail.)

to a small aperture in the center of the crystal face, we see the effect of the radial field falling off at the edges. This can be due to the crystal geometry and will be discussed in Section 5.2. Figure 2.18 shows the effect on the C Kα photopeak of reducing the field. The effect of wider collimation will be a superposition of such spectra with the outer weaker field region shapes dominating. This will occur when the field is very low or generally at very low energies even at normal bias.

These effects result in peak shifts and broadening at low energies. The radial superposition of such shapes probably accounts for the excess broadening of the C Kα peak with a downward peak shift in energy that has been observed [11]. An alternative explanation is the 'patchiness' of the surface barrier heights (lateral and depth variations), which seem to be a characteristic of Schottky contacts. A number of groups have calculated the scale of this patchiness [51], and with the CKα x-ray range in silicon of ~0.13 µm a high proportion of the interactions could be produced in reduced barrier height material.

2.7.7 General Radial Dependence

The condition $\mu R \geq 1$ may also represent the case at higher energies when the field is weakened, particularly at the edges of the front contact, allowing R to inflate considerably. This would also be the case where the detector is degraded, say, by contamination (ice, oil, etc.). Such spectra as those shown in Figure 2.15 are not unknown in practical situations.

2.7.8 Bulk Trapping

This was discussed in Sections 1.7.1 and 1.9.1 (Hecht and Trammell–Walter equations). We saw that hole trapping was generally more common than electron trapping in semiconductors, and the consequence was a long truncated quasi-exponential low-energy tail on the photopeak (see Figure 1.33). The quality of detector-grade silicon and germanium is such that trapping is only likely to be a problem after severe bulk radiation damage, and this is unlikely to occur in x-ray detectors during their normal use. Bulk trapping can also be present in Si(Li) detectors that are not properly compensated, and this will be discussed in Chapter 5. It can be a problem in very large HPGe gamma ray detectors (both hole and electron traps may be present and can be significant over distances of several centimetres), but for our purposes it is restricted only to compound semiconductors (CdTe, CdZnTe, HgI_2, GaAs, etc.). The situation in these will be discussed separately in Chapter 10. For low-energy x-rays, by the Shockley–Ramo theorem, only electron trapping will affect the spectrum and, because of the symmetry of the Hecht equation, shows itself as a low-energy tail that is the 'mirror image' of that for holes for the corresponding drift length, that is, increasing quasi-exponentially in magnitude toward a cut-off at low energy as depicted in Figure 2.19.

In the past, germanium was also lithium-drifted, and de-drifting on warm up was a major problem. As Ge(Li) detectors have not been commercially manufactured since the early 1970s (see Chapter 11), they will not be considered here.

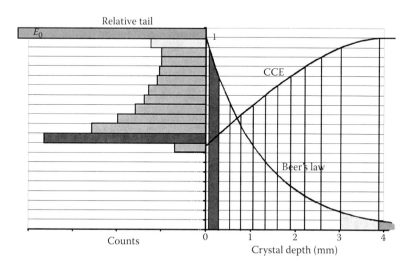

Figure 2.19 (See colour insert.) Response function for 22-keV x-rays with an electron-trap drift length of 2 mm.

2.8 Charge Gains

So far, we have discussed mechanisms whereby charge may be lost from the active volume of the semiconductor crystal. X-rays can also add an effective contribution to the response function from field free regions adjacent to the active volume through either energetic photo- and Auger electrons or x-ray fluorescence.

2.8.1 Energetic Electrons from the Contact

The effect of electrons from the contact has been discussed by a number of authors [3,10,11,13]. For example, Campbell included photoelectrons from a nickel contact of 10 nm thickness in his Monte-Carlo calculation. He was able to extract this contribution for a 5-keV x-ray (which will be the same for both Si(Li) and HPGe detectors) as shown in Figure 2.20 (C.M. Heirwegh and J.L. Campbell, private communication, 2012).

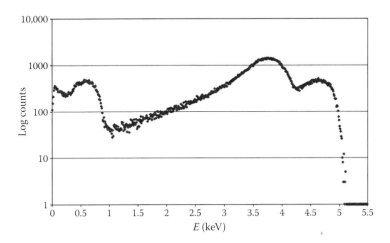

Figure 2.20 Results from a Monte-Carlo simulation using a 5-keV x-ray and a 10-nm nickel front contact showing electron contributions from contact.

When the 5 keV x-ray generates a photoelectron from the Ni L-shell (binding energy ~1 keV), the 4 keV photoelectron loses energy in the metal by the same processes described for loss from the silicon *to* the metal. The spectral shapes are therefore similar to Figure 2.8 but a 'mirror image' of them. The result is a broad hump at about 4 keV with a low-energy tail. The smaller hump at higher energy in Figure 2.20 represents photoelectrons from the Ni M-shell and that below 1 keV from Auger electrons. Figure 2.21 shows the calculated contributions from various contacts.

The general shapes can be explained as follows. If the metals were infinitely thick, the shapes would follow the inverse stopping powers in the metal and have mirror images of the 'flat' characteristic shown in Figure 2.8. For thicknesses d less than the range of the photoelectrons, there is obviously no contribution from depths greater than d, and this results in the rapid falloff at lower energies as shown. The result of going to thinner and thinner contacts is that although the intensity of photoelectrons is reduced, very little energy is lost in the metal, and we see sharp peaking at the corresponding electron energy. These effects are illustrated schematically in Figure 2.22.

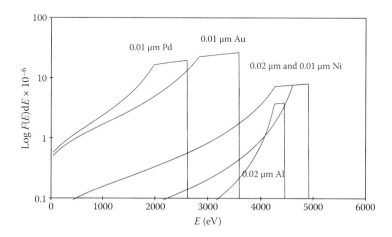

Figure 2.21 Calculated electron contributions from different metal contacts for 5.9-keV x-rays at 10 eV/channel. (From Lowe, B.G., *NIM A*, 439, 247, 2000. With permission.)

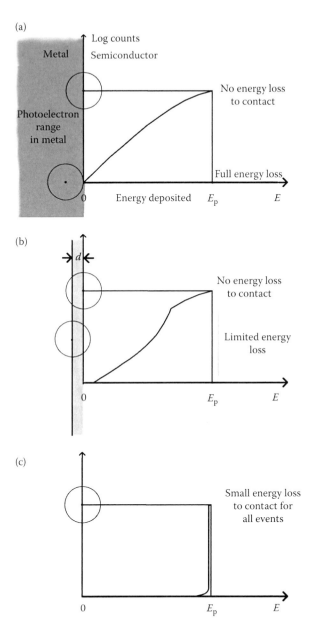

Figure 2.22 Contribution from different thickness metal contacts. (a) Thick metal contact, (b) thin metal contact and (c) very thin metal contact.

2.8.2 Choice of Metal Contact

From the above discussion, it is clear that we have some strategic control over the contribution to the response function from the front contact. Up to the 1980s, this was invariably a ~20 nm gold evaporated contact that, after the growth of a native oxide beneath it, is relatively robust. The lower the atomic number Z (photoelectric absorption varies as ~$Z^{4.5}$; see Section 1.2), the lower the contribution from the metal. In this respect, aluminium would seem a good choice (see Figure 2.21), but this makes an n$^+$ contact and would therefore require a reverse bias (i.e., positively biased window contact). This, in turn, requires a low-noise p-channel FET (which is not as available as n-channel FETs) or a more complex charge amplifier. Aluminium contacts on a Si(Li) detector were in fact realised [52] and due benefits were noted. One spin-off benefit was that the aluminium film (20 nm) has high emissivity and is relatively light-tight. Obviously, the thinner the metal contact, the less the contribution, but there is a limit set by the onset of noise due to the contact resistance. The formation of islands in thin evaporated gold contacts on silicon and germanium limits the thickness to greater than about 20 nm. Figure 2.21 shows that for gold (and palladium, another metal used particularly on germanium), this puts background counts into the low energy region below ~3.5 keV. This was particularly unfortunate as the IEEE standard [53] specified a background to be measured at around 1 keV. Peak-to-background (see Section 1.14.3) was usually around 3000:1 or less for Si(Li) detectors. The introduction of nickel contacts (see Section 11.20) increased this to more than 9000:1 by putting the contribution into higher energy regions, where the background levels were already high. It is also clear that any interface layer under the metal (dead layer or oxide) will reduce the contribution of electrons from the metal even further.

Ion implantation would at first sight seem to be an ideal solution. Boron makes a p$^+$ contact and has very low Z value. Great strides have been made in recent years to achieve very shallow implants (see Section 4.5), and such contacts are frequently used in HPGe, p–i–n diode, CCD, and SDD technology. Often, a coating of aluminium is deposited on the implant [12].

It has not, however, been found to be generally justified for Si(Li) detectors. Contacts for these will be discussed in more detail in Section 4.6.

In practice, the background is often dominated by other sources such as Bremsstrahlung radiation, but in energy dispersive x-ray fluorescence (EDXRF) and particle-induced x-ray emission (PIXE), for a 10 nm thick nickel contact (the practical limit) the contribution from the contact is just detectable particularly at higher energies.

2.8.3 X-Ray Fluorescence

XRF is, of course, nothing more than the occurrence of the photoelectric effect in the surrounding materials rather than in the semiconductor detector itself. In this case, the 'escaping' fluorescence x-ray may be detected. It will be strongest for high Z materials where the x-ray energies are just above the absorption edges of the atomic shells. Electrons—either from an electron microscope (EM) or photoelectrons from the surrounding materials or the semiconductor itself—can also be a source of fluorescence x-rays. Several candidates that may contribute to the response function are listed below:

(1) Dead layers in the semiconductor. For example, dead silicon or a thick oxide, can contribute a Si K x-ray in a silicon based detector.
(2) The metal window contact deposited on the semiconductor. Commonly Au, Ni, or Pd. These are normally <20 nm thick, and contributions can usually be neglected.
(3) The vacuum window material and the structure supporting it. Impurities in beryllium films, silicon, nickel, etc., supporting grids for boron nitride, silicon nitride, or polymer membranes.
(4) Intermediate films used for protection from contamination (cold traps), light, or infrared (particularly in HPGe detectors). Commonly aluminium films on meshes of nickel, steel, etc.
(5) Collimator materials. Usually, basically aluminium with carbon coating, but for TEM and other high

energy work, heavy metals such as Ta, W, or Pb may be involved.

(6) Electron traps. These, commonly used in front of the crystal in SEM work, are ceramic magnets (Fe, Ni, Co, etc.).

(7) Passivation or protection coatings on the crystal side walls. These are usually low-Z materials such as polymers, silicon, and silicon oxides, but may involve heavy elements such as germanium or polymers doped with lead or iodine. These conformal coatings can be excited by both high-energy x-rays and the photoelectrons from them.

(8) Materials at the rear of the semiconductor crystal, particularly solder joints (tin, lead, cadmium, and indium), indium used as a contact pad and the copper cold-finger. This will apply to high-energy x-rays, but in Si(Li) detectors (and other diffused contact detectors) the diffusion layer acts as a protecting screen. Thin detectors such as p–i–n diodes, bump-bonded PAD detectors, and SDDs require special consideration. Aluminium and gold are usually tolerated for XRF.

References

1. Phillips G.W. and Marlowe K.W., *NIM* 137, 525, 1976.
2. Campbell J.L., *XRS* 30, 230, 2001.
3. Inagaki Y. et al., *NIM* B27, 353, 1987.
4. Joy D., *X-ray Spectra in Electron Beam Instruments*, 53, Plenum, 1995.
5. Geretschlager M., *NIM* B29, 289, 1987.
6. Campbell J.L., *XRS* 30, 230, 2001.
7. Goto S., *NIM A* 333, 452, 1993.
8. Llacer J. et al., *IEEE* NS-24, 53, 1977.
9. Kalinka G., *NIM* B88, 470, 1994.
10. Campbell J.L. et al., *NIM A* 418, 394, 1998.
11. Lowe B.G., *NIM A* 439, 247, 2000.
12. Eggert T. et al., *NIM A* 568, 1, 2006.
13. Scholze F. and Procop M., *XRS* 38, 312, 2009.
14. Lepy M. C. et al., *NIM A* 439, 239, 2000.
15. Owens A. et al., *NIM A* 491, 437, 2002.

16. Zulliger H.R. and Aitken D.W., *IEEE* NS-17, 3, 187, 1970.
17. Fraser G.W. et al., *NIM A* 350, 368, 1994.
18. Papp T. et al., *XRS* 34, 106, 2005.
19. McCarthy J.J., *X-ray Spectra in Electron Beam Instruments*, Plenum, 67, 1995.
20. Pehl R.H. et al., *NIM* 59, 45, 1968.
21. Lowe B.G. and Sareen R.A., *NIM A* 576, 367, 2007.
22. Scholze F. et al., *NIM A* 439, 208, 2000.
23. Bilger H.R., *Phys. Rev.*, 163, 238, 1967.
24. Palms J.M. et al., *NIM* 76, 59, 1969.
25. Lowe B.G., *NIM A* 399, 354, 1997.
26. Papp T. et al., *XRS* 34, 106, 2005.
27. Pehl R.H., *NIM* 59, 45, 1968.
28. Roosbroeck W. van., *Phys. Rev.* 139, 5a, A1702, 1965.
29. Alig R.C., *Phys. Rev. B* 27, 968, 1983.
30. Lechner P.S. and Struder L., *NIM A* 377, 206, 1996.
31. Mazziotta M.N., *NIM A* 576, 367, 2007.
32. Kraft R.P. et al., *NIM A* 361, 372, 1995.
33. Alig R.C. et al., *Phys. Rev. B* 22, 5565, 1980.
34. Ryan R.D., *IEEE* NS-20, 1, 473,1973.
35. Smith R.A., *Semiconductors*, Cambridge University Press, 1959.
36. Varshni Y.P., *Physica* 34, 149, 1967.
37. Groom D.E. et al., *SPIE* 6068, 101, 2006.
38. Bloomfield D.J. et al., *XRS* 13, 2, 78, 1984.
39. Kalinka G., Atomki Annual Report, 91 (Debrecen, Hungary), 1991.
40. Van Espen P. et al., *XRS* 9, 3, 126, 1980.
41. Papp T. and Campbell J.L., *XRS* 30, 77, 2001.
42. Fitting H.-J., *Phys. Stat. Sol. (a)* 26, 525, 1974.
43. Iskef H., *Phys. Med. Biol.* 28, 535, 1983.
44. Ashley J.C. et al., *IEEE Trans.* NS-23, 1833, 1976.
45. Matthews J.L. et al., *NIM* 180, 573, 1981.
46. Batra, R.K., *NIM B* 28, 195, 1987.
47. Henke B.L. et al., Atomic and Nuclear Data Tables 27, 1, 1982.
48. Schmidt V., *Electron Spectroscopy of Atoms Using Synchrotron Radiation*, Cambridge Univ. Press, 1997.
49. Caywood J.M. et al., *NIM* 79, 329, 1970.
50. Kalinka G., *NIM B* 88, 470, 1994.
51. Monch W., *Semiconductor Surfaces and Interfaces*, 354, Springer, 1993.
52. Jaklevic J.M. et al., *NIM A* 266, 598, 1988.
53. IEEE Standard 759, 1984 (approved 1981), Standard Test Procedures for X-ray Detectors, New York IEEE, 1984.

3

Detector Artefacts

The contributions to the response function $F(E)$ are all of the type that one would expect to find in any semiconductor x-ray detector, however well manufactured. Obviously, these contributions will vary relatively from one type of detector [Si(Li), high-purity germanium (HPGe), compound and wide bandgap semiconductor, p–i–n diode, pixellated x-ray detector, silicon drift detector (SDD), charge-coupled detector, etc.], but the physics and the general nature of the effects will remain the same. How $F(E)$ varies for each of these will be discussed in later chapters, where each type will be considered in turn along with the processes that influence them. However, there are phenomena that distort $F(E)$ that are largely idiosyncratic to the manufacturing processes, the crystal environments, and sometimes the electronics. Artefacts remain part of $F(E)$ for a particular detector, and the definition in the end is rather arbitrary. For example, the external x-ray fluorescence (XRF) effects (Section 2.8.3) could be regarded as artefacts arising from the crystal housing materials used. Artefacts can vary from detector to detector and manufacturer to manufacturer and even among a manufactured batch. Not all detectors with artefacts are rejected during manufacture and indeed, some artefacts are impossible to eliminate. Users should be aware that they could influence their x-ray spectra.

3.1 Field Distortion

In the ideal planar detector, the field is high and uniform throughout the active volume and the boundary into field-free regions is abrupt. This criterion is seldom met, and we can list some of the characteristics that influence this outcome.

(1) *Crystal edge effects.* The use of contouring (grooves or shelves) in thick detectors and the lack of field confinement in thin planar wafers causes distortion and weakening of peripheral fields. Central anodes and integrated field effect transistors (FETs) in SDDs can also cause distortion (see Section 9.4). The use of close collimation and guard rings (see Section 1.16.3) can reduce this effect at the cost of crystal active area and electrical capacitance. We saw that contouring (Section 1.16.2) with a weakly n-type surface charge on the crystal walls pinches off the surface leakage currents in thick detectors. Field lines at the periphery can go to the surface and, as a consequence, signal electrons in these regions will be trapped at the surface and not reach the back contact resulting, by the Shockley–Ramo theorem, in counts going into the spectrum background. One consequence, for example, is that the background can be influenced by magnetic fields due to the Hall effect [1], and this is a consideration, for example, in transmission electron microscopy (TEM) applications. Where the surface positive charge on the walls is considerable, or the walls are 'chamfered', the electrons may go to the surface throughout the crystal depth, resulting in high flat background spectra. The charge may de-trap slowly in the sense that it flows along the surface and is eventually 'collected'. The evidence for this is that the detector dispersion (as well as, of course, the noise) increases at shorter processing times. This 'slow charge' may take many microseconds for it to be integrated into the signal. This leads to incomplete charge collection (ICC) and $F(E)$ being dependent on the signal processing time.

(2) *Graded contacts.* No contact, even the surface barrier, is truly monatomically abrupt. For example, gold can diffuse into silicon to some depth, giving rise to ICC in the diffused layer. This is a particular problem when high temperatures are involved as in lithium drifting

(as we will discuss in Section 5.1.6). Similar effects can be the result of contamination, for example, by vacuum pump oils (which are to be avoided). In Si(Li) and HPGe detectors, the n-type contact is made by diffusing lithium into the surface at high temperature. This makes a robust ohmic or junction contact, but achieving an abrupt contact is virtually impossible. Where high-energy x-rays penetrate into this diffusion tail, we can expect artefacts such as low-energy tails and shoulders on the photopeaks. In extreme cases, extended volumes of undepleted semiconductor can give rise to well-defined peaks ('ghost peaks'; see Section 3.4). This was also initially a problem with non-fully depleted p–i–n diodes. Ion-implanted contacts have been improved in recent years to make them almost as abrupt as surface barrier contacts.

(3) *Surface irregularities.* Damage (during machining, etching, or scribing) on the surface walls can distort the field locally. Even dust particles on the surface or in the passivation coating can charge up to give local field distortions and artefacts. At high temperatures, metals such as copper (which has a diffusion coefficient an order of magnitude higher than lithium) and gold can diffuse into the wall surfaces. If this occurs, local distortions and trapping can also give rise to 'ghost peaks'.

3.2 Spur

The term 'spur' was originally coined by Bloomfield and Love [2] to describe a background shape in some Si(Li) detector spectra that increase quasi-exponentially toward low energies (see Figure 1.53). This was observed even with a beryllium window, and so was clearly an artifact and was particularly troublesome for the analysis of light elements. It was noted that the intensity was proportional to the total number of counts in the spectrum but did not depend on the energies of the peaks in the spectrum. Bloomfield and Love attempted to explain this in terms of a window layer of ICC in the Si(Li). It is better explained by pockets of ICC in the bulk of the Si(Li) crystal. For example, the spur background dominates Si(Li) crystals that have not yet undergone, or not sufficiently undergone, a clean-up lithium drift. Geretschlager [3] postulated a similar idea to explain low-energy backgrounds. He suggested that electron–hole recombination takes place in 'clusters' of centres (such as poorly compensated regions of lithium-drifted silicon). Recombination is encouraged to take place at such 'defects' but must occur before the electrons and holes separate in the field. Because the background spur can occur even for very low energy x-rays, where the absorption takes place in a very small volume of the crystal, this is an unlikely explanation. The effect can be shown to be strongest at the edges of the crystal as collimation to the center reduces it considerably. We will discuss this in more detail in Section 5.1.8.

3.3 Non-Linearity

Elsewhere, we have discussed the non-linearity due to ICC by bulk trapping (extreme cases in compound semiconductors) and due to ICC at the window contact (Section 2.7.5) that affects linearity at low x-ray energies. However, we will point out here that very careful measurements, for example, in particle-induced x-ray emission (PIXE) studies [4], have shown the former non-linearity to occur even in Si(Li) detectors up to 30 keV (but not in HPGe detectors). This is probably attributable to recombination of carriers at the higher energies.

3.4 Ghost Peaks

3.4.1 What Constitutes a Ghost Peak?

Ghost peaks (also sometimes referred to as 'satellite', 'anomalous', or 'multiple' peaks) are peaks occurring in the x-ray spectrum of the detector that do not correspond to any lines, which can be explained as coming from the radioactive source or from any fluorescence or scattering. They are, in fact, *artefacts* of the detector itself. That they do not represent real incoming x-rays is verified by the fact that they are only a characteristic of some detector crystals and with no changes to the x-ray source or geometry they can, under certain conditions such as change of bias or temperature, vary in 'energy' (signal amplitude) and 'intensity' (signal frequency). Sometimes they can also be time dependent, only growing in time after application of bias. They are always independent of the electronics used. A test pulse through the crystal capacitance does not reproduce them, so they have a physical rather than an electronic origin. These facts identify them as essentially an ICC phenomenon, and bias change can be used as a diagnostic to identify them. They do, however, have a *very exact linear* energy relationship with the genuine x-ray lines present (the 'source' or 'parent lines') and are a reduced amplitude and frequency replica of them. In this respect, the term 'doppelgangers'—rather than 'ghosts'—might be a more appropriate term. In Si(Li) detectors, their 'energy' is usually in the region of 0.5–0.8 of the source line energy and their intensity is usually less than 1% of it, but they typically increase with increasing source line energy. They are attenuated in the same way as the source line when an absorber is placed between the source and the detector. Although they represent a small fraction of the parent peaks, they are significant for applications such as XRF, total reflection x-ray fluoresence (TXRF), and PIXE, and their presence has to be detected and corrective action should be taken during the manufacturing process, especially for these applications. Figure 3.1 shows an example of a Cd-109 spectrum taken with a particularly 'haunted' Si(Li) detector. The ghost peaks are indicated at 500 V (dots) and 800 V (line).

3.4.2 Probable Cause of Ghost Peaks

It is often instructive to observe how the applied field in a semiconductor detector influences $F(E)$. The effect is very variable, but an example was shown earlier (Figure 2.15). A further example is shown below in Figure 3.2.

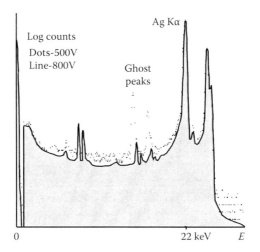

Figure 3.1 Ghost peaks observed in a Si(Li) detector at 500 V (dots) and 800 V (line) bias when exposed to silver x-rays from Cd-109.

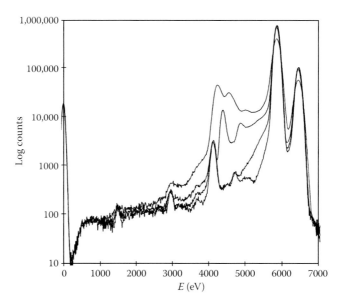

Figure 3.2 Effect of applied voltage on a Si(Li) detector. Curves top to bottom for 25, 50, 100, and 500 V. (From Lowe, B.G., *NIM A*, 439, 247, 2000. With permission.)

In both cases, we see the growth of the truncated tail and the appearance of 'ghost peaks'. Both effects can be explained by Equation 1.26 (Section 1.7.1), where it was shown that the radius R of the charge diffusion sphere inflates as the inverse of the field E

$$R \sim 1/E.$$

This results in a $1/E$ dependence in the depth of ICC first noted in Si(Li) detectors by Jaklevic et al. [6] and shown in Figure 3.3.

The tail truncation at roughly half the photopeak energy (see Figures 2.15 and 3.2) is a result of events taking place at or near the metal interface. As the bias is lowered and R at a given penetration depth expands more and more, events tend to a 50% charge collection efficiency. Because the relative magnitude of the tail is μR, a peaking takes place at roughly half the photopeak energy. This is schematically illustrated in Figure 3.4.

Continuing to reduce the bias will eventually result in events taking place in undepleted regions where we might expect little field and R approaching the dimensions of the

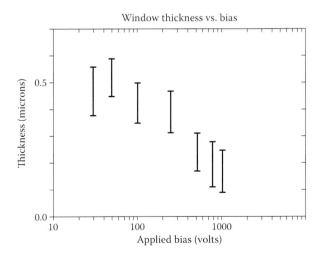

Figure 3.3 Drop in window thickness as a function of applied bias. (From Jaklevic, J.M. et al., *IEEE*, NS-18, 1, 187, 1971. With permission.)

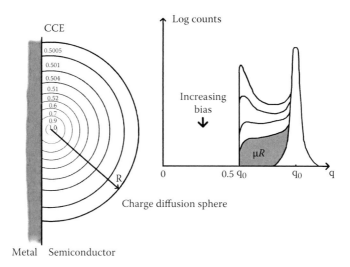

Figure 3.4 Shift from photopeak to tail with decreasing bias.

crystal. Figure 3.5 illustrates this schematically for Fe-55 x-rays in a Si(Li) detector at very low bias.

Events in the depleted region will give a 'normal' spectrum. Those in the undepleted region, where half the carriers on average diffuse toward the depleted region, will register

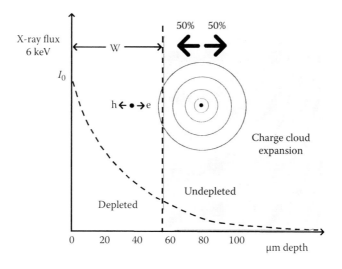

Figure 3.5 Events in undepleted region diffuse into depleted region and cause 'ghost peaks'.

as an x-ray event at half the energy. This will produce a well-defined 'ghost peak'. Note that the ghost peak 'signal' is generated almost entirely by the motion of holes in the depleted region.

The result will be a superimposed replica spectrum at ~0.5 times the parent energy, as shown schematically in Figure 3.6.

This artifact is basically a 'geometric' phenomenon. If the base material is of sufficiently high resistivity there will be some field in the undepleted region, and there will be a part polarisation of the charge cloud. Some holes in the undepleted region moving parallel to the boundary will be attracted back into the depleted region. This results in more charge being collected and the ghost peaks shifting up in energy relative to the parent. As the bias is increased and the depletion moves deeper into the crystal, the ghost peaks diminish and eventually disappear. However, if there are undepleted 'pockets' (usually near the back Li diffused contact), a higher energy source, such as Cd-109 (22 keV), may still show them.

Figure 3.7 shows how such pockets can form. With a strong n-type surface, the field at the corners of the mesa can be very weak and even reverse at full bias.

This can further be exacerbated by the removal of lithium from around the edges by its outdiffusion and by the

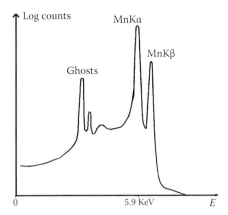

Figure 3.6 Ghost peaks resulting from interactions close to boundary between undepleted and depleted silicon and resulting mainly from 'hole' collection.

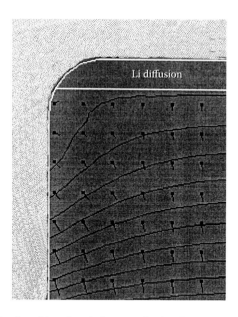

Figure 3.7 Undepleted 'pockets' close to the back contact.

preferential etching of the diffusion (undercutting) at the edges. Note also that the mesa is normally held at a small negative potential, the gate bias of the FET. The result can be a suppression of field in such regions where depletion is not established. It is clear that such regions will only give rise to ghost peaks if the x-rays penetrate into them at high energies. In practice, they are normally located in one or two regions of the crystal at most and are more likely to occur at extended mesa edges (large area detectors). Collimation to the central regions can sometimes help. It can be demonstrated that if the x-rays are directed directly into the rear (mesa) of the crystal, the ghost peaks are in the same position but are much more intense, as expected from this model. This supports the fact that these regions are close to the mesa and near the edge. If the silicon surface states change toward p-type, the peripheral fields will become stronger. This probably accounts for the time-dependent effects seen when the temperature is raised or the crystal is heavily irradiated.

Manufacturers take great pains to detect ghost peaks. They are sometimes cured by suitable collimation or a re-etch. The latter may, however, just produce a different ghost peak!

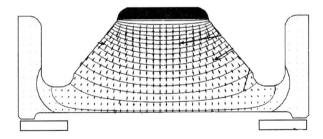

Figure 3.8 Example of a 'chamfer' geometry. (From White, G., US Patent 2007/0290141A2, 2006. With permission.)

In the case of Si(Li) detectors, extra cleanup drift can also be effective. The somewhat drastic step of removing the edge regions completely using a chamfer geometry [7] can also be taken as shown in Figure 3.8. The only real cure is test and rejection at the manufacturing stage.

Note that p-type high-purity silicon (HPSi) and HPGe planar detectors deplete from the lithium contact and ghost peaks are not observed at full bias.

3.5 Compton Scattering

We introduced Compton scattering in Section 1.2. The energy of the scattered x-ray and the recoil energy of the electron and the scattering angle between them can be treated semiclassically a shown in Figure 3.9.

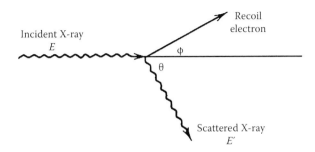

Figure 3.9 Compton scattering.

The energy E' of the scattered x-ray is given by

$$E' = E/(1 + E/mc^2 (1 - \cos\theta)), \qquad (3.1)$$

where mc^2 is the rest-mass energy of the electron (511 keV). Above about 15 keV, and increasing with energy, Compton scattering must be taken into consideration. The scattered x-ray is usually of high enough energy to escape in a silicon detector, but the energy E_e ($=E - E'$) of the recoil electron will be detected. From 3.1, it can be deduced that if $E \ll 1000$ keV, the recoil electron will have a continuum of energies up to a maximum (corresponding to directly backscattered x-rays) of approximately $E^2/1000$ keV (E in keV). Thus, for the x-ray energies, we are usually concerned with (<30 keV) the 'Compton edge' is less than 1 keV. However, one consequence is that *in the presence* of high-energy x-rays, this low-energy background can be troublesome. This is illustrated in Figure 3.10, which shows the spectra from an Am-241 source (which has a 60-keV x-ray) taken with a Si(Li) and a HPGe detector of similar geometry and under identical conditions.

The HPGe detector detects both the recoil electron and the scattered x-ray simultaneously and so removes the low-energy

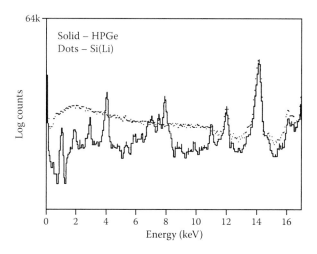

Figure 3.10 Improved performance of an HPGe detector compared to a Si(Li) resulting from absorption of both the recoil electron and the scattered x-ray when using Am-241. (From Cox, C.E. et al., *IEEE*, NS-35, 28, 1988. With permission.)

background observed in the Si(Li) detector. Another consequence of Compton scattering is that x-rays can be scattered from the surrounding material into the active volume and be detected. Even with good collimation, the backscattered x-ray from material at the back of the crystal (lithium-diffused silicon, metallisation, FET gate pin, copper cold-finger, etc.) can give humps in the background depending on the scattering angle. From Equation 3.1, a 'backscatter peak' may be expected in the region of E minus the Compton edge energy.

3.6 Self-Counting

Insulating materials, such as boron nitride, polytetrafluoroethylene (PTFE or 'Teflon'), and epoxies, used to house the crystal in the cryostat, can charge up, particularly in high-voltage TEM applications. Boron nitride is a commonly used dielectric, which at liquid nitrogen temperatures is particularly prone to charging, and the discharge of electrons into the cryostat vacuum can give rise to spurious signals and even fluorescence x-rays from surrounding material such as aluminium. This 'self-counting' effect can last many months. On warming the cryostat, the self-counting rate of the detector increases and at room temperature the dielectric is usually totally discharged. HPSi does not escape the effects of this 'self-counting', or in fact 'ghost peaks'. Both have been described by Japanese researchers [9,10] using deep detectors made from this material. They observed 'ghost peaks' from an α-particle source and a 'ghost' Compton edge from a gamma ray source as well as self-counting from the PTFE mount. This self-counting mechanism was first proposed by Chevalier [11]. The history of the 'self-counting' and 'ghost peak' phenomenon is discussed further in Section 11.19.4.

3.7 Absorption Edges

Any material through which the incoming x-ray passes will produce a discontinuity in the absorption curve and detector efficiency at the binding energy of the atomic shell of the material (see Section 1.1). This can be detected in the spectrum background continuum and is known as an 'absorption edge'. For example, the K-shell absorption edge of silicon is at 1.84 keV. Because the penetration depth of the x-ray decreases abruptly at this energy, it can have an adverse effect on ICC at the surface of silicon detectors, and the step height of the edge in a continuum spectrum can be used as a measure of the thickness of the silicon 'dead layer' [6,12]. In good-quality modern silicon and germanium detectors, the absorption edges of the materials are not usually detectable in the spectrum continuum. The absorption edge can obviously be important where vacuum windows are used and high-Z materials are to be avoided. However, the supporting structure material for the window (commonly silicon) may produce an edge as well as fluorescence. Such windows are discussed in Section 11.21, but an example of the transmission in a state-of-the-art thin window is given in Figure 3.11.

The absorption edges resulting from some of the materials used (coated polymer) are observed, and these will dominate the efficiency of the detector at low energies. It is usually advantageous to spread the absorption edges over several elements. It is obvious that the efficiency of the detector must be

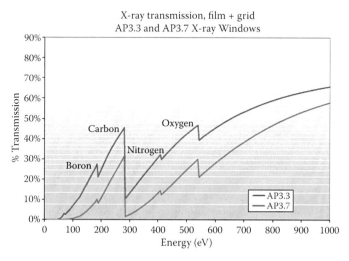

Figure 3.11 Transmission of x-rays through a state-of-the-art thin window. (From Moxtek UTW AP3.3 literature: Moxtek, Provo, USA [now part of a Japanese company Platechno.] With permission.)

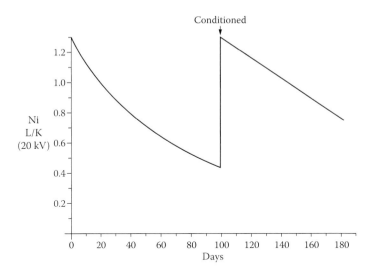

Figure 3.12 Ice removal as observed by the Ni L/K ratio measurement. (From Lowe, B.G., *Ultramicroscopy*, 28, 150, 1989. With permission.)

known accurately for any of the so-called 'standardless' analytical work [14]. A more insidious source of absorption edge is attributable to an ice layer [15] that can form on any cooled semiconductor surface. The oxygen absorption edge at 543 eV can have adverse effects for low energy efficiency [16], and regular warming ('conditioning') is necessary. Figure 3.12 shows the effect of such a conditioning on the Ni L/K ratio in an SEM at 20 kV. An even more sensitive test for an ice layer is provided by the Cr L-lines that straddle the oxygen absorption edge.

3.8 Sum Peaks

Although pile-up rejection (see Section 1.9) can electronically veto the signals from consecutive x-ray events that overlap each other in time, they cannot of course achieve this if they arrive within the resolving (shaping) time of the discriminator. Short-time constant noise is therefore an important parameter. Such events will add the two amplitudes together to give a 'sum peak', and this artifact must be recognized as such. The magnitude of the sum peak will increase with counting rate and the discriminator resolving time [17]. The problems are thus best avoided by operating detectors at low to medium counting rates. Although it is not difficult to recognize sum peaks from the summed energies of the principle peaks present, they can, of course, often obscure other peaks. For example, $CaK\alpha$ plus $SiK\alpha$ energy is close to $CrK\alpha$ energy. At low energies, the pileup rejecter must discriminate between a genuine signal and a random noise excursion; otherwise, pileup on noise will occur. A further consideration is the loss of counts from a peak into the sum peak. For a discussion on pileup, including pileup in SDDs, see Statham [18].

3.9 Electron Contamination

In EM applications using thin window detectors, a magnetic electron trap is usually effective in removing electrons. However, in some applications where very high energy electrons are also present e.g., scanning TEM (STEM) at high kV and high take-off angles, these may enter the detector and result in a background added to the already present Bremsstrahlung. Figure 3.13 illustrates this for a 100 keV electron beam on a copper TEM specimen grid bar in a Be window Si(Li) detector.

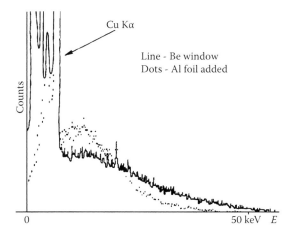

Figure 3.13 Extra background events from energetic electrons entering the detector on an STEM at 100 keV. (From Lowe, B.G., *Ultramicroscopy*, 28, 150, 1989. With permission.)

The dotted spectrum was taken with a thin aluminium foil interposed. The electron contribution is decreased (but the extra Bremsstrahlung generated in the foil is also observed).

References

1. Craven A.J. et al., *Ultramicroscopy* 28, 157, 1989.
2. Bloomfield D.J. et al., *XRS* 13(2), 78, 1984.
3. Geretschlager M., *NIM B* 28, 289, 1987.
4. Maenhaut W. et al., *Bull. Soc. Chim. Belg.* 95, 407, 1986.
5. Lowe B.G., *NIM A* 439, 247, 2000.
6. Jaklevic J.M. et al., *IEEE* NS-18, 1, 187, 1971.
7. White G., US Patent 2007/0290141A2, 2006.
8. Cox C.E. et al., *IEEE* NS-35, 28, 1988.
9. Shiraishi F. et al., *IEEE* NS-29; 1, 775, 1982.
10. Shiraishi F. and Takumi Y., *NIM* 196, 137, 1982.
11. Chevalier J.P., NSF/CNRS Workshop, Aussois, France, 1988.
12. Craven A.J. et al., Inst. Phys. Conf. 78, Ch7, 189, 1985.
13. Moxtek UTW AP3.3 literature: Moxtek, Provo, USA (now part of Japanese company, Polatechno).
14. Alvis M. et al., *Microsc Microanal* 12, 406, 2006.
15. Cohen D.D., *NIM* 193, 15, 1982.
16. Lowe B.G., *Ultramicroscopy* 28, 150, 1989.
17. Statham P.J., *XRS* 6, 94, 1977.
18. Statham P.J., *Microsc. Microanal.* (suppl. 2), 1428, 2007.

4
Contacts

We have seen how the response function $F(E)$ is influenced to an inordinate degree by the physics taking place at the front contact of the semiconductor x-ray detector. Incomplete charge collection (ICC) results from the escape to the contact of what may be termed 'hot' electrons (primary Auger and photoelectrons and their immediate progeny), 'warm' electrons (the high energy tail to the distribution at the end of the cascade process), and 'thermal' electrons in energetic equilibrium with the crystal lattice after the outward diffusion of the charge cloud. Furthermore, 'hot' electrons from the metal contact are sprayed into the semiconductor adding background events. All these effects are undesirable from a spectroscopic perspective, so we must address the topic of how to eliminate or reduce them. There have been many theoretical discussions on semiconductor surfaces and interfaces (e.g., see Monch [1]), and the topic of contacts can be as complex and deep as you wish to make it. And yet, we still believe that certain aspects are not well understood. We therefore make the caveat at this point that, in what follows, some of the explanations proposed are the authors' own ideas based on many years of experimentation with various contacts to silicon, germanium, and CdZnTe.

4.1 Metal

The choice of metal has already been discussed (Section 2.8.2). We saw that, taking into account the bias polarity and electronegativity of the metal (see Table 5.1), light elements such as Al and Ni were to be preferred to the more 'traditional' Au and Pd. It only remains for us to add that the interposing of a low-Z intermediate layer between the metal and the semiconductor, as we will discuss below, would have the advantage of absorbing a proportion of the electrons emanating from the metal. Obviously, we need to achieve a balance between reducing this component of background while not adversely affecting other parameters such as noise, x-ray absorption, fluorescence, and absorption edges.

4.2 Parameters Influencing ICC

We saw in Chapter 2 that the relative magnitude of the integral tail on the photopeak is given by μR, where μ is the linear x-ray absorption coefficient and R is loosely defined as the 'radius' of the charge cloud before it is arrested by the field as described in Section 1.7.1. We also saw that R is given by Equation 1.26

$$R = D/\mu_{\text{mob}}E,$$

where μ_{mob} is the mobility of electrons in the applied field E and is constant up to the saturation velocity v_s. It gives the inverse field dependence of R mentioned in Section 3.4.2. D is the diffusion coefficient of the thermal (or near thermal) electrons at the appropriate temperature T. At normal applied detector bias, the drifting electrons are near saturation velocity, which is not a strong function of temperature [2]. D was given by Equation 1.27

$$D = kT\mu_t/e,$$

where μ_t is the mobility of the thermal (or near thermal) electrons and decreases strongly with increasing temperature (see Figure 1.13). Fred Goulding [3] at Berkeley initially discussed this in 1976. We conclude that R can be reduced proportionately by

(1) Decreasing μ_t
(2) Increasing E
(3) Increasing T

or any combination of these. The last dependency (3) has consequences for detectors manufactured using silicon lithography (discussed in Chapter 7) that usually operate at higher temperatures than Si(Li) and HPGe detectors.

4.3 Reflection

The idea that charge could be reflected back from the contact was first suggested by the Berkeley group [4] in 1990, but no mechanism was proposed. In Chapter 2, we assumed a passive nature for the window contact of the semiconductor as depicted in Figure 2.12. We saw that the effect of reducing the radius of the charge sphere R in the bulk of the semiconductor was to reduce the relative magnitude μR of the peak tail as was shown in Figure 2.14. However, the contact itself (the surface or near-surface layer) is never as simple as being an abrupt transition from semiconductor bulk to pure metal. It can be highly complex and is, in many cases, not fully understood. This is related to semiconductor surfaces and interfaces, a topic we will visit later. For the time being, we will simply state that the contact can have physical characteristics that aid charge collection in a nonpassive manner. For example, if this layer has decreased charge mobility μ_t or enhanced field E, we can regard a fraction of the charge (usually electrons in practice) as being 'reflected' back at the surface into the bulk of the semiconductor. This is illustrated in Figure 4.1.

A fraction r of the charge in the shaded volume is 'reflected' *at the interface*, and we can rewrite the charge collection efficiency $f'(Z)$ given by Equation 2.12 as

$$f'(Z) = f(Z) + r[1 - f(Z)], \tag{4.1}$$

where $Z = -z/R$.

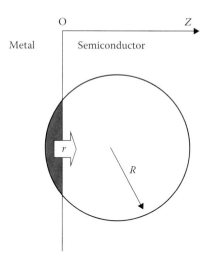

Figure 4.1 Reflection of charge at contact.

When this form is used in the simple uniform charge diffusion model discussed in Section 2.7.4, the tails of the photopeaks are modified as shown in Figure 4.2.

The total relative magnitude of the tail is unchanged by the reflection (it is still μR), but the counts are drawn more and more into the peak as r increases. Ultimately, this leads to an asymmetry that will affect the full-width half-maximum of the peak and the effective Fano factor measured. It can be

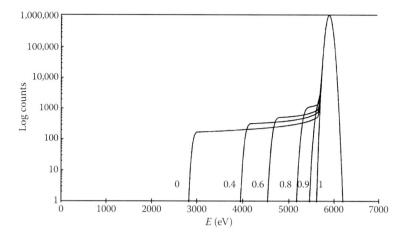

Figure 4.2 Effect of reflection ($r = 0$–1) on 'tail'. (From Lowe, B.G., *NIM A*, 439, 247, 2000. With permission.)

shown that the point of truncation E_C (corresponding to charge generation at $z = 0$ in the crystal) increases according to

$$E_C = (1 + r)E_0/2. \tag{4.2}$$

The values of E_C observed are discussed in Section 4.6.2, but for Fe-55 x-rays for most modern Si(Li) detectors it is virtually merged into the photopeak.

At low energies, the x-rays are absorbed more and more in the layer of ICC at the surface, and the tail begins to dominate the peak. The reflection thus has even greater effect on the low energy photopeaks and is capable of restoring the energy linearity to some extent as discussed in Section 2.7.6. Figure 4.3 shows the calculated effect on the Kα x-ray lines of B (183 eV), C (277 eV), and O (525 eV) as r is increased.

The correlation between low energy nonlinearity and photopeak tail at 5.9 keV is demonstrated in Figure 4.4 for a large number of Si(Li) detectors. The tail factor T is the ratio of the tail counts to fitted Gaussian peak count (see Figure 1.53) and in practice includes contributions from the SiKβ escape peak, radiative Auger bands, and fast electron losses. ΔE is the shift of the C Kα peak from its true position (277 eV). The

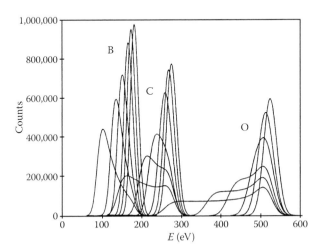

Figure 4.3 Effect of reflection on low energy x-ray photopeaks for same values of r as in Figure 4.2. (From Lowe, B.G., *NIM A*, 439, 247, 2000. With permission.)

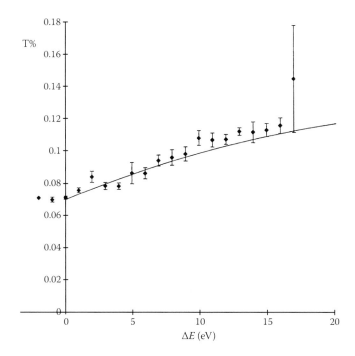

Figure 4.4 Correlation between tail and nonlinearity at CKα as measured in a large number of Si(Li) detectors. (From Lowe, B.G., *NIM A*, 439, 247, 2000. With permission.)

line is their expected dependence taken from Figures 4.2 and 4.3, assuming R and r do not change with energy (see Section 4.6.3). As they both refer to conditions at the end of the cascade process, this is not unreasonable. For a more detailed explanation, see Lowe [5].

Alternatively, the same concepts can be described in terms of a surface recombination velocity, s [6]. If the carrier density is n (charge carriers/cm^3) in the region of the contact, the loss at the contact is described in terms of a current density j (charge/cm^2/s) where

$$j = sn,$$

where s (expressed in cm/s) is the surface recombination velocity. The best characteristics of a junction (low leakage current and high breakdown voltage) correspond to the lowest values of s. For example, a freshly etched surface has s ~50–100 cm/s, whereas a lapped surface has s ~10,000 cm/s. Thus, it can be regarded as an 'inverse' of the reflection coefficient r; $s = 0$ for perfect reflection and $s > 0$ for contacts that have charge loss to the surface.

Let us look at the various contacts to semiconductors used and analyze their qualities in these terms. The formation of junction contacts to semiconductors was discussed briefly in Sections 1.5 (Schottky barriers) and 1.6 (diffused p–n junctions). In semiconductor detectors, the deposition of metals on semiconductors rarely results in a simple Schottky barrier but more usually, owing to interfacial layers, in a very complex not well understood 'surface barrier'.

4.4 Diffused Junction Contacts

Diffused junctions were the first junctions to be manufactured in semiconductors for rectifiers and later transistors in the early 1950s and were first used for photodiodes around 1955. These were also shown to be sensitive to ionising radiation around the same time. Their use as a window contact for low-energy particle and x-ray detectors, however, was limited by the thickness of the diffusion (usually >200 nm), and the high temperatures required made them completely unsuitable for lithium-drifted detectors, which were far better served by the intrinsically shallow surface barrier contacts. However, the lithium drifting process uses a lithium diffusion, at around 400°C, which acts as the reservoir for the drifting process and this is usually left in place as a robust noninjecting ohmic back contact. This temperature is not high enough to affect the charge lifetime of carriers in the p-type silicon (usually a few ms), but temperatures above ~200°C even for a short time *after* drift will subsequently give rise to lithium precipitation and trapping–detrapping noise. The depth of the lithium diffusion is not critical and, of course, the diffusion is undertaken before the drifting. However, we saw in Chapter 3 that the uniformity and abruptness of the diffusion has a bearing on artefacts for high-energy x-rays.

Diffused junction p–i–n photodiodes manufactured with n-type silicon and a boron diffusion at around 1000°C as a window contact are still manufactured. Boron diffusion makes a good p-type contact and phosphorous a good n-type contact. They are still also used for low-energy x-ray detectors commercially, especially as the low leakage currents make liquid nitrogen unnecessary. Being manufactured from standard HPSi wafers by simple single-sided photolithography, they are much less expensive than Si(Li) or HPGe detectors [7], and the quality of the response is adequate for many undemanding x-ray fluorescence (XRF) applications. They can be encapsulated and cooled by Peltier devices, and their reliability has led to them being used for many portable and remote applications. They are even capable of detecting carbon (277 eV), albeit with considerable ICC (photopeak at 230–240 eV).

Boron and phosphorus diffusions are now commonly achieved by depositing from the gas phase (B_2H_6 and PH_3). An added bonus of the phosphorus diffusion is that it also acts as a 'getter' for impurities [8] and is sometimes diffused and subsequently removed for this purpose. For better x-ray detection performance, ion-implanted p–i–n diodes have replaced diffused junctions, and we will now discuss these.

4.5 Ion-Implanted Contacts

Experiments with ion-implanting silicon also started in the early 1950s at Bell Laboratories, and it was soon realised that it was a good alternative to a diffused junction for a particle detector. In 1962, the first implant of phosphorous on p-type silicon was used followed soon after by a boron implant on n-type silicon. Ion-implanted contacts never caught on with Si(Li) detectors, as the surface barrier were less expensive and more straightforward to apply. Also, the damaging annealing process (>600°C) required for silicon after implant was an undesirable feature. Apart from this, the few experiments carried out seemed to yield inferior results particularly on spectral background and linearity at low energy. The ion implantation of HPGe x-ray detectors has been more successful [9], and they have been commercially available since 1990 (e.g., from Canberra Instruments) [10–12]. The implantation of boron on n-type HPGe gamma ray detectors is now routine.

Ultrashallow junctions are a requirement not only for x-ray detectors but also for IC production, as the dimensions of these are scaled down. Ion implantation lends itself well to these photolithographic processes. The ion implantation of HPSi for x-ray detectors such as p–i–n, silicon drift detectors (SDDs), and charge-couple devices (CCDs) is also now done routinely. These have sometimes shown poor results at low energies in the case of SDDs [13], but techniques have been used to make these implants shallower (<60 nm) [14,15]. These techniques include:

(1) Low energy (~5 kV) implant from BF_2 gas phase.
(2) Implant angle selection (to avoid channelling).
(3) Orientation selection of silicon ($\langle 100 \rangle$ for minimum channelling).
(4) Implant through a silicon oxide or nitride layer (0–50 nm thick). This technique was first used to form a graded implant that reduces edge breakdown [16].
(5) Implant through a-Si (to avoid channelling and to make the implant shallow). This also suppresses the diffusion current [17].
(6) High dose rate (B diffusion is retarded for higher dose rates).
(7) Formation of an amorphous layer. Tasch et al. [14] used transmission electron microscopy (TEM) analysis to show a sharp smooth amorphous–crystalline interface for high doses ($10^{15}/cm^2$).
(8) Annealing at high temperatures (>900°C) [18] or in a two-step anneal: 600°C, 30 min, +1050°C, 10 s [15]. The temperature must not be so high that diffusion changes the doping profile.
(9) Laser annealing is effective for shallow ion-implanted junctions [19]. Excimer lasers are usually preferred because of their short wavelength.

A great advance came with the introduction of 'engineered' ion implanted contacts in the form of a 'double implant' by Hartmann et al. [20] (discussed in the next subsection).

4.5.1 Double Ion-Implanted Contacts

By using various combinations of the above techniques, a boron p+ implant into n-type silicon can be achieved roughly 40 nm below the surface. By carrying out another implant on this of the *opposite polarity* (phosphorous n+) of lower doping and slightly deeper profile, this junction can be tailored to considerably enhance the local field as shown in Figure 4.5.

This 'compensated' layer can support fields (full curve in Figure 4.5) of 60 kV/cm and will therefore act as a reflecting barrier for back-diffusing electrons. It is also possible that some amorphisation might be relevant to the contact. Any amorphous layer in a region where the field is not zero will reduce the mobility μ_t, suppress carrier diffusion, and enhance reflection. The double-implant technique can, of course, be applied to any p–n junction device such as a p–i–n diode, back-illuminated fully depleted CCD, or HPGe detector. The contact is completed with a metallisation of aluminium that also shields the semiconductor from light and IR radiation. The x-ray spectra from modern ion-implanted p–i–n diodes, SDDs, and HPGe detectors testify that the reflection coefficients are very close to unity in these cases.

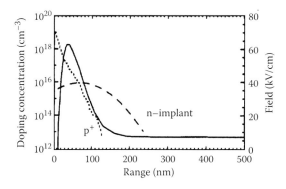

Figure 4.5 Mixed implants (n and p) used to maximise field. (From Hartmann, R. et al., *NIM A* 387, 250, 1997. With permission.)

4.6 Surface Barrier Contacts

Surface barrier contacts were first made in 1950 being a natural 'area contact' development from the metal point contact, or cat's whisker. Their development for radiation detectors ran parallel to that of diffused junctions, being first used as photodiodes and then as radiation detectors around 1955, but they were more difficult to make and were more fragile. Up to 1984 (when Ni contacts were introduced), all commercial Si(Li) detector manufactures used a simple evaporation of gold (typically 20 nm thick) onto the clean etched surface. There was usually a delay introduced in a dry atmosphere before use because the barrier needed to 'harden' or 'age' (in fact, to allow the native oxide to grow beneath the gold layer, thus reducing leakage current). It is interesting to note that although a similar procedure was used on germanium and the material characteristics at low temperatures are not that different from silicon, the quality of x-ray spectra, particularly at low energies, was far inferior in the former case. No one gave a satisfactory explanation for this difference at the time. This is exemplified by spectra taken at Berkeley [21] in 1977 with a HPGe detector (Figure 4.6).

This should be compared with low-energy x-ray spectra taken with a Si(Li) detector using a similar technology using a nickel contact [22] from 1985 as shown in Figure 4.7.

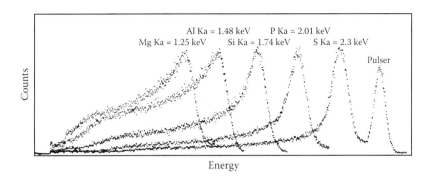

Figure 4.6 'State of play' showing a 1977 HPGe detector's response to soft x-rays. (From Llacer, J., *IEEE* NS-24, 53, 1977. With permission.)

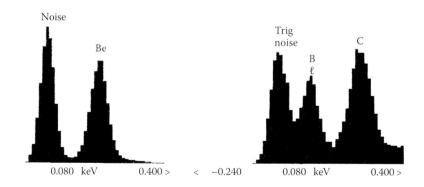

Figure 4.7 Low-energy response of a Si(Li) detector. (From Statham, P.J., Assoc. Nat. de la Res. Techn. (ANRT), *Microanalyse et Microscopie Electr. A Balayage*, Paris, Nov. 1985. With permission.)

Armed with this evidence, the Berkeley group condemned the use of HPGe detectors for x-ray spectroscopy despite just falling short of an explanation (reflection at the native oxide of silicon) for the difference.

4.6.1 Amorphous Interface

The deposition of an amorphous layer of semiconductor under the metal contact significantly improves the charge collection. This is true for Si(Li) detectors [23] but is most dramatic for HPGe detectors. By depositing an amorphous layer of germanium (~10 nm) before the metal contact is applied (nickel is preferred to gold, as discussed in Section 2.8.2), the reflection coefficient is enhanced. This is demonstrated in Figure 4.8, which reproduces spectra using a planar HPGe detector with a-Ge contact. Figure 4.8a shows a Fe-55 spectrum. The Al peak is from the internal IR shield (see Section 6.2.8). Figure 4.8b shows an AlKα peak in detail. This is just above the Ge L absorption edge and so is prone to ICC because of surface effects. This should be compared with Figure 4.6, noting that it is a linear plot.

The prime mechanism for this effect is not clear. Amorphous semiconductors (a-Si, a-Ge), like their crystalline counterparts, have bandgaps and are capable of being depleted and supporting considerable fields (resistivity ~1.6×10^8 Ω cm for a-Ge). Indeed, they have been used for radiation detection

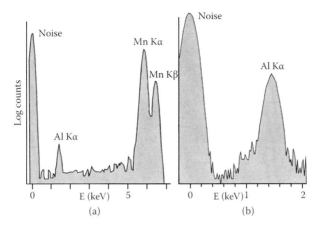

Figure 4.8 Spectra obtained using an amorphous germanium layer under front metal contact. (a) Fe-55 spectrum and (b) Al K spectrum.

[24], especially for x-ray imaging, but their random structure gives rise to low electron lifetime and mobility. The bandgaps, however, are not abrupt, and it is possible that this allows the electrons to continue to lose energy by ionisation longer, resulting in a higher degree of thermalisation at the expense of diffusion.

A 'passivation' effect (see Section 4.6.2) and reduction in mobility may also play a part. The difference between silicon and germanium is that silicon has a natural stable interfacial layer (the native silicon dioxide), whereas germanium does not (the germanium oxides are unstable and water soluble). This property of silicon is fortunate but unfortunately unique among the semiconductors. We will return to a discussion of oxides below.

4.6.2 Interface States and Recombination

Fixed charged interface states are a source of charge recombination and therefore ICC. We saw in Section 1.11.3 that these can give rise to surface leakage currents if present on the semiconductor walls. If present under a metal contact, the diffusing charge cloud that impinges such a layer can undergo some loss of carriers by recombination. The density of these states is dramatically reduced by applying an effective passivation.

Various methods of surface passivation are discussed in Section 5.1.10. The method most often applied in general for surface passivation is a layer of high-quality silicon oxide and another, a layer of amorphous semiconductor. The relative importance of the application of such films for electron reflection and reduction of interface states under contacts is debatable, but no doubt both aspects contribute to reducing ICC. In the case of the ion-implanted contacts discussed above, both a thin layer of high-quality oxide [25] and hydrogen passivation [26] have been used.

4.6.3 Dielectric Interface

Let us first restrict ourselves to silicon. For many years, it was recognized that one of the most important roles of the SiO_2 layer in electronic devices was, in general, to maintain the carriers that participate in normal device operation inside the silicon. If the native oxide is responsible for some reflection and improved charge collection in x-ray detectors, then perhaps it can be improved upon, for example, by making it thicker? Barrier heights are modified by interface states with an oxide present (see Section 1.5.3), and it was shown very early on that when oxygen was deliberately excluded, for example, by contacting cleaved crystals in vacuum [27], barriers were very poor or nonexistent. Any thermal or near-thermal electrons penetrating into the dielectric from the semiconductor will find themselves in the large (~12 eV) forbidden energy gap of the dielectric and consequently, the wave functions decay rapidly with depth of penetration. In other words, we will have near-perfect reflection [28], aided of course by the applied field. The situation is as depicted in Figure 4.9.

This shows the band diagram for silicon with a thin oxide layer and a negative bias applied to the metal contact. Hot primary electrons and 'warm' electrons, with energies greater than the oxide barrier height, may move against the field into the metal and be lost (injection over the barrier as shown in Figure 4.9). However, if the oxide is sufficiently thick, the lower energy electrons, including thermal ones, will be reflected back into the bulk of the silicon. With a native oxide (thickness ~1–3 nm), some tunnelling and trapping of such electrons does take place, and tail cut-off energies E_C (Figure 4.2) have

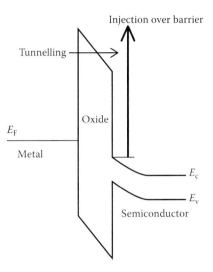

Figure 4.9 Large barrier height in oxide reduces loss of thermal electrons.

been reported for 'conventional contact' Si(Li) detectors from 0.5 E_0 [29] to 0.78 E_0 [30] corresponding (by Equation 4.2) to reflection coefficients r of 0 to 0.56. For any given detector, any dependence of these ratios on energy appears to be weak, as expected, and this is true at least between 1 and 6 keV [31,32]. If thicker oxides are engineered, as discussed below, r approaches unity [5]. Figure 4.10 demonstrates this for light

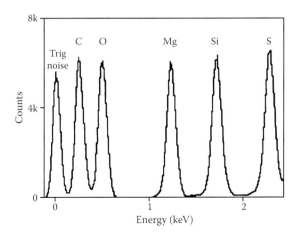

Figure 4.10 Light element spectra using a HPGe detector. (From Cox, C. et al., *IEEE* NS-35, 28, 1988. With permission.)

elements and HPGe using a dielectric contact [33]. Again, these can be compared to Figure 4.6.

As a side effect of the oxide interface layer, a proportion of the electrons injected from the metal are absorbed before they reach the active volume of the semiconductor. For example, the stopping power for 1-keV electrons in SiO_2 is ~20 eV/nm [34]. This can have a significant effect in reducing the background contribution to the response function at low energies.

4.6.4 Equilibrium and Instabilities

For photoelectric events in the bulk, the electrons move to the back contact and holes toward the surface barrier contact and quickly spread out over it. For any electrons diffusing to this contact as described above, recombination at the interface after an ionising event is very unlikely. The mobility of electrons in SiO_2 is about 40 cm^2/V s for temperatures below 150 K [35]. Hole mobilities are lower by factors of at least 7 orders of magnitude in the same temperature range [36], so that the holes accumulate at the interface enhancing the field in the oxide. This field builds up because of ionising radiation and quiescent bulk leakage current until field-assisted tunnelling (Fowler–Nordheim current) (see Figure 4.9) by electrons from the metal causes recombination with these holes and an equilibrium condition is set up. The tunnel current increases exponentially with the field in the oxide and therefore provides negative feedback and stability. Another way of describing the requirement for stability is that the oxide layer must be sufficiently thick to give high electron reflection (r ~ 1) but be thin enough for the tail of the metal wave function to penetrate it and allow an electron tunnel current to recombine with holes trapped at the interface. In practice, thicknesses of 4–9 nm have been shown to be effective. If the oxide is too thick, the accumulation will cause a screening of the negative bias on the metal (and reduction of the bulk field) or in the extreme, a breakdown of the dielectric. The nature of an x-ray entrance window with an oxide under the metal contact is shown schematically in Figure 4.11. As the holes accumulate at the interface the field across the oxide grows, and field-

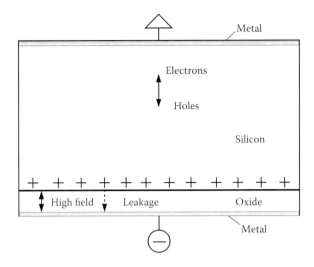

Figure 4.11 An oxide under the metal contact gives rise to hole accumulation.

assisted tunnelling of electrons from the metal increases to maintain an equilibrium condition.

   Instabilities manifest as the growth with time of low-energy tails on the photopeaks, and usually this recovers if the bias is removed for a certain period. Presumably, this allows the holes time to disperse from the interface and leak away. The quiescent tunnel current appears as a detector leakage current (contributing to noise), which can also become unstable. Note that the dielectric interface represents a capacitance of the order of 0.1 μF even for small area x-ray detectors, so that recovery times can be long and may be extended if long lifetime interfacial charge states are involved. A sudden discharge of the capacitance can damage the dielectric [and incidentally, the field effect transistor (FET)], and this has implications when disconnecting the bias (often built into the preamplifier).

   Thin oxide layers are capable of supporting very high fields before the onset of dielectric breakdown. The field can be as high as 3×10^7 V/cm in SiO_2 at room temperature [2]. It increases slightly at lower temperatures [37], and the dielectric breakdown field is also reported to increase as the film becomes thinner [2]. This means that a film of 5 nm can support ~15 V and a field many thousands times greater than the average field in the bulk. The enhanced field will cause the reflection of electrons that penetrate the dielectric, contributing to the reflection coefficient, a mechanism first proposed by Scholze [30].

Contrary to some beliefs [38], the charge reflection phenomenon described above is effective for both rectifying and ohmic detector window contacts. The same structures work, for example, on Si(Li) and n-type HPGe detectors (where it is a rectifying contact) and HPSi and HPGe p-type detectors (where it is an ohmic contact). In thick detectors, such as Si(Li), HPSi (2–3 mm), and HPGe detectors, charge collection at the back contact is not usually an issue, but in thinner detectors (<1 mm) such as SDDs, p–i–n diodes, and CCDs, it can be. Of course, in the case of SDDs (see Section 1.17 and Chapter 9) that deplete from both the front and back of the wafer, the same implantation techniques can be used for both. In this case, the small central anode is an ohmic contact (also usually ion implanted), and injection from it will not be significant compared with the large area of the rectifying contact.

These dielectric contacts are more stable at lower temperatures where the quiescent hole generation and tunnel current is low. Obviously, poor thermal contact (caused, e.g., by loose coupling or by ice formation in the bottom of a liquid nitrogen dewar) may also be the cause of instability. It should be pointed out that instabilities in standard surface barrier contacts (native oxide) are almost as common as in the modified ones. At higher temperatures (–90°C and above, in the range of Peltier coolers), the hole accumulation at the interface may cause instability and an implanted contact, as used for p–i–n diodes and SDDs, is more appropriate. Very high energy-rate product ionising radiation will give the same effect, but this is also usually reversible. Another cause of instability is contamination, for example, by ice or vacuum pump oil (oil-free cryopumps are always preferred). Any testing regime must include a period (many hours) of exposure to a high x-ray energy-rate product either with a source (e.g., Cd-109) or from an x-ray tube, together with close monitoring of the peak tailing.

4.6.5 Methods of Applying a Dielectric

There have been a number of attempts to 'engineer' better surface barrier contacts. A 10-nm tungsten oxide interface film was shown [39] to improve the spectral performance and stability of gold-n-type silicon surface barrier detectors operated for both particles and x-rays at room temperature in 1986. The film was evaporated by heating the tungsten wire in an atmosphere of oxygen, and various species of oxide were deposited in a rather uncontrolled manner. Much research has gone into the methods of growth of high-quality SiO_2 films because it is especially important in metal–oxide–semiconductor (MOS) technology. They usually involve the growing of the oxide in wet oxygen at high temperatures (>1000°C), and although appropriate for HPSi such as used for SDDs, p–i–n diodes, PADs, and CCDs, etc., are inappropriate for Si(Li) detectors and, of course, HPGe detectors. For Si(Li) detectors, a 'cold' silicon oxide deposition method must be used so as not to affect the lithium compensation. Because of the lack of a suitable native germanium oxide, silicon oxide or some other dielectric must be used on HPGe. The deposition by vacuum evaporation of SiO_2 starts when the heated source is at 1300°C, which does not make it practicable. Deposition is usually carried out by RF sputtering, electron beam, or chemical vapour deposition. However, the monoxide, SiO, has almost as good properties (dielectric strength of 3 MV/cm, resistivity 4×10^{12} at room temperature). The conduction properties have been studied, for example, by Adachi et al. [40]. It sublimes during evaporation in a 10^{-4} Torr vacuum at a modest source temperature of 850°C. Furthermore, the monoxide maintains its composition on deposition. It may be remarked that even for a grown or deposited SiO_2 interface layer, the layer immediately in contact with silicon is likely to have characteristics (such as bandgap) of SiO_x, where $x \sim 1$. The ease of application of SiO has led to its use as one of the alternatives as a semiconductor surface passivation [41] and a contact interface for Si(Li) detectors [23]. David Joy (University of Tennessee) and Jon McCarthy (Noran Instruments) carried out a study [42] in which the influence of various contact manufacturing parameters on ICC in a Si(Li) were reported. Their results of the dispersion measured for low-energy x-rays are illustrated in Figure 4.12.

The upper curve (Figure 4.12a) compares the photopeak dispersions measured for the 'old' technology with that fitted from Joy's Monte-Carlo model. The lower curve (Figure 4.12b) does the same for the 'new' technology. The fits with the model were given in terms of the normalised recombination velocity s and diffusion length D_L. The improvements are:

D_L = 1.75 mm (old), 1.5 mm (new)
s = 35 cm/s (old), 0.1 cm/s (new)

There was no reduction in the depth of the ICC layer itself. This work represents an improved recombination velocity or

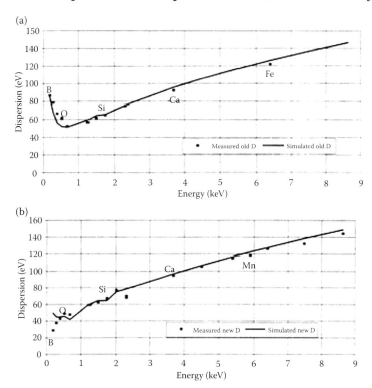

Figure 4.12 Plots of dispersion vs. energy for so-called 'old' technology vs. 'new' technology. (a) 'old' technology and (b) 'new' technology. (From McCarthy, J., *X-ray Spectroscopy in Electron Beam Instruments*, Williams, D.B. (Ed.), 67, Plenum Press, 1995. With permission.)

reflection coefficient, but it is not exactly clear what processes gave the improved performance. Device 'aging' is mentioned in the oxide growth process. Their model underestimates the dispersion for the lightest elements. This was also found by Lowe [5], and both concluded that the anomalous broadening could be due to the diffusion of the 'warm electron' component of the charge cloud toward the contact.

In the MOS structure used in the manufacture of CCDs, the electrons are held under each pixel by the application of a positive charge on the aluminium deposited on the SiO_2 layer, which acts as a near-perfect reflector, being thick enough to prevent any tunnelling. The charge is prevented from diffusing laterally by both this field and implanted channel stops (see Section 1.15 and Chapter 9). The trapping in the silicon bulk is reduced to such an extent that in clocking from one pixel to the next down the chain, very little charge is lost. The substrate of CCDs can also be etched back to form an entrance window (back-illuminated CCDs) where ion implant technology can be used (see Chapter 8). As a result, the spectral performance of scientific CCDs, as for SDDs, approaches that of Si(Li) detectors.

4.6.6 Quality of Dielectric

It is not necessary for the dielectric film to be continuous as the reflection of electrons is localised. In fact, on HPGe it has been found that discontinuous films can avoid excessive hole accumulation and instability (see Section 6.2.7). Presumably, recombination takes place at the discontinuities. In grooved detectors, problems arise—as we shall discuss later—in the peripheral regions of the front contact where the field is reduced and equilibrium conditions may not prevail. Either this region can be ion implanted, or the metal contact masked to confine the field to the central regions. Suitable collimation prevents excessive x-ray–induced charge from causing further instability.

The quality of the dielectric is as important for the detector contact as it is for MOS technology, and much research has been carried out in this respect. Full treatments can be found, for example, in the work of Barbottin and Vapaille [37].

The following can cause poor dielectric properties and instabilities:

(1) Holes embedded in the oxide. These act as electron traps for the tunnel current. For example, they can be embedded in the oxide during deposition or possibly by any sudden discharge of the bias capacitor, causing dielectric breakdown. Recovery normally takes place on warming the detector to room temperature.

(2) Contamination by ions. H_2O, Na^+, Li^+, etc., in the oxide reduce the dielectric strength and act as traps. Water dissolves and attacks SiO_2. It should be noted that the mobility of Na^+ ions is even greater than that of Li^+ ions. Both of them will 'poison' the dielectric. For example, lithium can migrate from the compensated material into the oxide if the temperature is raised sufficiently for a long period (>50°C over many hours). This has consequences for the vacuum baking of cryostats of finished detectors. Techniques using chlorine ions [37] during deposition have been used successfully for MOS structures since the early 1970s to leach out positive ion impurities present. For example, Kemmer [43] used HCl during oxide growth.

4.7 Low Injection Contacts

One of the main requirements of ohmic contacts, as well as the rectifying contacts, is that they should not generate excessive leakage current. The injection of electrons or holes should be suppressed to form a good 'blocking' contact. Although, as we saw in Section 1.11.3, surface leakage currents usually dominate and thermally generated currents can be reduced to acceptable levels by cooling, diffusion current and localised injection across the barrier must be considered. The latter is usually attributable to the quality of the surface barrier (etching conditions and cleanliness) or ion implant and to any nonuniformity or physical damage to the fragile junction. Let us consider a shallow p-type window contact on an n-type

semiconductor with an n+ ohmic contact on the back and sufficient reverse bias to give over depletion (i.e., a high field at the
back contact). We previously looked at the origin of the diffusion current density in Section 1.11.2 and saw that it was due
to the diffusion of electrons from the front and holes from the
back. We saw that it depends on the concentration gradient of
minority carriers at the depletion region edges in both cases.
Note that if the p-type side has high doping and is shallow, the
current is mainly due to holes injected from the n+-type contact side. Thus, the current per unit area is given by Equation
1.45 (Chapter 1).

$$I_d = ep_0 \times D_p/L_p$$

Note that diffusion length L_p is $\sqrt{(D_p\tau_p)}$, where τ_p is the hole
lifetime.

We also had Equation 1.44

$$np_0 = n_i^2.$$

Thus, because p_0 is inversely proportional to n at a given
temperature, this current is inversely proportional to both the
doping concentration N_D and $\sqrt{\tau_p}$, so both of these parameters
should be high for low leakage. The carrier lifetimes may be
reduced by any high-temperature processes during manufacture such as thermal growth of oxide and annealing after ion
implantation. Gettering effects of chlorine ions can alleviate
this problem to a certain extent. A more detailed analysis of
diffusion currents has been carried out [17] where it was concluded that 'high–low' doping concentrations in contacts suppress the diffusion currents, and the depth of the highly doped
region should exceed the minority carrier diffusion length. In
practice, the concentrations and depths that can be achieved
by lithium diffusion and ion implanting are adequate in this
respect. The implanting techniques of Joseph Kemmer at the
Technical University of Munich [43] were instrumental in his
achievement of very low leakage currents (<1 nA/cm² at room
temperature) for his planar devices. Modern techniques have
extended this to <0.6 nA/cm² [44].

In the case of Si(Li) detectors, p$^+$ surface barrier contacts should contribute little injection provided precautions are taken (see Chapter 5). The n$^+$ lithium diffusion is the remnant of the reservoir for lithium drifting and as such has very high doping levels, but it is difficult to control its abruptness. This also will be discussed further in Chapter 5.

4.7.1 Use of Amorphous Layers in Ohmic Contacts

There is a long history of the use of amorphous layers under metal contacts to semiconductor radiation detectors to provide good blocking ohmic contacts [45–48]. These have mostly been quite thick layers (100–500 nm) of a-Ge. Presumably the amorphous layer acts as a region of short diffusion length (low μ_t) for the minority carriers in the junction, and this acts to suppress the diffusion current significantly below that which would be achieved by a simple thin metal layer on a clean etched surface. In amorphous germanium, the Fermi level is pinned close to mid-gap, and it therefore acts as a blocking contact for holes and electrons. This has been taken advantage of in 'bipolar' germanium detectors [49] that use a-Ge for both contacts as well as the sidewall passivation.

4.8 Edge Termination of Contacts

The abrupt termination of a contact at its edge, particularly a rectifying contact, will give rise to a very high field region. This can cause current injection or in the extreme complete electrical breakdown. The surface barrier, being the shallowest of the junctions, is the most problematic in this respect. In the Si(Li) detector, there is a natural solution in the form of a graded junction, in that in the grooved or top-hat geometry the surface barrier contact can extend beyond the drifted profile and terminate on relatively low-resistivity material (see Figure 1.60). The latter has a shallow depletion depth. In fact, the metal usually terminates on a degenerate lapped surface that serves as a pressure contact for the applied bias. Thus, the groove, introduced for diverse reasons, facilitates the edge termination of the surface barrier contact and also makes the crystal easy to handle. The same argument applies to the planar HPGe detector where the full depletion region is confined by the action of the groove itself. This is one reason for the manufacturers' reluctance to move away from these geometries (see Section 5.2).

In the case of diffused contacts, the outdiffusion of the doping gives rise to a naturally graded termination. In the case of lithium diffusion this is at the edge of the mesa, and in other diffusions (boron, phosphorus), used in planar technology, the diffusion goes under the oxide mask that passivates and protects the edge naturally.

In ion implanting, graded contacts at the edges can be achieved by

(1) Increasing the angle of implant at the edges
(2) Implanting with a tapered silicon oxide or nitride mask [16]

A further technique is to use guard rings to reduce the radial field at the edge. These can be 'floating' or biased, and many such rings can be used [44].

4.9 Radiation Damage

Semiconductor x-ray detectors are used in many radiation environments, and separate considerations are required regarding their survival for each application. As they have been designed to be sensitive to ionising radiation, it is not surprising that there is no perfect solution and both temporary (degradation often mitigated by annealing) and permanent damage can occur. The problem increases in magnitude when moving from applications such as total reflection XRF (TXRF; low x-ray fluxes), XRF (moderate but controllable x-ray fluxes), through to scanning electron microscopy (possibility of high electron fluxes), TEM (high energy and high fluxes of x-rays and electrons), particle-induced x-ray emission (PIXE; possibility of proton damage), x-ray astronomy (cosmic ray backgrounds), up to high energy physics (particle tracks, heavy ions, neutrons, etc.), and military applications.

There are many treatises on radiation damage, for example, those presented by Van Lint [50,51], and there is also good coverage in the work of Spieler [52], so we will not dwell on the topic here, but merely raise it as appropriate, for the various types of detectors discussed later. The most common form of radiation effects concerns dielectric charging, and we have already seen that artefacts can be caused by this (Section 3.6). It is therefore also clear that where thick oxides are used in the manufacture of the detectors (or FETs), they will be more prone to damage. When taken out of equilibrium conditions by very high fluxes (even of low energy x-rays), oxides trap holes and disturb the field distribution. It is not surprising that MOS structures suffer from this more than p–n junction structures, and a great deal of effort and money has been invested in mitigating these problems, not least for military applications. In the case of energy dispersive x-ray microanalysis on electron columns, it is prudent to give the semiconductor crystals protection (good collimation, retraction, shutters, etc.) as it is rarely the case that analysis is required during 'overload' high dead time conditions. 'Conditioning' (controlled annealing—see Section 3.7) can also solve temporary charging problems.

References

1. Monch W., *Electronic Properties of Semiconductor Interfaces*, Springer, 1993, ISBN 3-540-20215-3.
2. Sze S.M., *Physics of Semiconductor Devices*, Wiley, 1981.
3. Goulding F.S., *NIM A* 142, 213, 1977.
4. Madden N.W. et al., *IEEE* NS-37, 171, 1990.
5. Lowe B.G., *NIM A* 439, 247, 2000.
6. Joy D.C., *Rev. Sci. Instrum.* 56, 9, 1772, 1985.
7. Nowotny R. et al., *NIM* 147, 477, 1977.
8. Evensen L. et al., *NIM A* 326, 136, 1993.
9. Slapa M. et al., *NIM* 196, 575, 1982.
10. Canberra HPGe detectors, Canberra, Ultra-Low Energy (ULE), USA.

11. Tench O., Advances in X- and gamma radiation detectors, *Trans. Am. Nucl. Soc.* 79, 104, 1996.
12. Streli C. et al., *NIM A* 334, 425, 1993.
13. Newbury D.E., *Scanning* 27, 227, 2005.
14. Tasch A.F. et al., *NIM B* 112, 177, 1996.
15. Hartmann R. et al., NIM A377, 191, 1996.
16. Zandfeld P., *Solid State Electron.* 19, 659, 1976.
17. Rawlings K.J., *NIM A* 245, 512, 1985.
18. Maisch T. et al., *NIM A* 288, 19, 1990.
19. Bailey P. et al., *SPIE* 1344, 356, 1990.
20. Hartmann R. et al., *NIM A* 387, 250, 1997.
21. Llacer J. et al., *IEEE* NS-24, 53, 1977.
22. Statham P.J., Assoc. Nat. de la Res. Techn. (ANRT) *Microanalyse et Microscopie Electr. A Balayage*, Paris, Nov. 1985.
23. Cox C. et al., *NIM B* 241, 436, 2005.
24. Kaplan S.N. et al., *IEEE* NS-33, 1, 351, 1986.
25. Hartmann R. et al., *NIM A* 439, 216, 2000.
26. Struder L., *XRS—Recent Tech. Adv.*, Chapter 4, Van Grieken R., Tsuji K., Injuk J. (Eds.), Wiley, 2004, ISBN 0-471-48640-X.
27. Siffert P. et al., *IEEE* NS-11, 3, 244, 1964.
28. Polushkin V. and Sharp S., IEEE NSS Conf. Rec.1; N14-97, 396, San Diego, 2006.
29. Campbell J.L., *NIM B* 49, 115, 1990.
30. Scholze F. et al., *NIM A* 339, 49, 1994.
31. Inagaki Y. et al., *NIM B* 27, 353, 1987.
32. Kalinka G., *NIM B* 88, 470, 1994.
33. Cox C. et al., *IEEE* NS-35, 28, 1988.
34. Ashley J.C. et al., *J. Electron Spectrosc. Relat. Phenom.* 24, 127, 1981.
35. Sah C.T. et al., Properties of silicon, *EMIS Data Review Series No. 4*, p. 613, 1987, ISBN 0 85296 475 7.
36 Sah C.T. et al., Properties of silicon, *EMIS Data Review Series No. 4*, p. 620, 1987, ISBN 0 85296 475 7.
37. Barbottin G. and Vapaille A. (Eds.), Instabilities in silicon devices, Vol. 1, *Silicon Passivation and Related Instabilities*, North Holland, p. 348, 1986, ISBN 0 444 87944 7.
38. Eggert T. et al., *NIM A* 568, 1, 2006.
39. Ishizuka Y. et al., *IEEE* NS-33, 1, 326, 1986.
40. Adachi H. et al., *J. Phys. D Appl. Phys.* 4, 988, 1971.
41. Dinger R.J., *J. Electrochem. Soc.* 123, 9, 1398, 1976.
42. McCarthy J.J., *X-ray Spectroscopy in Electron Beam Instruments*, Williams D.B. (Ed.), 67, Plenum Press, 1995.
43. Kemmer J., *NIM* 169, 499, 1980.

44. Evensen L. et al., *NIM A* 337, 44, 1993.
45. Bramley A., *Bull. Am. Phys. Soc. Ser. 2*, 5, 355, 1960.
46. England J.B.A. and Hammer V.W., *NIM* 96, 81, 1971.
47. Hansen W.L. et al., *IEEE* 24, 1, 61, 1977.
48. Luke P.N. et al., *IEEE* NS-41, 4, 1074, 1993.
49. Amman M. et al., *Proc. SPIE 4141*, 144, 2000.
50. Van Lint A.J. et al., *Mechanisms of Radiation Effects in Electronic Materials*, Vol. 1, Wiley, 1980.
51. Van Lint A.J., *NIM A* 253, 453, 1988.
52. Spieler H., *Semiconductor Detector Systems*, Oxford University Press, 2005, ISBN 0-19-852784-5.

5

Si(Li) X-Ray Detectors

An introduction to Si(Li) detectors was given in Section 1.16. Here, we will discuss the manufacture of these detectors and various related topics in further detail. The previous comments on the response function, spectral artefacts, and contacts are, of course, general, but we will elaborate on these where they specifically apply to this type of detector. The Si(Li) detector has been the workhorse of x-ray spectroscopy since the late 1960s, and the literature on them is extensive, written mainly by the pioneers in government laboratories and large well-established companies under government contracts, mainly in the United States. The progress since the middle to late 1970s has been made inside the relatively small commercial companies that took over the role of manufacturing, and these innovations have not been well documented. This fascinating development will be discussed in Chapter 11, and here we will endeavour to describe the most state-of-the-art devices.

5.1 Manufacture of Si(Li) Detectors

The basic process was developed and described in 1965 by the Berkeley Group [1] and has not changed radically, although these researchers would have been amazed at the performance of the modern-day Si(Li) detector compared to those early efforts. In 1979, Jim McKenzie (Chalk River, Bell, and Sandia Laboratories) [2] said, 'There are almost as many recipes for making semiconductor counters as there are silicon and germanium crystals or, at the least, detector fabrication facilities.' He was referring to the variations in the basic process, and we would add now that each 'detector fabrication facility' probably has more than one process that they use, for example, as insurance when one fails or for particular applications where the standard one cannot be used.

In Section 1.15, we classified Si(Li) detectors as *charge induction detectors*. Being of a classic planar structure, the motion of the charge generated by the x-ray induces a signal at the field effect transistor (FET), and this charge has to be completely cleared from the bulk before another x-ray can be analyzed. They can also be categorised in terms of their x-ray absorption characteristic as 'thick' detectors being 2.5–5 mm thick in contrast to 0.27–1.0 mm detectors where the starting material is a silicon 'wafer' as processed in more or less standard semiconductor 'foundries'. As a consequence, the starting material is not a wafer but a 'slice' from an ingot (or boule), usually 4–5 mm thick and 50–100 mm in diameter.

5.1.1 Selection of Material

Boron-doped single crystal silicon is the starting material. As lithium is an n-type dopant the starting material must be p-type, and as the silicon is to be lithium drifted in order to increase its resistivity, very high purity is not a critical starting criterion. However, if the lithium drifting process is to be omitted, as in the HPSi detector (which we will discuss in Chapter 6), a resistivity of >20 kΩ cm (all resistivities quoted at room temperature) is necessary. Although this higher purity material can be lithium drifted, for Si(Li) detectors p-type material of resistivity >0.1 kΩ cm can be drifted in reasonable time (days to weeks) at reasonable temperatures (<150°C). The lower limit on resistivity is attributable to the fact that at higher temperatures the conduction may approximate to 'intrinsic' (see Section 1.4) and the reverse bias junction required for drifting may be lost. From our earlier discussions on the theory of detectors, we saw that we want good charge lifetime (>2000 μs). Usually, 1–3 kΩ cm 'dislocation-free' (<10^4 cm^{-2}) silicon is selected for drifting, and it is more important that the resistivity and lifetime be uniform throughout the ingot. This represents boron concentrations in the range 4×10^{12} to 10^{13} cm^{-3}. Float-zone purified material is preferred because of its lower oxygen content, which must be $<10^{15}$ cm^{-3}. Oxygen contamination, because it forms Li–O donors, can impede lithium drifting and is generally to be avoided. Zone refinement uses one or more molten zones, which are moved through the ingot at about 10 cm/h, and sweeps the impurities along with them. The segregation between melt and crystal can reduce the impurity level to below 1 part in 10^{10}. An improvement to the zone refining was made in 1952 [3] at Bell Laboratories by using a vertical silicon ingot supported at both ends and scanning a single floating molten zone down it. This is known as float-zone silicon (FZ-Si). The surface tension of the molten silicon is sufficient to stop it from running away, and hence it never comes in contact with the outer vessel that had previously been the source of contaminants. In practice, it is usually purified in a helium gas atmosphere. The float-zone refining method is not very effective for boron (the segregation coefficient 0.8 is too close to unity), and this remains as an almost uniform acceptor impurity naturally (and fortuitously for lithium drifting) produces p-type material. The phosphorus donor impurities have a much lower segregation coefficient, and slight variations in temperature during growth can give rise to n-type striations in p-type material [4,5]. It has been shown that the phosphorus concentration in FZ-Si can be highest at the center of the ingot [5] because of evaporation from the molten zone. Vacancy clusters are only present in FZ-Si and have been shown to cause lithium precipitation in drifted material. They may therefore be the cause of spectral artefacts because of electron trapping in some cases. They also appear to be more mobile than *individual* vacancies and can be gettered out using phosphorus [6]. This technique is used fairly routinely now for detector-grade HPSi wafers [7] as noted in Section 4.4. A high concentration of diffused phosphorus at the silicon surface is a source of silicon interstitials that combine with the vacancies. The vacancies tend to be concentrated at the ingot center [8] in some material. The present authors, however, have not noted any drift or performance-related dependence of crystals cut from the center of FZ-Si ingots [9]. The orientation of the silicon crystal ingot is usually available as $\langle 111 \rangle$. This choice for detectors is historical and was chosen because it presented the greatest density of atoms/unit area for radiation. Moreover, the channeling of electrons through the crystal, reducing the stopping power, is less likely. However, the fixed charge at the silicon oxide interface is about an order of magnitude less on $\langle 100 \rangle$ than $\langle 111 \rangle$ silicon, which is why the former is usually used in the semiconductor industry. In practice, no difference has been found with both orientations as far as Si(Li) or SDD detectors [10] are concerned. There are only a small number of crystal growers, such as Topsil (Denmark), Wacker (Germany), and Komatsu (Japan), which supply the 'detector-grade' silicon both for lithium drifting, and HPSi and detector manufacturers are rather vulnerable in this respect. The situation for HPGe is even worse, and detector companies have mainly chosen to grow their own material in this case.

5.1.2 Lithium Diffusion

Lithium diffuses readily into silicon interstitially. By Fick's law, the diffusion rate is proportional to the concentration gradient, and the constant of proportionality is called the diffusion constant D. This diffusion rate has been studied using p–n junctions since 1953 [11] and increases rapidly with temperature as $D = D_o \exp(-E/kT)$ over a wide range of temperatures. The experimental values are $D_o = 0.0025$ cm^2/s and the activation energy $E = 0.655$ eV [12,13]. The diffusion constant D is related to the mobility of the lithium by the Einstein diffusion equation, $D = \mu kT/e$. The temperature of the diffusion is limited by the fact that we want to preserve the carrier lifetime of the material. After diffusion, the lithium concentration N_D follows an error function with depth x. The initial drift process described in Section 5.1.7 relies on the diffusion of lithium into the depleted region and therefore on the gradient $\delta N_D/\delta x$. Pell [12] shows that this gradient at the junction is given by

$$\delta N_D/\delta x = -N_D c/2Dt, \tag{5.1}$$

where c is the depth of the junction and t is the diffusion time. So, to promote the initial drift, we need to limit the time of diffusion t and quench the silicon quickly.

The ingots are sliced to the thickness desired using a diamond saw, and the faces are lapped flat and parallel and then cleaned. The lithium metal can be painted on to one face (either previously etched or left finely lapped) as a suspension in oil, but it is more usual to evaporate it in a vacuum chamber dedicated for this purpose. This also usually acts as the diffusion furnace in one process without breaking the vacuum. It should be noted that lithium must be regarded as a contaminant for all subsequent processes and should be isolated from them as much as possible. The slice is placed on a suitable clean isolation sheet (aluminium foil or mica sheet) on a hot plate in the vacuum chamber. The hot plate is brought up to temperature (~400°C) and a generous evaporation of lithium is applied. After closing a shutter to stop the deposition, the slice is left at this temperature for a further 10–25 min depending on the depth required. This is usually in the range of 0.2 to 0.3 mm. Care should be taken that the silicon is not exposed to any hot surfaces that could contaminate it (copper, brass, gold, etc.). Variants of this procedure do exist. For example, the diffusion can be carried out under an argon atmosphere [14–16], and some believe that argon 'wets' the silicon surface aiding lithium adhesion. In order to arrest the diffusion quickly to avoid a deep lithium concentration tail, the slice can be cooled rapidly using water cooling or flipping the slice onto a previously cooled heat sink. This helps later to initiate drifting and will provide a reasonably low injection junction when reverse biased during the initial lithium drift. The excess lithium is removed and the sheet resistivity, which should be <2 Ω cm, is checked. The depth of diffusion can be measured by lapping an edge of the slice and copper staining. As copper staining is commonly used throughout manufacture of Si(Li) detectors to delineate the p–n junction, we will now discuss this topic further.

5.1.3 Copper Staining

Copper staining was introduced by Whoriskey in 1958 [17] as a technique for visually indicating the position of the p–n junction in semiconductors. The technique, as now used, involves the preparation of a copper sulphate solution with a

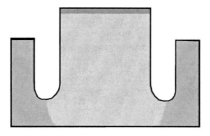

Figure 5.1 Results of copper staining.

drop of hydrofluoric (HF) acid added to remove the oxide film on the silicon. The solution is spread onto the surface (preferably finely lapped to prevent 'draw back' of the solution at the edges) and illuminated with a bright white light so that the photovoltaic effect biases the p–n junction in the reverse direction and aids the reaction. The Cu^{2+} ions preferentially decorate the n-type surface. The higher the concentration of free electrons, the brighter is the copper plating. The sample should be washed in deionised (DI) water to fix the stain. If a drifted crystal is lapped transversely to show the profile after copper staining, the lithium diffusion and the compensated volume are clearly visible, as shown in Figure 5.1.

The process relies on the deposition rate of copper from the staining solution onto the silicon being greater for n-type silicon than for p-type. The reaction is

$$Si + 4HF + 2\,CuSO_4 = SiF_4 + 2Cu + 2H_2SO_4.$$

The copper ends in solution from this reaction as Cu^{2+} ions that require two electrons to plate out. The plating therefore is faster on n-type silicon than on p-type. Note that Li does not enter the reaction, and indeed Whoriskey demonstrated staining on gallium-diffused silicon. The technique is extremely versatile and can be used, for example, to show any junctions in as-grown HPSi and HPGe ingots.

5.1.4 Machining of Si(Li) Crystal Profile

The reason for contouring with a groove or 'top hat' profile is to reduce surface leakage current as was previously discussed

(Section 1.16.2). The lithium-diffused face provides a rugged contact for the final diode and apart from metallisation needs no further processing. The metallisation can be applied before machining (which has the advantage that no masking is required), but because the machining process is—by its nature—dirty, it is usually carried out as a final stage before drifting. The profile is first machined (usually by a computer controlled machine) and then the detector crystal diced out from the slice and lapped to the desired thickness. Many different machining methods, such as diamond drilling, carborundum slurry grinding, ultrasonic drilling, and spark erosion, have been used with success and, provided precautions are taken (such as avoiding drill chatter, excessive speed, overheating, and chipping on removal of drill), they do not induce damage deep into the crystal. No one method can be particularly recommended. Subsequent silicon etching removes damaged material preferentially, and this is usually sufficient to leave a good surface. Drills will naturally need to be profiled to produce the desired shape (groove, top hat, chamfered), and side walls are normally at least partially removed (see Figure 5.1). The latter allows better flow of etch and better inspection, and lowers the electrical capacitance of the device.

5.1.5 Etching

Etching to remove damage and leave a quality 'polish' or 'mirror' surface is the most important part of all the processing steps. The effect of whatever is done to the surfaces after etching depends on quality of the surfaces and the *repeatability* of the process. This is as true of side-wall etching as it is of the final entrance window etching. For example, the 'smoothness' of the etched window surface has been correlated with charge injection in surface barrier detectors [18]. Each manufacturer will have their own preferred recipe that suits them, and all we can do here is emphasise some basics and the need for repeatability. A 'loss of process' is often caused by one seemingly insignificant change that creeps in unnoticed. Any 'development' or variations of the process is best undertaken by those normally running the production line or at least very familiar with it so that only *one* change is made at a time, and this is well documented.

The basic silicon etchant is a mixture of concentrated HNO_3 acid and HF acid and is carried out in polytetrafluoroethylene (PTFE; 'Teflon') beakers. The former oxidises the surface and the latter dissolves this oxide. A 3:1 ratio of these has been found to work well and is known in the industry as CP4 (see Section 11.11 for derivation), although other definitions are used. The reaction can be buffered, usually with glacial acetic acid, in various ratios and is known as CP4A. Such mixtures can be purchased ready mixed, but most would prefer to make their own 'fresh' mixes. A relatively long etch (~6 min) is required after machining with a good method of masking the surfaces that do not require etching (the front face, the mesa, and the outer walls). This can be done by clamping these surfaces in PTFE jigs and etching in a PTFE basket, a procedure now more commonly used for large-volume germanium detectors. PVC tape can also be used with care. A more convenient method for small crystals is to use a masking wax, such as black 'picein' (e.g., 'Apiezon-W') wax dissolved in trichloroethylene (TCE) or a related (less toxic) solvent. This commercial wax, also known as 'black wax', is very pure and adheres well to both silicon and PTFE. This enables the crystal to be attached to a PTFE 'spatula' (as shown in Figure 5.2) or some sort of etching machine.

It is also possible to etch a serial number onto the crystal edge during this procedure. The walls and groove should be blow dried with nitrogen and inspected thoroughly after masking. Particles or wax (which can be attracted electrostatically to the silicon) will cause pits that cannot be removed by subsequent etching.

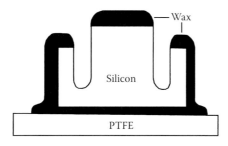

Figure 5.2 Method of masking silicon in preparation for etching of groove.

Care must be taken in order to achieve the required mirror finish. In particular, the following points should be noted:

(1) The etchant must not become overheated. For a long etch (~3 min or more) the etchant can be precooled. In the extreme, a hot etch leaves an 'orange peel' appearance.

(2) Air must not get to the surface being etched. Beads of isolated etch will heat. A steady even flow of fresh etchant across the surface in a rotary action is recommended without any excessive turbulence.

(3) The quench must be rapid, again not allowing air to come into contact with the surface. Usually, a quick quench in DI water is followed by methanol. A variation is to omit the DI water and flood wash with methanol.

The wax is removed by immersion in TCE and blow drying with nitrogen. Initial drift can be carried out without a passivation being applied. A metal contact can be evaporated onto the mesa. This is usually a thick coating of aluminium with an option of another metal (e.g., Ni), to prevent oxidation. Gold on aluminium is likely to give the 'purple peril' and should be avoided. Nickel can also be used as a front contact. However, there needs to be some method of stopping the initial drift as soon as the lithium has reached the front face ('punch-through'; see Section 1.16.1); otherwise, there would be an accumulation of lithium and a transverse spreading of the compensated profile. Note that even if batches of crystals are drifted together, they are likely to punch-through after different drift times. There are various options available such as measuring the change in resistivity [19], in which case the metal needs to be applied as two electrodes on an etched surface or the detection of light or short-range ionising radiation, in which case a p-type surface barrier (e.g., Ni or Au) needs to be applied, again on an etched surface. The capacitance can also be monitored because this decreases and will level off on punch-through. The authors have found that a reliable indication of punch-through is the rapid increase in bulk current when the compensated volume reaches a degenerate lapped surface coated with nickel. The p–n junction formed by the uncompensated p-type silicon then

Figure 5.3 (**See colour insert.**) Copper staining revealing 'punch through'.

breaks down, and the high current (a few milliamperes) can be used to terminate the drift. This method was first used by Geoff Dearnaley [20] at Harwell, UK, in 1964. The completion of the initial drift is confirmed by lapping the metal off the crystal face and copper staining. If the stain indicates the presence of compensated silicon (usually in the center), the crystal can be lapped back further until the stain covers the required area that is largely defined by the bottom of the groove (see Figure 1.57c). Figure 5.3 shows the copper stains on a group of crystals after punch-through.

It is essential that the punch-through profile be known, because it will represent the surface junction area that must be protected from pressure points and damage. This is achieved by masking the crystal periphery (the dark areas in Figure 5.3) and etching a shallow recess (see Figure 1.60). It is also essential that all traces of copper be removed with nitric acid before cleanup.

5.1.6 Lithium Drifting

An introduction to lithium drifting was given in Section 1.16.1. The process had its origins in the diffusion work of Erik Pell and others at General Electric Research Laboratory in 1960 and is now unique to Si(Li) technology. Contrary to popular belief, it is not a difficult process to carry out, and yields are generally very good. The lack of supply of very high purity material, which would not require drifting, is not the great handicap that it was for HPGe (because of the dedrifting of Ge(Li) crystals at room temperature). Apart from the extra time needed to drift the material, it has few disadvantages and indeed has the advantage of clearly defining the active volume of the crystal independent of applied bias. If the resistivity of the starting material is not too high, there will be little depletion beyond the compensated volume. The drifting is carried out in two stages for practical reasons: an 'initial drift' at a high temperature, which can be carried out relatively quickly and which largely defines the active volume, and a 'cleanup' at a lower temperature during which the bulk leakage current is much lower and therefore leads to much more complete compensation. This is because the Li^+ ions also compensate the negative space charge of the bulk current during initial drift leading to 'overcompensation' at the higher temperature. Ideally, we would want to compensate at very low temperature and current, but the drift times would be unacceptable. A compromise is an initial drift at 110–120°C and a cleanup at 85–90°C. Although a direct association of the positive Li^+ ion and a negative acceptor such as B is not necessary for bulk charge compensation in the crystal [21], the Coulomb force between them will result in pairing in a localised complex, particularly at low temperatures. This pairing reduces the scattering of charge carriers in the Si(Li) crystal and increases their mobility.

5.1.7 Initial Drift

The activation energy for the ionisation of the lithium atom is very low (33 meV in silicon and 9.3 meV in germanium), and remembering that the thermal energy at room temperature is about 25 meV, moderate temperatures only are needed to give the singly charged anion mobility in an electric field. The mobility μ_L of the Li^+ ion in a field E, for example, the field of the reverse-biased p–n lithium-diffused silicon junction, is related to the drift velocity in the field by the usual equation

$$\mu_L = v/E, \qquad (5.2)$$

where v is the ion drift velocity.

Pell [22] gave the following empirical formula for the Li⁺ ion mobility

$$\mu_L = (26.5/T) \exp(-7500/T).$$

This agrees well with that derived from the values of D given in Section 5.1.2.

Figure 5.4 shows the mobility of Li⁺ ions in silicon as a function of temperature.

Note that the mobility at 120°C is ~500 times that at room temperature. Si(Li) detectors do not degrade significantly if kept for a time (<1 year) at room temperature, contrary to many reports.

When a reverse bias V is applied to p-type silicon, a depletion layer of depth W is formed supporting a field E. The depletion depth W is given approximately (see Section 1.6.1) by

$$W \sim \sqrt{(\rho V)}/100, \tag{5.3}$$

where ρ is expressed in kΩ cm, V in volts (V), and W in mm.

Thus, with a 500-V bias, the depletion depth at the start of drift is roughly of the same magnitude as the Li diffusion depth. We also saw that the field has the form

$$E(x) = E(0)[1 - x/W], \tag{5.4}$$

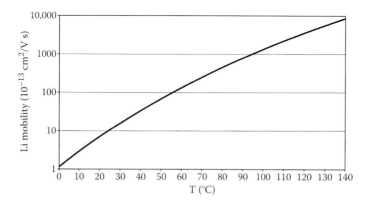

Figure 5.4 Lithium ion mobility in silicon as a function of temperature.

where x is measured from the p–n junction into the depleted region. Thus, the field is a maximum at the junction and falls off linearly to zero at the depletion depth W as shown in Figure 1.23. If the bias is kept constant, at the start of drifting this field is at its maximum, and both surface and bulk leakage currents can be excessive. Therefore, at this critical stage, it is usual to ramp up both the bias and the temperature over 10–15 h or so. The computer-controlled drifting developed at Berkeley [19] monitored the current, waiting an hour between incremental changes. By this means, a 'well behaved' crystal could be ramped up to initial drift conditions (110°C at 1 kV in their case) in ~3.5 h. Great care should be taken that there is no damage to the mesa edge and the field tangential to the surface can be reduced by a slight chamfer, although the undercutting effect of the etchant (caused by the lithium and local turbulence) normally provides this naturally. Note that the field in the heavily lithium diffused region is always zero, so that the driving force behind lithium drifting at the start is mainly *diffusion* (aided by temperature) from this reservoir. The field is only high in the depletion region, which progresses as it becomes compensated. In this region, the Li$^+$ ions are available to compensate the acceptor impurities, thus levelling the field and advancing the depletion layer into the silicon as shown in Figure 1.58. A Japanese group [23] cleverly showed this progress using the technique of electron beam–induced drift current (EBIC) in an SEM as shown in Figure 5.5.

Thus, the process takes place at the compensated/uncompensated boundary layer aided by the diffusion of the Li$^+$ ions. In a simple treatment, if a constant voltage V is applied and a compensated depth W has been formed, we can write the Li-drift velocity as

$$v = dW/dt.$$

Assuming perfect compensation

$$E = V/W. \tag{5.5}$$

Then Equation 5.2 gives

$$WdW = \mu_L V dt \tag{5.6}$$

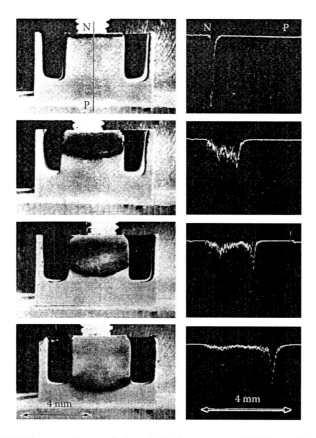

Figure 5.5 Electron beam–induced drift current (EBIC) applied to partially drifted Si(Li)s. (From Watanabe, E. et al., *IEEE* NS-36, 1, 117, 1989. With permission.)

which when integrated gives

$$W^2 = 2\mu_L Vt. \tag{5.7}$$

Equations 5.5 and 5.7 show that, for a fixed bias V, the field across the compensated layer decreases slowly as the inverse of W and the inverse square root of the drift time t. To a first approximation, the compensated depth W increases as the square root of time. The relationship between W and drift time t for a constant bias of 500 V for various temperatures is shown in Figure 5.6.

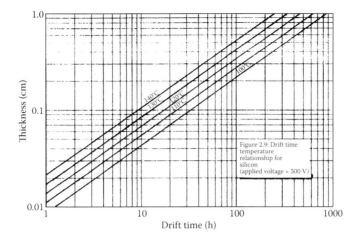

Figure 5.6 Drift times for different drift temperatures.

As an example, we can consider the time to complete the drift of a 3-mm crystal at 500 V and 110°C. Figure 5.6 gives ~113 h or ~4.7 days. In practice, the drift times may exceed this because of the effect of the contouring. Furthermore, the Li⁺ ion trapping and detrapping process puts the drifting into the 'subdiffusion' [24] regime, in which the diffusion times follow a 'Pareto' law rather than Fick's law. As a result, the drift times are longer and much more variable than expected.

As first pointed out by Jim Mayer [25], the field distortions due to the space charge current during initial drift are minimised by drifting at high fields. Also, a high field at the start of the drift process is desirable because it helps 'harden' the diffused junction profile [12]. This effect is shown in Figure 5.7.

For a drift field E, the lithium tail can be shown [12] to have a gradient

$$\delta N_D/\delta x = -E\mu_L N_D/D.$$

As the drift progresses, the field decreases and the sharp transition region profile is frozen in. This hardening of the junction has a number of benefits. It minimises the layer of series resistance, minimises the depth giving incomplete charge collection (ICC), and gives a 'high–low' junction, thus reducing the diffusion current in the final device (see Section 4.7). A

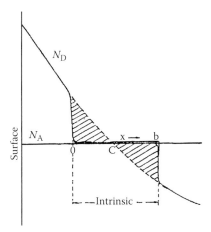

Figure 5.7 Impurity profile after drifting. (From Pell, E.M., *J. Appl. Phys.*, 31, 291, 1960. With permission.)

constant drift bias of 600–800 V is common and leads to an initial field at the junction approaching 10 kV/cm.

Let us now look at the detail of the drift process at the compensated/uncompensated p–n junction interface as it progresses from the crystal mesa toward the front contact.

If we take the junction as our origin $x = 0$, the field is obtained by integrating Poisson's equation:

$$\varepsilon\varepsilon_0\,\mathrm{d}E/\mathrm{d}x = e[(N_A - N_D) - n(x) + p(x)], \tag{5.8}$$

where e is the electronic charge, $\varepsilon\varepsilon_0$ is the dielectric permittivity of the silicon, N_A and N_D are the acceptor and donor concentrations, respectively (in this case, represented by the impurity B and lithium ion), and $n(x)$ and $p(x)$ are the mobile electron and hole densities, respectively. Neglecting the presence of mobile charge, the compensation is perfect if $N_D = N_A$ and there is no space charge present, and by 5.8 field E is uniform. In a more rigorous treatment [26], the space charge due to the current flow of holes and electrons (the difference in hole and electron densities) is accounted for. Assuming the currents are thermally generated in the bulk, the density of electrons and holes increases linearly across the drifted volume. This is an approximation to 'intrinsic' conduction. For negligible carrier recombination in this volume, the space charge $(N_D - N_A)$

after compensation is linear, and under these circumstances we might expect the p–n junction to be halfway across. More precisely, it is in the ratio of the two carrier mobilities and therefore roughly 1/3 distance from the lithium diffused n⁺ side. The higher electron mobility means that they, on average, travel roughly twice as far as the holes from their origin in a given time in the same field. Hence, the plane in which their densities are equal is twice as far from the front contact as it is from the lithium diffused back. It is this uneven distribution of free charge in the volume during drift that distorts the field and gives rise to an uneven compensation or 'graded junction' [25]. The practical solution to this problem, first suggested by Lauber [26], is to redrift (or 'cleanup') at a lower temperature to 'level' this distribution after the initial drift.

However, 'intrinsic' conduction does not apply at normal drifting temperatures (<150°C), and taking into account the minority carrier concentrations due to the reverse bias diode we note that this will be mainly an electron current from the p-side. This is especially true when the original resistivity of the silicon is reasonably high as is usually the case. The existence of this current will cause a shift of the p–n junction toward the p-type side, and in practice the overcompensation persists throughout the depth and the p–n junction ends up at the front even after the levelling effect of clean-up. Experimentally, in the finished device, x-ray signals are observed at very low bias (5–20 V), indicating that it is at (or at least very close) to the front p⁺ contact and the whole bulk can be regarded as very slightly n-type (ν-type).

Following Lauber [26], we consider the flux $f(x)$ of Li⁺ ions during the drift process. The two elements driving this flux are the concentration gradient of the Li⁺ ions $\delta N_D/\delta x$ and the field E resulting in the drift velocity given by 5.2. We can thus write

$$f(x) = D\delta N_D/\delta x + \mu N_D E. \qquad (5.9)$$

If the flux changes by δf because of compensation in the depth element δx, this must be exactly balanced by the accumulation $\delta N_D/\delta t$ of ions in this element in order to satisfy continuity.

$$\delta f = (\delta N_D/\delta t)\delta x \tag{5.10}$$

Combining 5.9 and 5.10 gives the change of lithium ion concentration N_D at x with time as

$$\delta N_D/\delta t = D\delta^2 N_D/\delta x^2 + \mu\delta(N_D E)/\delta x. \tag{5.11}$$

In principle, Equation 5.11 and Poisson's equation (Equation 5.8) with suitable expressions for $n(x)$ and $p(x)$ can be used to give the lithium ion density N_D and field E at any point and time in the drifted layer. Lauber [26] solved this problem for a number of specific cases, but a simple argument in accordance with Goulding [21] shows that the compensation process is 'self-regulating' in the sense that it works toward uniformity. Suppose, for example, there is an island of above average concentration of lithium in the 'compensated' layer as shown schematically in Figure 5.8a. Then, the resulting nonuniform field is as shown in Figure 5.8b.

The higher field on the right of the excess lithium results in a higher current of lithium ions out of the island than the current of lithium ions into it from the left. Thus, uniformity is eventually restored.

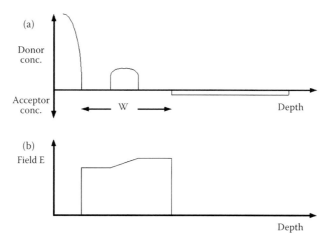

Figure 5.8 Lithium excess will be reduced by field gradient. (a) Excess lithium at a depth in the crystal and (b) non uniform field acting in a direction to reduce this excess.

In practice, the drifting can be carried out with high yields in air [27] or liquid [25], but both have to be carried out in darkness to reduce photocurrents. Any temperature-controlled chamber can be used, but a clean environment is essential. As noted earlier, particulates of copper, brass, gold, lithium, etc., should be specifically excluded. A passivation that is temperature resistant (such as a SiO coating) can be applied, but under these conditions bulk currents tend to dominate and the groove will not pinch off surface currents until the drifted volume has reached it. In some cases where the silicon surface has gone p-type, the lithium can spread laterally at the groove and stall the drift. Liquid drifting is normally carried out in inert fluorocarbons. These can be mixed to give the required constant boiling temperature. Liquids are thought to give more uniform temperatures, enable drifting at higher power, and keep the surfaces cleaner and more stable. However, contamination can still be a problem. Computer control [19] can be used but is not essential if currents are monitored on a daily basis. The bulk current increases steadily during drift because it is proportional to the volume of the compensated region. Typically, for 4 mm deep 10–50 mm^2 crystals at 110°C and 800 V bias, the current increases from <100 to ~500 μA after 18–20 days. The current then increases rapidly over a day or so. There can be an automatic cut-off of the bias when a preset current (~4 mA) is exceeded. This guards against irreversible dielectric breakdown and signals the lithium punch-through to the degenerate front face.

5.1.8 Cleanup

The processes taking place during cleanup are not well understood. Reducing the temperature from 110°C to 90°C (say) reduces the bulk current and the lithium mobility by a factor of ~2.8. Cleanup has even been carried out at room temperature [28]. In order to clean up the lithium drift, the crystal must support a bias of 600–800 V at the lower temperature and lower reverse leakage current (less than 100 μA) for a time, which is learned by experience but is typically 16–30 days depending on the thickness and area of the crystal. The compensated profile should not alter significantly during the cleanup operation. A relatively rugged p$^+$ junction must be made on the front face, and in the past diffused or ion-implanted contacts have been tried with a view to further drifting toward the surface to obtain a thin window [28] but the usual choice is to make a surface barrier contact by the evaporation of gold onto a clean etched surface. It is probable that gold diffuses into the silicon during cleanup as it is found necessary to lap the surface back before remaking the contact to avoid ICC in the final device. The type of surface barrier contacts that give the high reflection coefficient described in Section 4.6.2 are found not to be suitable at elevated temperatures. This is probably because of dielectric breakdown in the contact. Usually, the gold surface barrier is recessed to protect the surface as mentioned. During cleanup, a number of changes take place that can, if desired, be monitored. After initial drift, compensation is not complete and we cannot treat the planar device as a simple capacitor with parallel plates and a silicon dielectric (dielectric constant 11.7). A capacitance plot against bias (C–V plot) does not give a flat characteristic because we still have significant space charge present. In fact, it was shown by Mayer [25] that for the linear 'graded junction' described above, Poisson's equation in one dimension gives the depletion region a $V^{1/3}$ dependence or the C–V plot a $V^{-1/3}$ dependence. As the cleanup levels the lithium distribution, so the C–V plot is also 'levelled'. Figure 5.9 shows the original C–V plots of Mayer as a result of drifting at various temperatures.

In practice, even after cleanup there is usually a very slight decrease in capacitance with bias that is due to incomplete compensation. The excess lithium also causes electron trapping and trapping–detrapping and its associated noise. Figure 5.10 shows the effect of cleanup upon the noise at a processing time $\tau = 40$ μs.

The same effect on noise is noticed if the crystal temperature is raised above ~200°C even for a short time but can be restored by a repeated cleanup. Before cleanup, x-ray spectra exhibit symptoms of electron trapping as shown in Figure 5.11 that compares Fe-55 spectra at 100 and 500 V bias before cleanup in a log plot.

This exponential 'spur' in the low-energy background (see Section 3.2) can be explained in terms of electron trapping centres as described below.

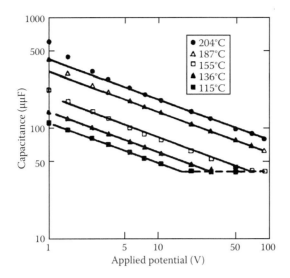

Figure 5.9 Mayer's original capacity vs. voltage plot for drifting at different temperatures. (From Mayer, J.W., *J. Appl. Phys.*, 33, 2894, 1962. With permission.)

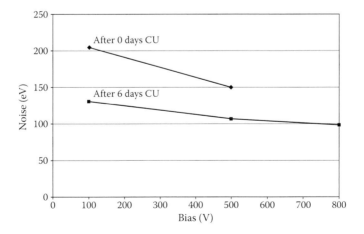

Figure 5.10 Effect of 'cleanup' on noise measured at 40 μs processing time.

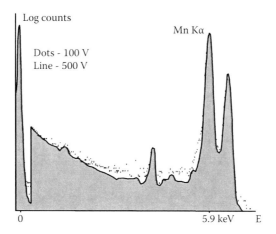

Figure 5.11 Evidence for electron trapping before 'cleanup'.

5.1.8.1 Electron Trapping for Low-Energy X-Rays

For low-energy x-ray spectra using thick detectors of depth D, we can assume that all charge is generated at a depth $x \ll D$, and the signal is virtually entirely due to electrons. We will consider two cases. The first case involves a uniform distribution of individual traps that will reduce the signal from all x-rays to some extent by the incremental trapping of electrons. The second case refers to a relatively sparse distribution of strong trapping *centres* that will trap the majority of the charge generated by any x-ray event that encounters it. The first case is described by the Hecht equation (see Section 1.7.4), which, putting $x \sim 0$, reduces to

$$\eta \sim q/q_0 \sim L_e/D(1 - \exp[-D/L_e]),$$

where L_e is the electron drift length. The charge collection efficiency (η) is constant for all x-ray events, so the prediction is that the photopeak is shifted down in energy from E_0 with no other effects except for statistical broadening. Indeed, this shift is often used to measure the drift length of electrons in compound semiconductors. This is clearly not the case for Figure 5.11, where there is no such peak shift.

Considering the second case, we can define a new mean free path length λ for the charge cloud as a whole and its encounter with a trapping center. For simplicity, we will assume that the charge cloud is completely removed and no longer contributes to the signal. If we have N_o charge clouds generated at $x \sim 0$ after travelling a distance x, we will have

$$N = N_o \exp(-x/\lambda).$$

By Ramo's theorem, the equivalent background energy E after travelling distance x is given by

$$E = E_o(x/D).$$

The background spectrum thus has the form

$$N = N_o \exp - (E/E_o) (D/\lambda)$$

$$\ln(N) = \ln(N_o) - (E/E_o) (D/\lambda).$$

A logarithmic plot of the energy spectrum will show a photopeak at the correct position with a background of slope $-D/E_o\lambda$.

Figure 5.12 shows a spectrum with no trapping superimposed on a simulated background for $\lambda \sim 0.9$ mm. This can be compared to Figure 5.11.

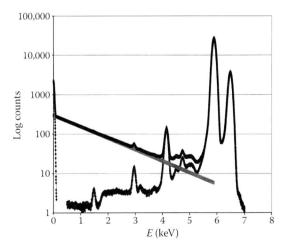

Figure 5.12 Comparison of a real spectrum (trap free) of Fe-55 with a simulated spectrum with trapping assuming $\lambda \sim 0.9$ mm.

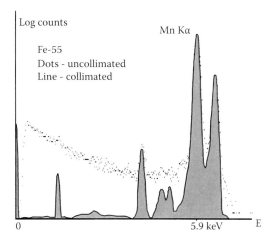

Figure 5.13 Fe-55 spectrum from a detector that had no cleanup showing that the center of crystal (collimated spectrum) did not require cleanup and presumably had a low concentration of trapping centres.

However, it appears that a larger-than-expected proportion of charge clouds traverse the crystal without being trapped, implying that the distribution of traps is not uniform. This is confirmed by collimation as displayed in Figure 5.13, which shows Fe-55 spectra before cleanup, collimated and uncollimated.

It is not clear why the trapping centres have this radial distribution. It may be due to the outdiffusion of lithium to the crystal side walls (this is apparent in Figure 5.5) during initial drift. As cleanup progresses, the trapping centres are dispersed and the background spectrum throughout the crystal levels, leaving that depicted by the line in Figure 5.13.

5.1.9 Passivation Theory

After the drifting process, the front surface barrier contact is made as described in Chapter 4, and it is usual practice to apply a surface passivation to the exposed walls of the crystal. The role of the passivation is to stabilise the surface in a state that also minimises the surface leakage current during operation at low temperature. The passivation can be so effective that the FET gate leakage current sometimes dominates the crystal leakage current. The topic of surface states was introduced in Section 1.11.3 and the effect of contouring on the pinch-off of surface currents in Section 1.16.2. We will first consider the conditions for minimising surface currents on a flat semiconductor surface. The passivation of Si(Li) detectors is specialised because high-temperature processes such as thick oxide growth that are standard in silicon foundry planar processes are ruled out because of the effect on lifetime and lithium compensation of the silicon. Before discussing the practicalities of various passivation techniques, let us look a little closer at the theory.

Unlike metals, semiconductor surfaces and interfaces can naturally acquire large layers of space or surface charge. This is possible because of the low densities of free carriers that result in much larger *screening lengths* (Debye lengths) and to surface states that are energetically within the bandgap of the semiconductor. These states can be charged, and it is these large space-charge regions that complicate and dominate device operation. The surface states on clean (e.g., vacuum cleaved) semiconductor surfaces are attributable to the unsatisfied 'dangling bonds'. This deficiency of electrons leads to a strongly n-type surface. These bonds can be partially satisfied by distortion of the lattice at the surface but can also become bound to *adatoms* that provide covalent bonds to the surface, giving rise to adsorbate-induced states. So, we have to consider first the energy levels of the semiconductor surface atoms or dangling bonds on clean surfaces and then those of the adsorbate-induced states. Figure 5.14 shows the electron probability distributions (wave functions) for electrons in these two cases.

The wavefunctions of the surface states decay into vacuum on one side and into the bulk semiconductor on the other. We therefore have surface-localised states in a complex band structure rapidly decaying into the vacuum and semiconductor. These surface states exist in a *virtual energy gap* and are sometimes called *virtual gap states*. All atoms (with the exception of rare gases) can form chemical bonds with the semiconductor surface and are then called adatoms. Provided they are sparse, they interact with their nearest neighbour semiconductor atom only, forming a heteropolar 'molecule'. The adatoms themselves are electrically neutral, but the center of gravity of the electronic charge making up the covalent bond may

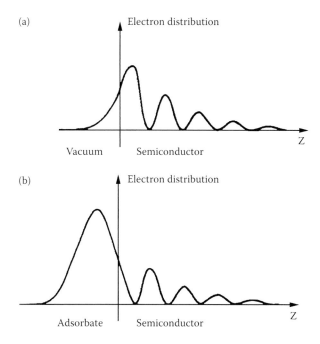

Figure 5.14 Electron wave functions at surface. (a) Electron probability distribution at the surface with just a vacuum present and (b) electron probability distribution when the surface is rich in adatoms.

be shifted toward or away from the surface. In other words, they may be considered as polarised, and they form donor or acceptor-like states in the virtual gap at the surface of the semiconductor. The adatoms can thus become charged positive or negative when the virtual gap states are occupied depending on their donor or acceptor nature. In order to maintain charge neutrality, they induce free electrons or holes beneath the surface of the semiconductor, and conducting n-type or p-type 'channels' are formed. The polarity of the surface states depends on the *electronegativity* of the adatom relative to the semiconductor as explained below.

We can regard the mobile surface charge carriers in the semiconductor contributing to the surface conductivity as sitting in a channel or potential well. This can be modelled as triangular (field constant) and infinite at the surface to prevent electrons from escaping. Such models result in quantised energy bands, and these drastically change the mobile charge distribution so that it peaks *below* the surface (at ~2 nm at 150 K), not *at* the surface as classically predicted. The carrier mobilities at the surface are modified by the fact that there will be scattering at the surface and the sides of this potential well.

Diatomic heteropolar molecules, formed of atoms A and B (where B may be the semiconductor), have dipole moments. This gives the covalent bond between the atoms their partial ionic character. Adsorbates on a semiconductor surface will therefore also induce surface dipoles. The amount of ionic character (termed the 'ionicity') of bonds in diatomic molecules AB is correlated with the *difference* in electronegativities of the atoms A and B. The ionicity is the difference in their electronegativities. The electronegativity increases from ~2 for the elements central in the periodic table increasing toward the right (toward halogens) and decreasing toward the left (toward the alkyl metals). The electronegativity of molecules bound to the semiconductor is the geometric mean of its constituents A and B. For the semiconductors in group IV (e.g., silicon and germanium) being in the middle of the periodic table, they have values ~2. This is also true of 3–5 and 2–6 compound semiconductors, and in fact the electronegativity for all semiconductors is 2 ± 0.1. The more electronegative atoms are larger and therefore generally more polarisable. The electronegativities of some of the important elements and their ionicities with respect to silicon are given in Table 5.1.

Note that these are purely surface effects and not the same as the doping in which contacts are made by the diffusion or implanting of the atoms into the semiconductor bulk from the surface. Under normal conditions, semiconductors are inert to molecular hydrogen and nitrogen, for example. Ionic compounds can be used to supply effective atomic adsorbates for passivation. Hydrogen *atoms* from HF acid treatments will terminate the surface dangling bonds and oxide and hydrogen passivations dominate silicon processing. The Si–H covalent bond energy is much higher than the Si–Si bond energy. This means that hydrogen will saturate the dangling bonds protecting the surface against chemical reactions such as oxidation. The halogens, Cl, Br, F, and I are expected to behave in a similar manner. H-coverage, for example, decreases the initial sticking coefficient of oxygen on the surface of silicon by

TABLE 5.1
Electronegativity of Elements and Ionicity with Respect to Silicon

Element	Electronegativity	Ionicity
F	3.98	−2.08
O	3.44	−1.54
Cl	3.16	−1.26
Br	2.96	−1.06
I	2.66	−0.76
Au	2.54	−0.64
Pb	2.33	−0.43
Pd	2.2	−0.3
H	2.2	−0.3
P	2.19	−0.29
As	2.18	−0.28
B	2.04	−0.14
Ge	2.01	−0.11
Ni	1.91	−0.01
Si	1.9	0
Co	1.88	0.02
Cr	1.66	0.24
V	1.63	0.27
Al	1.61	0.29
Mg	1.31	0.59
Li	0.98	0.92
Na	0.93	0.97
K	0.82	1.08
Cs	0.79	1.11

a factor of ~10^{13}. It was first thought that fluorine, not hydrogen, was the active element in terminating the dangling bonds by HF dips [29]. Calcium fluoride quenches [30] and evaporations [31] have also been used in the past. Conformal coatings can also be doped with compounds of fluorine, chlorine, iodine, arsenic, or lead, for example.

If both acceptor and donor type surface states are present, one of them will always get charged resulting in band bending regardless of whether the bulk semiconductor is n-type or p-type. As the surface charge or density of surface states increases, the band-bending *saturates*. This is termed the *pinning* of the Fermi level to mid bandgap at the surface. This saturation level occurs at lower surface charge densities in 'intrinsic' (high-purity) semiconductors. Under these conditions of high density surface charge states, the band bending (or position of the Fermi level, which is the same), are pinned in energy to within ~2.3 kT eV per decade of the change in the surface charge density: a very weak dependence on temperature and surface state density. This means that surface states in excess of this critical density will have no further effect. This critical density is small (~10^{12}/cm^2) compared to the number in a complete monolayer of adatoms (~10^{15}/cm^2). The bulk volume of a Si(Li) or HPGe crystal is close to intrinsic so that the Fermi level is very close to mid-gap (see Section 1.4). Almost flat-band conditions are achieved when the Fermi level at the surface is also around mid-gap. Two classes of material come close to satisfying this, dielectrics and amorphous semiconductors. If their Fermi levels differ from those of the semiconductor surface states, we will have some band bending, surface charge, and a conducting surface channel. Amorphous semiconductors or oxides may also be doped to control this.

To summarise, the surface states, like bulk states, have either a donor or an acceptor character, that is, they can become either positively (donor) or negatively (acceptor) charged, or be neutral. The mobile carriers produced in the semiconductor channel beneath can dramatically change the conductivity of the surface (see Figure 1.42) and generally dominate the leakage currents. It is important that any adatoms or molecules introduced for passivation are not themselves mobile (water is an example). Si(Li) surfaces are very sensitive to ambient atmospheres, the problem being that the substrate type is not well defined (being very close to intrinsic) and if the lithium compensation is poor may even run from n-type to p-type along its length.

5.1.10 Passivation Practice

We can categorise the types of practical passivations available as follows:

(1) A surface 'treatment' by various chemicals and etches
(2) A conformal coating by application of a fluid that is subsequently cured
(3) A vacuum deposition, usually by thermal evaporation or sputtering

As we have seen, the subsurface channels of mobile electrons or holes (n-type or p-type surfaces) induced by adatoms or molecules determine the surface leakage currents and noise. The optimum condition for an n-type bulk and a flat surface would seem to be slightly p-type to produce depletion or 'inversion'. However, a well-compensated Si(Li) crystal can be regarded as 'intrinsic' (technically ν-type), and we recall (Section 1.16.2) that with a groove or top-hat geometry the objective is to fix the surface (at least at the bottom curve of the contour) with a moderately n-type surface. In practice, a combination of usually proprietary etches, treatments, and coatings that yield low leakage and high stability are found by experience.

A freshly etched surface tends to be strongly n-type (because of the deficiency of electrons in the dangling bonds at the surface) moving toward p-type in a dry atmosphere. A thick thermally grown oxide gives a slightly n-type surface, by virtue of the embedded positive charge. This is used extensively in semiconductor device manufacturing but is unsuitable for Si(Li) and HPGe detectors. The early work at Bell Laboratories [32], Raytheon [33], and elsewhere classified the effects of exposure to various ambients such as water vapour, HF vapour, ammonia, methyl alcohol, acetone (all driving surfaces n-type) and oxygen, ozone, chlorine, and iodine (all driving surfaces p-type), whereas TCE, toluene, and xylene left the surface fairly neutral. Etches and quenches can also be doped, for example, with iodine. The earliest treatments usually involved the exposure of etched surfaces to HF, methanol vapour (to remove water vapour), exposure to a dry atmosphere,

ending with a TCE rinse while monitoring the room temperature leakage current, and relied on rapid transfer to a vacuum
environment in a cryostat. Hydrogen peroxide with acids (usually HF) that clean the surface and passivate it with a surface
oxide [34–36] has also been used.

Hydrogen passivation is a more recent innovation, although
the uniqueness of HF-treated silicon surfaces was recognized
by Buck [37] at Bell Laboratories back in 1958. Dilute HF
treated silicon surfaces have a 'paraffin wax–like' hydropho-
bia, and over the years it was appreciated that in the process
the dangling bonds at the surface were terminated by hydrogen and this constituted a form of passivation. The techniques
of H-passivation have been reviewed by Kern in *Handbook of
Semiconductor Wafer Cleaning Technology* [38]. The growth
of a native oxide in ambient air following treatment reduces
defects and impurities and reduces the surface interface
charge down to levels of $6 \times 10^{11}/cm^2$. The water rinse and drying process requires great care and attention to cleanliness.
A series of 'dump-and-rinse' cycles ending with pure water
cleans the surface and brings the surface charge to even lower
levels. The surfaces are then relatively stable.

The use of a conformal coating was an early form of passivation, applied after a conventional etch. Such coatings are
still in use. Photoresists and polymers, such as polyimide [39]
and paralene C, have been successfully used [40]. One problem with these thick organic coatings is the mismatch of the
temperature coefficient of expansion to silicon. This can lead
to cracking on thermal cycling.

Low-temperature glass coatings also have a long history.
For example, Mayer (Hughes Research Laboratories) and
Williams (RCA) [41] were using arsenic glass on silicon detectors in 1960, and coatings by the decomposition of siloxanes
were patented [42] in 1962. Hughes Laboratories took out a
patent in 1965 [43] for glass passivation of a silicon junction
diode. In 1984, Honeywell commercialised siloxane-based
'spin on glass' materials. These are applied in a liquid carrier (ethanol, butanol, etc.) spun to obtain a thin even coating
before curing to drive off the organic solvent and water. On a
silicon crystal they leave a deep blue sheen at the bottom of
the groove, indicating a thickness of around 0.3 μm. If applied

thinly they do not suffer from cracking on thermal cycling and can be cured in a short time at relatively low temperatures (~200°C) [44]. Later, phosphosilicate glass was introduced. This is widely used in lithographically patterned planar detector technology [10] and has the added advantage in that it getters sodium from the underlying native or thermally grown oxide.

Vacuum-evaporated coatings such as CaF_2 have already been mentioned, and Dinger [45] (who was working at Chalk River)—after assessing many dielectric evaporants including MgF_2, TiO_2, Al_2O_2, and some glasses—came to the conclusion that SiO reduced the number of surface states to ~10^{11}/cm^2 on HPGe. It is likely that the final film is SiO_x, where $x \sim 2$. A coating of ~0.2 μm gives a light blue colour and can also be applied to silicon, for example, during cleanup drift. The SiO form of passivation on HPGe was used effectively for many years (see Section 6.2.6) until workers at Berkeley [46] introduced sputtered amorphous germanium (a-Ge) that they considered to be superior. The sputtering was carried out in argon, and the rate and coating uniformity were better using AC sputtering. They found that, by adding some hydrogen to argon (up to 17.5%), a-Ge:H coating could give flat-band conditions. The coating thicknesses could typically be 0.03 to 0.3 μm. Other workers at Berkeley [47] later applied the same process (but replacing a-Ge with a-Si) to Si(Li) detectors. The results were not as successful. This can probably be explained by the surface free energy of germanium being lower than that for silicon, so the germanium wets the surface of silicon but not the other way around. However, in the case of the contoured Si(Li) crystals, a straightforward vacuum thermal evaporation of germanium of about 14 nm is found to leave an effectively passivated surface. The coatings, although thin, are very hard and chemically stable.

To summarise, the above discussion demonstrates that there are many options available for passivation of the semiconductor surface, and these are settled upon by trial and error and are usually proprietary. Crystals without any coating other than some sort of 'wet chemistry' treatment and quench can be very stable, but they are more of an 'art' and reproducibility is difficult. Apart from hydrogen passivation, it is probably correct to say that most commercial manufacturers would opt for a rugged protective coating that can be removed if reprocessing is necessary. Another consideration is that materials in the passivation itself, including dopants, may be fluoresced either directly or by photoelectrons. This may be a problem where especially low levels of elements are to be analyzed and the solid angle for x-rays is very large as, for example, in total reflection x-ray fluorescence (TXRF) or particle-induced x-ray emission (PIXE).

5.2 Legacy of Contouring

The arguments for contouring in Si(Li) detectors (and HPSi and HPGe detectors) can be summed up as follows:

(1) It confines the compensated volume during a lithium drift.
(2) It facilitates the pinching off of surface channels that makes passivation less critical and increases the yield of passivated crystals with respect to their reverse leakage current.
(3) It provides a rugged edge termination of the surface barrier contact avoiding high fields.
(4) It provides a reverse bias p–n junction at the groove surface that is a barrier to hole current.
(5) It facilitates crystal handling and labelling.

'True planar' Si(Li) (and HPGe) crystals in the form of right circular cylinders or cubes *can* be manufactured, but the surface processing and passivation is then much more critical and not commercially attractive. One cost of contouring is that the detector capacitance is increased (by typically 0.2–0.3 pF), which is significant for small area detectors. As a rule of thumb for 3-mm-thick detectors, the capacitance is 0.3 pF for every 10-mm^2 active area plus 0.2–0.3 pF because of the periphery. However, a greater cost is paid in that the field at the entrance window is no longer uniform and has a radial dependence as Figure 5.15 shows.

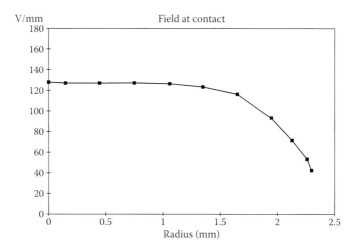

Figure 5.15 Radial dependence of field at contact.

This shows the front contact field on a simple electrostatic modelling for a 20-mm² active area Si(Li) crystal with groove surface charge. Clearly, the field is influenced over a distance of ~1 mm by the groove and collimation to 10 mm² is necessary for good charge collection. Because of the 30% difference in dielectric constant, this effect will be enhanced for grooved HPGe crystals. This radial dependence has two consequences. First, as noted in Section 4.6.5, it affects the stability of the contact in terms of charge collection, and second, quite severe collimation may be required particularly for large solid angle applications such as transmission electron microscopy (TEM) and TXRF. Of course, in a grooved geometry the field will become more uniform as the groove to front contact distance decreases. It is noticeable that successive etches improve charge collection, but this is difficult to control practically and eventually makes for a very fragile crystal. The effect of the weakened field at the contact diminishes with increased x-ray energy, and large area crystals are less affected and tend in general to have better charge collection characteristics. The radial dependence of ICC has been well documented both in terms of efficiency [48] and effective dead layer [49]. Because magnetic fields can deflect primary electrons into peripheral regions, they can also influence the charge collection, for example, in TEM applications [50].

It is important to note that lithographic planar devices such as p–i–n diodes are not immune from such peripheral effects and require collimation or guard rings for quality x-ray spectra. Side grooves [51] have been used on HPSi detectors as shown in Figure 5.16b and might improve the situation, but these have not been reproducible in Si(Li) detectors and there will still be regions of poor charge collection.

A group at Berkeley [39] attempted to overcome the problems by removing the groove after drifting and bevelling the corners to form a 'bullet' shape as shown in Figure 5.17.

This, for a given active area, also reduces the capacitance, and they referred to them as 'large area low capacitance' (LALC) Si(Li) detectors. The LALC detectors required very high bias (>2 kV) to fully deplete (with no bevel, they showed a pronounced 'ghost peak' at half energy because of the undepleted regions in the corners; see Section 3.4.2). These high biases with no groove gave high leakage currents, and although one company took them up commercially for a while they could not compete with the superior charge collection of conventional groove or top hat geometries.

An early Si(Li) geometry [52] is shown in Figure 5.18, where the lithium drift was arrested ~1 mm before punch-through. This was the work of the Yale group lead by Al Bromley and is of particular note. They ground a recess through the uncompensated material into the drifted material and made a gold surface barrier contact.

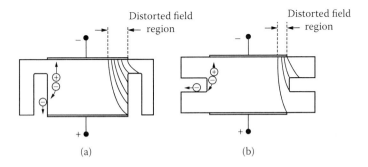

Figure 5.16 Side grooves. (a) Conventional grooved detector and (b) side groove detector. (From Ohkawa, S. et al., *NIM*, 226, 122, 1984. With permission.)

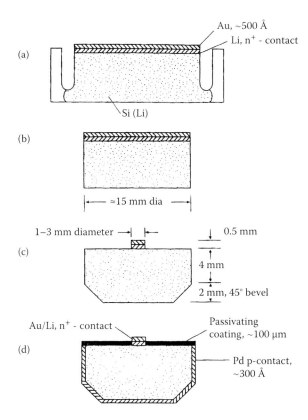

Figure 5.17 'Bullet'-shaped structure. (a) Grooved structure, (b) outer walls removed, (c) front face bevelled and (d) contact and passivation applied. (From Rossington, C.S. et al., *IEEE* NS-40, 354, 1993. With permission.)

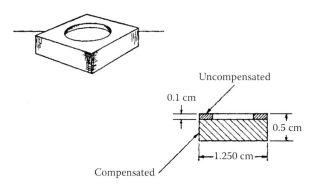

Figure 5.18 Interesting early geometry with a recessed surface barrier contact. (From Chasman C. et al., *NIM*, 24, 235, 1963. With permission.)

These detectors gave good reliable β spectra at −80°C, and the geometry avoids the weak peripheral field regions associated with the grooved geometry. Such a device, perhaps with a side or shallow mesa groove, might have been a model worth pursuing.

Another, more easily solved, problem associated with the groove is the tendency for dust or loose particles to drop and be trapped in them. High takeoff angle detectors can be particularly prone to this. These particles, depending on their nature, can cause field distortion, electrical breakdown, or horrendous microphony.

5.3 Losing the Process

From time to time in the manufacture of Si(Li) detectors, the yields can fluctuate. When a process is 'lost' it is probably for more than one reason, and unravelling the cause can be difficult and time consuming. The following questions need to be asked:

(1) Has cleanliness, temperature, or humidity changed significantly?
(2) Has a new procedure crept in (possibly with a change of personnel)?
(3) Can we identify when the yield fell and which part of the process is responsible?
(4) Have material specifications (silicon and chemicals) been maintained?

It is sometimes an advantage to have alternative processes (contacts, passivation, quenches) that might indicate the stage at which things are going astray. Processing silicon slices that have yielded good results in the past might indicate poor current material. Processing a HPSi slice to make a HPSi detector (see Chapter 6) might indicate some drifting problems. The most common problem encountered is increased reverse leakage current, which unfortunately may be attributable to the front face contact, the side walls, or both. The same can be said for poor backgrounds. It is not unknown for the yield to rise again to normal levels for no explicable reason!

References

1. Lothrop R.P. et al., UCRL-16190, 1965.
2. McKenzie J.M., *NIM A* 162, 49; *Detectors in Nuclear Science*, Bromley D.A. (Ed.), North-Holland, 1979 ISBN 0 471-05661-8.
3. Theuerer H.C., US patent 3060123, Bell Labs, 1952.
4. Watanabe E. et al., *IEEE* NS-35, 1, 33, 1988.
5. Glasow P.A. et al., *MRS* 16, 17, 1983.
6. Walton J.T. et al., *IEEE* NS-41, 1031, 1994.
7. Evensen L., *NIM A* 326, 136, 1993.
8. Abe T. et al., *Semiconductor Silicon*, 105, Electrochemical Society, 1990.
9. Topsil Semiconductor Materials A/S, Frederikssund, Denmark and Wacker Chemie AG, Germany.
10. Spieler H., *Semiconductor Detector Systems*, Oxford University Press, 2005, ISBN 0-19-852784-5.
11. Fuller C.S. et al., *Phys. Rev.* 91, 193, 1953.
12. Pell E.M., *J. Appl. Phys.* 31, 291, 1960.
13. Canham L.T., *Properties of Silicon*, INSPEC, London, 455, 1980.
14. Stab L. et al., *NIM* 37, 113, 1965.
15. Bertolini G. and Coche A. (Eds.), *Semiconductor Detectors*, American Elsevier (John Wiley), 1968, ISBN 044410142X.
16. Protic D. et al., *IEEE* NS-49, 1993, 2002.
17. Whoriskey P.J., *Appl. Phys.* 29, 867, 1958.
18. Kim H. et al., *NIM A* 579, 117, 2007.
19. Landis D.A. et al., *IEEE* NS-36, 185, 1989.
20. Dearnaley G. et al., *NIM* 25, 238, 1964.
21. Goulding F.S., *NIM* 43, 1966.
22. Pell E.M., *Phys. Rev.* 119, 1014, 1960.
23. Watanabe E. et al., *IEEE* NS-36, 1, 117, 1989.
24. Klafter J. et al., *Physics World*, Vol. 18, No. 8, p. 33, IOP UK, 2005.
25. Mayer J.W., *J. Appl. Phys.* 33, 2894, 1962.
26. Lauber A., *NIM* 75, 297, 1969.
27. Blankenship J.L. et al., *IRE Trans. Nucl. Sci.* NS-9, 3, 181, 1962.
28. Protic D. et al., *IEEE* NS-50, 4, 1008, 2003.
29. Weinberger B.R. et al., *J. Vac. Sci. Technol. A* 3, 887, 1985.

30. deWit R.C. et al., *IEEE* NS-15, 352, 1968.
31. Sher A.H. et al., *NIM* 53, 341, 1967.
32. Buck T.M., *Semiconductor Nuclear Particle Detectors*, Dabbs J.W.T. and Walter F.J. (Eds.), NAS Pub., 871, 1961.
33. Statz H., *NAS Nucl. Sci. Rep.* 32, 99, 1961.
34. Llacer J., BNL 8442, 1964.
35. Llacer J., BNL-7853 and *IEEE* NS-11, 3, 221, 1964.
36. Chabal Y.J., *Properties of Silicon*, 211, INSPEC London, 1999.
37. Buck T.M. et al., *J. Electrochem. Soc.* 105, 709, 1958.
38. Kern W. (Ed.), *Handbook of Semiconductor Wafer Cleaning Technology*, Noyes Publications, 1993, ISBN 0-8155-1331-3.
39. Rossington C.S. et al., *IEEE* NS-40, 354, 1993.
40. Jantunen M. and Audet S.A., *NIM A* 353, 89, 1994.
41. Dabbs J.W.T. and Walter F.J. (Eds.), *Semiconductor Particle Detectors*, NAS-871; Nuclear Science Series, Report Number 32 (Asheville Conference, Washington), 1961.
42. Sandor J.E., Method of Placing Thick Oxide Coatings on Silicon, US Patent 3158505, Fairchild, 1962.
43. Jenny D.A., Making p–n Junction under Glass, US Patent 3397449, Hughes Aircraft, 1965.
44. Pai P.-L. et al., *J. Electrochem. Soc. Solid State Sci. Technol.* 134, 11, 2829, 1987.
45. Dinger R.J., *J. Electrochem. Soc.* 123, 9, 1398, 1976.
46. Hansen W.L. et al., *IEEE* NS-27, 1, 247, 1980.
47. Walton J.T. et al., *IEEE* NS-31, 1, 331, 1984.
48. Alfassi Z.B. et al., *NIM* 143, 57, 1977.
49. Pajek M. et al., *NIM B* 42, 346, 1989.
50. Craven A.J. et al., *Ultramicroscopy* 28, 157, 1989.
51. Ohkawa S. et al., *NIM* 226, 122, 1984.
52. Chasman C. et al., *NIM* 24, 235, 1963.

6

HPSi and HPGe X-Ray Detectors

The manufacture of high-purity silicon (HPSi) and high-purity germanium (HPGe) detectors does not differ basically from that of Si(Li) detectors described in Chapter 5, except of course that the lithium drifting stages are omitted. The geometries used are similar and the same contact technologies (Chapter 4) can be used. We will discuss in this chapter the details of the *differences* in each case and point out any differences regarding their operation.

6.1 HPSi X-Ray Detectors

The first point to make is that here we are considering *thick* detectors (2–5 mm) that are not accessible (at present) to lithographic wafer techniques, which are restricted to less than ~1 mm. The latter is the topic for Chapter 7 and is concerned with p–i–n diodes, pixel array x-ray detectors, silicon drift detectors, and charge-coupled devices that in general use resistivities less than a few kΩ cm. In this respect, the starting material for Si(Li) detector manufacture can be regarded as 'HP p-Si', but depleting a thick detector with reasonable bias (<1 kV), without lithium drifting would require resistivities of 20–100 kΩ cm. The next point to note is that such 'detector-grade' material is not as readily available as the lower resistivity material (and even this, as we noted in Section 5.1.1, is restricted to just a few suppliers). The Japanese company Komatsu has led the way with the production of ultra-high purity silicon, particularly that manufactured by the decomposition of mono-silane [1], and one Japanese company (Horiba, a subsidiary of Hitachi) has based its x-ray detector technology on this material. Without neutron transmutation doping (NTD), a low-temperature process that converts silicon to phosphorus and hence from p-type to n-type, the HPSi available is generally p-type and <100 kΩ cm.

The first obvious difference between the Si(Li) and a p-type HPSi detector is that the former remains slightly n-type after drifting and so depletes from the front contact, and the latter depletes from the back lithium diffused junction. It is noticeable that, for example, a Fe-55 source will not produce any spectrum until the bias is sufficient to fully deplete the device. On overdepletion, the spectrum improves and with an appropriate front contact the quality of the spectrum is as good as that of a Si(Li) detector. This is true even though on p-type material the ion implant or surface barrier window is ohmic and no longer the junction contact. In fact, the x-ray spectra from HPSi detectors are indistinguishable from those from Si(Li) detectors. Another difference is that the depleted volume is no longer bounded by the lithium compensated volume and at high bias depletion can go beyond the bottom of the top hat rim or the groove bottom. This is undesirable and is an argument for selecting a resistivity that is not *too* high. For a bias of 800–1000 V, resistivities in the range 20–50 kΩ cm are ideal. One of the advantages of using HPSi is that artefacts associated with drifting (see Chapter 3) are not present.

6.2 HPGe X-Ray Detectors

Because of its lower melting point (see Table 6.1) germanium is easier to purify than silicon, and HPGe detectors date back to 1970 [2] when they largely replaced Ge(Li) detectors. The history of their use for x-rays was a checkered one, as discussed in Chapter 4 and in more detail in Section 11.23. The reasons for the superior performance of the detectors are discussed in Section 6.2.7. Before discussing the details of the manufacture

TABLE 6.1

Differences in Physical Properties of Silicon and Germanium and their consequences

Property	Si	Ge	Units	Comment on Ge
Atomic number	14	32		Greater x-ray absorption
Density	2.3	5.3	gm/cm³	Greater x-ray absorption
Lattice constant	0.543	0.566	nm	Reasonable match to silicon
Knoop hardness	1150	780	kg/mm²	Machining, handling, and mounting
Shear modulus	64	41	GPa	Machining, handling, and mounting
Linear coefficient of expansion	2.6	5.8	μm/m/K	Crystal mounting
Thermal conductivity	10	2	W/cm K	Both good, copper is 5
Melting point	1412	937	°C	Easier to evaporate and purify
Oxide	Stable	Unstable		Unreliable for lithography and passivation
Dielectric constant	12	16		Capacitance, Debye length
Energy gap	1.1	0.67	eV	Lower operating temp (<100 K), higher sensitivity to IR radiation
Energy/electron–hole pair	3.74	2.96	eV	Lower noise; better energy resolution
Fano factor	0.116	0.106		Better energy resolution
Electron mobility	21,000	36,000	cm²/V s	Higher drift current and diffusion length
Hole mobility	11,000	42,000	cm²/V s	Higher drift current and diffusion length

Note: All measurements at liquid nitrogen temperature.

of HPGe x-ray detectors, let us first look at the differences in the physical properties of silicon and germanium that are likely to have a bearing. Table 6.1 lists some of these and their consequences.

6.2.1 Manufacture of HPGe Detectors

The first comment to make is in regard to the care of machining, handling, and mounting. Germanium crystals are hard but also very brittle. Corners of the mesa and front periphery are easily damaged, and unpadded contacts can cause pressure points giving rise to microcracking. The latter is not visible until the surface is etched, but they will give rise to leakage current or breakdown that will fail the detector. The relatively high thermal coefficient of expansion and the lower thermal conductivity dictate that the temperature should not be changed rapidly (<20 K/min) especially for large volume crystals.

6.2.2 Selection of Material

Germanium single crystal is usually Czochralski grown. For small area x-ray detectors, p-type material is chosen although n-type devices will work with identical processing. The material should have between 5×10^9 and 5×10^{10}/cm^3 electrically active shallow impurities. This range is dictated by the required depletion depth (typically ~3 mm) and the fact that as we decrease the acceptor impurities the minority carriers increase, and so too does the reverse bulk leakage current (see Section 1.5.1). The material is usually grown in the $\langle 100 \rangle$ direction. The charge lifetime should be >10 ms, and etch-pit density should be between 500 and 5000/cm^2. Some manufacturers of large-volume HPGe gamma-ray detector manufacturers (e.g., Ortec, Canberra) grow material for their own use but may also sell to third parties. Hoboken (Umicore) in Belgium is an independent supplier of detector-grade material.

6.2.3 Lithium Diffusion

The steps for making the lithium-diffused n-type contact are the same as for silicon except that the temperature and time of diffusion are lower—that is to say, ~9 min at 260°C or ~7 min at 280°C. The diffusion can be carried out under an argon atmosphere. This should give a diffusion depth of ~0.3 mm and a sheet resistivity of 2–4 Ω cm. Because x-rays are unlikely to interact at the back of the detector, the diffusion profile is not as critical as for a Si(Li) and, as mentioned, cooling after diffusion should not be too rapid.

6.2.4 Machining of HPGe Crystal Profile

Because of the brittleness of germanium, grinding is preferable to diamond cutting. The groove/top hat shaping and the dicing from the slice can be carried out by low-speed 'drilling' with a mild steel tool in a carborundum slurry. The front and mesa edges should be slightly bevelled to avoid chipping.

6.2.5 Etching

Masking is done by painting black wax (see Section 5.1.5) on areas to be protected as shown in Figure 6.1.

Notice that the outer edge of the groove (if used) is also etched because the proximity of a sharp edge of a charged intrinsic material could influence the field in the bulk when biased. The etchant for germanium differs from that of silicon by the addition of fuming nitric acid to the CP4. The proportions of *HNO$_3$/HF/fuming nitric* can vary from 3:1:1 to 7:2:1. This should give a 'polish' etch. The goal is to achieve an etch rate that is neither too fast (producing etch 'rashes') nor too slow (resulting in excessive unstable oxides). The etching

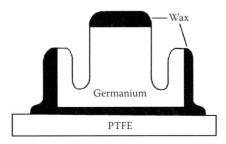

Figure 6.1 Masking a HPGe crystal before etching the groove.

procedure should follow the same rules as for silicon, but on a long etch (~6 min) to avoid overheating it can be cooled by placing the PTFE beaker in an outer basket of ice.

6.2.6 Passivation

The electro-negativities and ionicities of germanium are similar to silicon, and many of the comments on the passivation of the etched surfaces of silicon also apply to germanium. Wet chemical treatments have been used, but again these usually come under the category of 'black art'. The groove can be spray-etched and then quickly quenched in methanol and blown dry with nitrogen. The aim is to avoid water vapour, and this recipe should end with a slightly n-type surface. Providing the vacuum cryostat is water-free, it should be stable. The natural native oxides of germanium are porous, nonstoichiometric, and water-soluble. Nitric acid–based etch oxidations are also non-stoichiometric and do not give a stable passivation. However, a successful wet chemical method has been given by Gregory et al. [3]. This uses HF with trace (<1% in water) hydrogen peroxide that gives a water insoluble phase and leads to stoichiometric GeO_2. Hydrogen peroxide was first used by Llacer [4] and, more recently, by Chabal [5]. Some etches (e.g., KOH solution) will preferentially etch along specific lattice planes. These 'preferential' etches can be used to produce a rough surface on a microscopic scale on crystal walls and perhaps reduce surface leakage in germanium. Chlorine termination of dangling bonds is an equivalent to hydrogen termination for silicon. A treatment with dilute HCl on $\langle 111 \rangle$ germanium [6] appears to do this.

The evaporation of SiO, as noted in Section 5.1.10, was developed at Chalk River [7] and has been widely used by many workers since. After etching as above, ~50 to 200 nm of SiO is downward evaporated. The thickness is not critical and can be measured on a capacitance thickness monitor or merely stopped at 100 or 200 nm when the coating looks blue. A variation of this method is to add boron trioxide (B_2O_3) to obtain a borosilicate glass that can then be sputtered [8].

The Berkeley a-Ge:H sputtered coating [9] is also widely used for large-volume gamma-ray detectors, but the authors have found that on small grooved devices an evaporated film of a-Ge, 7–15 nm thick, also produces low leakage current at <100 K.

6.2.7 Contacts

As we have indicated, the back contact is usually made by lithium diffusion as for a Si(Li) detector. The front face can be recessed as in the Si(Li) detector, and the same ion implant or surface barrier techniques can be used (see Section 4.6) to make the front contact. This is the case even though the contact is not the rectifying junction. However, because of the higher dielectric constant, the groove has a larger influence on the peripheral field than in a Si(Li) crystal [10–12], and instabilities in the contact can be problematic [13]. These manifest as incomplete charge collection (ICC) or increased injection current. Correct formation of the contact can overcome this, but there are other methods—such as implanting the peripheral regions, confining the metallisation to the required active area in the center, or reducing the groove depth—that can be used. The latter will, of course, increase to some extent the capacitance and noise. When a dielectric is used, it helps if it is confined to the center uncollimated area or if it is discontinuous.

6.2.8 Thermal Screening

Besides having to operate at lower temperatures (~100 K, which is usually achievable using a liquid nitrogen cryostat), the germanium crystal must be screened so that it is not exposed to infrared (IR) radiation because the smaller bandgap extends the sensitivity into the IR spectrum. Using the approximation $E = 1.24/\lambda$ (where E is expressed in eV and wavelength λ is in μm), the energy gap of Ge puts the cut-off wavelength λ at 1.85 μm, which is into the mid-IR. This, unfortunately, includes the front entrance contact as the thin metals used are normally transparent to both light and IR. Aluminium has a high emissivity for both light and IR but would make an n-type contact (see Table 5.1) on p-type HPGe if used as a contact. Usually, an internally cooled thermal shield consisting of a 100-nm film of aluminium on a metal or silicon mesh support is used in front of the crystal. It is quite remarkable how much this reduces the leakage current. The same effect can be produced by cooling the detector external window (with care!). Because of this, HPGe detectors usually perform better in a vacuum environment (e.g., on a scanning electron microscope) than in ambient air.

6.3 Performance

The performance of HPGe x-ray detectors is well documented [14–17]. As noted previously, the low-energy x-ray performance above 1 keV is well characterised by spectra taken with a Fe-55 source. Figure 6.2 shows such spectra taken with a HPGe detector overlain with one taken from a state-of-the-art Si(Li) detector.

From the above spectra, it is clear that

(a) The resolution of HPGe detector is superior.
(b) The tail on photopeak is as good as a Si(Li).
(c) The noise is comparable.

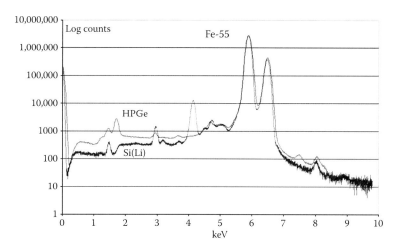

Figure 6.2 Comparison of response of a Si(Li) and a HPGe detector to Fe-55.

(d) The background around 1 keV and above is higher in the HPGe detector (peaks are Al and Si, from collimator and window support).
(e) The background below 1 keV is higher in the HPGe detector.

These characteristics are explained by

(a) ωF being ~28% less in germanium (see Section 1.9.1).
(b) The reflection coefficient is as high (or perhaps higher) than that of Si (see Section 4.3).
(c) The noise might be expected to be lower by ~21%, because the 'intrinsic gain' ω is lower. This is rarely achieved in practice because the detector capacitance C_d for equivalent geometry is higher as the dielectric constant ε is higher by ~30%. This affects the voltage and $1/f$ noise (as C_d) and the dielectric noise (as $\sqrt{C_d}$) (see Equation 1.43).
(d) The photoelectrons' escape to the front contact has a greater probability in germanium than silicon (see Section 2.7.3).
(e) There are added contributions from photoelectrons from the thermal screen (Al foil and supporting structure) in the case of the HPGe detector.

Note that items (a) and (b) have often not been generally recognized [18,19], and there may be technical solutions to item (e), such as a cooled atmospheric thin window or thermal screen deposited onto the germanium crystal itself. At higher energies, germanium has the advantage in that the stopping power is greater (this can just be seen by comparing the Mn K_β peak heights at 6.5 keV in Figure 6.2). A 3–4 mm deep HPGe crystal has good efficiency up to ~50–100 keV (see, e.g., Figure 1.2). In the presence of high-energy x-rays, the Compton background is less (see Figure 3.10) because the scattered x-ray is more likely to be stopped inside the crystal. Escape peaks, however, are more problematical especially above ~10 keV. There are more, and more intense, escape peaks because the fluorescent germanium $K\alpha$ x-rays (9.9 and 11 keV) are of higher energy than those in silicon (1.74 keV) and therefore more likely to escape. Physics also works negatively for energy dispersive x-ray microanalysis at high energies by reducing the ionisation cross sections and therefore the amplitudes of the K-lines of the heavier elements relative to the lower elements (the inner atomic shells are screened by the outer shells). However, their sensitivity to heavy elements is still much better, and there are distinct advantages in energy dispersive x-ray fluorescence. All these points must be balanced and HPGe detectors show real advantages at both high and low energies for energy dispersive x-ray spectroscopy.

The nonlinearity problems that dogged Si(Li) detectors for many years (see Section 11.20.2) are virtually absent in HPGe

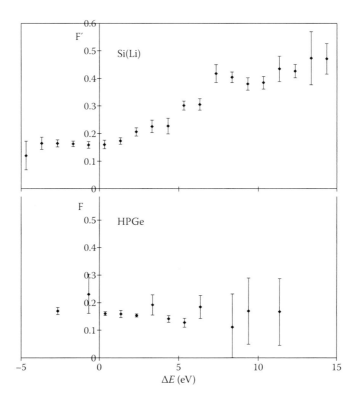

Figure 6.3 Effective Fano factor for a range of detectors against measured shift in energy at carbon K line position. Top: Si(Li) detectors; bottom: range of HPGe detectors. Scatter in ΔE on the latter is largely due to peak location uncertainty, and it is clear that ICC is less in evidence in HPGe at this energy. (From Lowe, B.G., *NIM A*, 339, 345, 1997. With permission.)

detectors (see Figure 4.8). This is partly because the L-shell absorption edges, above which we may expect poor charge collection because of the steep decrease in the x-ray range, is higher in germanium (L1 abs ~1.41 keV) compared to silicon (L1 abs ~150 eV). In the case of silicon, it can cause ICC at the B, C, N, and O photopeaks (nonlinearity, broadening, and tailing; see Figure 4.3). In the case of germanium, we might expect the Al photopeak (1.48 keV) to be broadened, but this is not generally observed. The lower ω in germanium causes the threshold energy at which ionisation is no longer possible to be extended to lower energy (the bandgap is 0.67 eV). The residual electrons (which can then only lose energy to phonons) have a lower energy distribution and, it is reasonable to suppose, a higher reflection coefficient at the contact. The effects of the loss of thermal and 'warm' electrons seen in silicon are not generally observed in HPGe. This is illustrated in Figure 6.3, which plots the effective Fano factors (increased by ICC) for the CKα peak for a large sample of Si(Li) and HPGe detectors against ΔE (the displacement from 277 eV).

References

1. Itoh D. et al., *MRS* 16, 39, 1983.
2. Baertsch R.D. and Hall R.N., *IEEE* NS-17, 3, 235, 1970.
3. Gregory O.J. et al., *J. Electrochem. Soc., Solid State Sci. Technol.* 135, 923, 1988.
4. Llacer J., BNL-7853: NS-11; 3, 221, 1964.
5. Chabal Y.J., EMIS Data Review Series 20, *Properties of Silicon*, INSPEC, 211, 1999, ISBN 0 950-1398.
6. Lu Z.H., *Appl. Phys. Lett.* 68, 520, 1996.
7. Dinger R.J., *J. Electrochem. Soc.* 123, 9, 1398, 1976.
8. Martin G.N. et al., *Monitoring Research Review: Ground-Based Nuclear Explosion Monitoring Technologies*, 787, 2008.
9. Hansen W.L. et al., *IEEE* NS-27, 1, 247, 1980.
10. White G. et al., *NIM A* 234, 535, 1985.
11. Burns P.A. et al., *NIM A* 286, 480, 1990.
12. Allsworth F.L. et al., *NIM* 193, 57, 1992.
13. Pugh N.D. et al., *Phys. Med. Biol.* 35, 625, 1990.
14. Cox C.E. et al., *IEEE* NS-35, 28, 1988.

15. McCarthy J.J. et al., *Proceedings of the XIIth Int. Congress for EM*, Vol. 2, 90 San Francisco Press, 1990.
16. Sareen R.A., *J. X-ray Sci. Technol.* 4, 151, 1994.
17. Lowe B.G., *NIM A* 439, 247, 2000.
18. Rossington C.S. et al., *IEEE* NS-39, 4, 570, 1992.
19. Scholze F., *Handbook of Practical XRF Analysis*, Chapter 4, Beckhoff B. et al. (Eds.), Springer, 2006, ISBN 10-3-540-28603-9.
20. Lowe B.G., *NIM A* 339, 345, 1997.

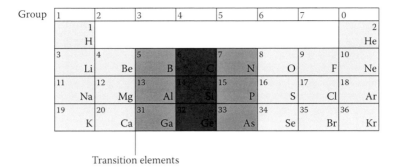

Figure 1.7 Part of the periodic table.

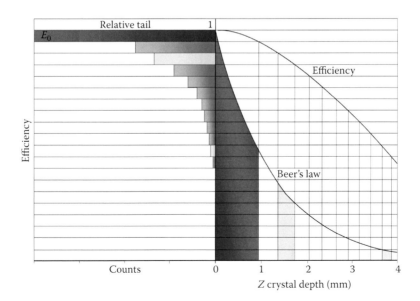

Figure 1.33 Generating a pulse height spectrum of a 22-keV x-ray incident on a 4-mm-deep detector in the presence of hole trapping.

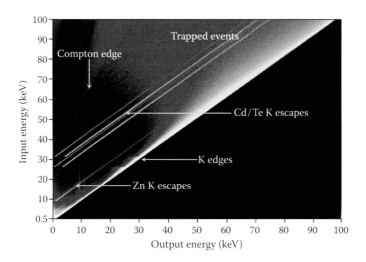

Figure 1.54 Simulation of the response function for a 2-mm CdZnTe detector.

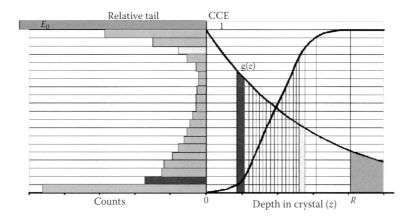

Figure 2.4 Spectrum generated from arbitrary charge collection efficiency (CCE) function g(z) on right-hand side of figure.

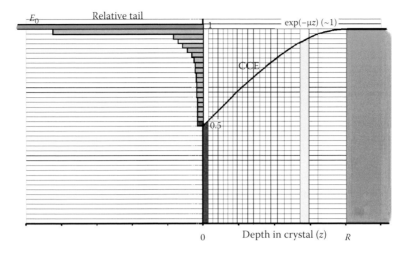

Figure 2.13 Populating bins or channels in a simulated spectrum.

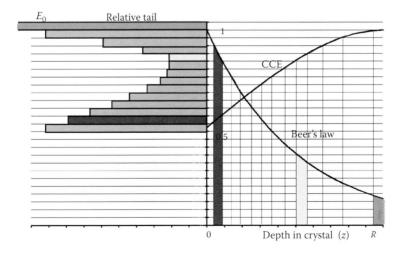

Figure 2.17 Spectrum resulting from low energy x-rays.

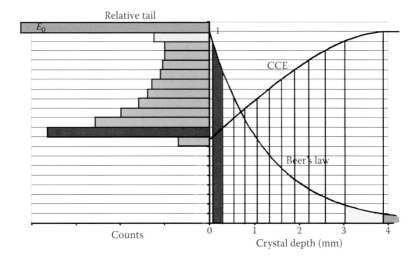

Figure 2.19 Response function for 22-keV x-rays with an electron-trap drift length of 2 mm.

Figure 5.3 Copper staining revealing 'punch through'.

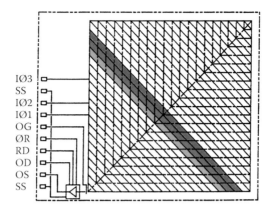

Figure 8.8 Movement of charge in diagonal areas to readout node.

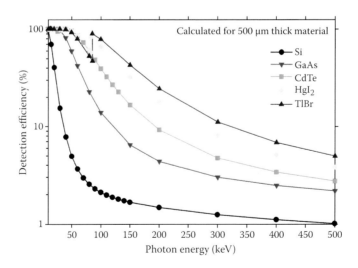

Figure 10.1 Detection efficiency of a range of WBGS compared to Si. (From Sellin, P.J., Development of Semiconductor Gamma-ray and Neutron Detectors for Security Applications (presentation pdf: Dept. of Physics, Univ. of Surrey, UK, 2009. With permission.)

Figure 11.48 Examples of the Logos of Detector Companies.

Figure 11.54 Detector geometries salvaged from redundant equipment.

Figure 11.57 Front face of a Kevex crystal illustrating their excellent etching technology.

Figure 11.62 Early EDAX front–end.

Figure 11.82 SDD assembly from Ketek.

Figure 11.83 SDD from XGLab.

7

X-Ray Detectors Based on Silicon Lithography and Planar Technology

In this chapter, we will discuss the types of x-ray detectors that are manufactured using more-or-less standard lithographic techniques (both single-sided and double-sided silicon wafer processes). The fact that they share this feature with field effect transistors (FETs) presents the technical possibility of integration with them. The important cases of charge-coupled devices (CCDs) and silicon drift detectors (SDDs) are not manufactured using standard integrated circuits (ICs) technology and will be discussed separately in Chapters 8 and 9, respectively. We will be considering only direct excitation x-ray spectroscopy in silicon and not the family of detectors that rely on intermediate media such as optically coupled scintillators, phosphors, and image intensifiers. We will also not be considering devices primarily designed for imaging.

Large silicon wafers and wafer technology are at present restricted to thin (<1 mm) slices but have the advantage of the many years of experience accumulated in silicon chip and IC development in many commercial silicon foundries. Furthermore, they use well-developed and well-understood SiO_2 passivation technology and allow for large-scale production once the architecture and masks are established. Generally, the leakage currents of diodes so manufactured are low even at moderate temperatures, and if cooling is necessary at all, simple TEC (thermoelectric or 'Peltier' cooler) devices usually suffice. All these considerations lead to convenience and considerable unit cost reduction. The first devices reported as manufactured using lithography and SiO_2 passivation specifically for x-ray spectroscopy (as opposed to merely 'photodiodes') were p–i–n diodes [1] at the Technical University of Munich, Garching, Germany, and were originally made for medical applications. The term 'p–i–n' (p-type–intrinsic–n-type) used here refers to contacts made on HPSi, and one must be aware that early Si(Li) detectors were also often called p–i–n (or even 'n–i–p' earlier) diodes. Individual p–i–n detectors were followed by arrays of p–i–n diodes or 'pixel detectors' (which we will call PXDs) with a 'hybrid readout' [2–4] and CCDs (the first practical demonstration of CCDs for x-ray spectroscopy was in 1979 [5]). SDDs, first conceived in 1983 [6], have since taken on a dominant role in x-ray spectroscopy (and will be discussed in Chapter 9).

7.1 Silicon p–i–n Diodes

Silicon p–i–n diodes represent the basic form of semiconductor x-ray detectors. Being a simple planar geometry, they are easy to understand and to simulate. They are abrupt p^+-high resistivity-n^+ reverse-biased junctions as described in Section 1.6.1. There we saw that they are characterised by relatively high fields (several kV/cm) near the junction contact even at moderate (<100 V) bias. Figure 7.1 shows a cross section of a single-sided lithography-based p–i–n detector manufactured on a 270-μm wafer of ~5 kΩ cm n-type silicon in its simplest form. N-Type silicon is preferred because it depletes at a lower bias voltage than p-type silicon (see Section 1.6.1).

The n^+ junctions can either be diffused (e.g., phosphorus) or implanted. The x-rays enter through the p^+–n junction so that in the case of low energies the signal is mainly from the transport of electrons. The p^+ diffused junctions at the window

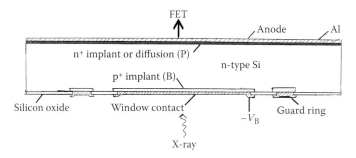

Figure 7.1 Cross section of a p–i–n detector with a guard ring around the cathode.

contact have now been replaced in most cases by a boron implant. The details of the lithographic processes involved can be found, for example, in Spieler's book [7]. The guard ring shown in the figure can be biased (e.g., $-V_B$) or 'floated' and has the effect of making the field in the central area more uniform and of reducing surface leakage (see Section 1.16.3). Collimation is necessary to reduce edge effects. One problem associated with all such 'thin' detectors with relatively shallow back contacts is the fluorescence of materials that are behind the detector. For example, fluorescence x-rays from Sn or In (solder), In, Ni, or Au (from contacts), Cu (from a cold finger), and Bi, Te (from TEC elements), can be detected when high-energy x-rays are present. This can sometimes be a serious problem in analytical work involving such elements or those close to them in the periodic table.

The analysis of these devices follows that outlined in the Section 1.6.1. Using the approximation for the depletion depth for n-type silicon $W \sim \sqrt{(3\rho V)}/100$ (where ρ is expressed in kΩ cm, V in volts, and W in mm), the example above should fully deplete at a bias of ~60 V. Leakage currents at 300 K are normally in the range of 1–120 nA/cm^2 and when moderately cooled ($-90°$C) they will usually allow an overbias of 3–4 times the depletion bias. Under these circumstances, for small-area (5 mm^2) devices used for x-ray spectrometry the capacitance is ~80 pF/cm^2 at full depletion, reducing to ~60 pF/cm^2 with over-depletion. For larger area (80 mm^2) devices, this can reduce to ~25 pF/cm^2. This does not compare favourably with the equivalent capacitances for a Si(Li) crystal of 3.5–5 pF/cm^2.

In the late 1970s, surveys were carried out [8–10] on the use
of commercially available photodiodes for x-ray spectrometry.
It was shown that x-ray resolutions (~200 eV) could be obtained
at temperatures of –70°C (in the range of a multistage TEC).
Commercial exploitation included Amptek (founded 1977) with
XR-100 launched in 1994. By hermetically sealing the cooler,
FET, and crystal, and backfilling with a suitable gas (such as
nitrogen or argon), the problem of outgassing is avoided and
temperatures of ~–30°C are maintained. These detectors are
reliable and relatively low cost for the less-demanding x-ray
fluorescence (XRF) applications. Figure 7.2 shows a comparison of a Centronic phosphorus diffused junction photodiode
and a Si(Li) detector for Fe-55 spectra.

Amptek now offers Peltier cooled p–i–n diodes with 145 eV
resolution (6 mm² area, 500 μm thick), and many companies
(e.g., Aracor, Moxtek, Oxford Instrument) have based both
tabletop and portable XRF analyzers on Peltier cooled p–i–n
diodes. Much progress on radiation hardness and leakage current (reduced to ~1 nA/cm²) [11] has come about because of the
requirements of high-energy physics experiments that require
similar technology for position-sensitive particle detectors.
Oxide passivation, although technically the best, cannot be
regarded as ideal. Fixed positive charges can lead to high field

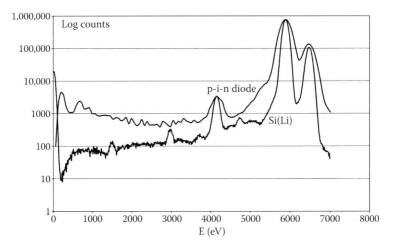

Figure 7.2 Fe-55 spectrum taken with a Si(Li) and a p–i–n diode.

regions near p$^+$ contacts on n-type silicon and being hydrophilic can be affected by environmental humidity [11,12].

7.2 Avalanche Photodiodes

We touched on the topic of 'avalanche ionisation' and avalanche photodiodes (APDs) in very high internal fields in semiconductors in Section 1.15. We saw in Chapter 6 that the reason for the better energy resolution and potentially lower noise in germanium compared to silicon detectors is that the 'intrinsic gain' (1/ω) is higher. More electron–hole pairs are generated for a given energy x-ray. The question that naturally then arises is, if we could increase the gain (and hence signal/noise ratio) using an internal multiplication or 'avalanche' process (as in an electron or photomultiplier), would there be any advantage? But before discussing the performance of APDs, let us look at the construction of such devices. The field must be sufficiently high (>200 kV/cm) for carriers to accelerate and then have enough energy to produce secondary carriers by further ionisation. This *impact ionisation* is avoided in all other types of devices because it may be self-sustaining and lead to electrical breakdown. It is one reason for limiting the bias on semiconductor diodes. But with care and very stable bias and temperature conditions, it can be controllable. Because the mobility of electrons is higher than that for holes (by a factor of 3 in silicon), electron multiplication normally dominates. And because the mobility of electrons is temperature dependent, APDs have to be temperature stabilised. The gain G for electrons increases exponentially with the distance x drifted by the primary electron

$$G = \exp(\alpha x),$$

where α is the electron ionisation coefficient and depends exponentially on the drift field in the avalanche region. APDs are normally operated with G having values between 10 and 100. The avalanche region is produced by a relatively shallow high field region (~300 kV/cm) created by judicious doping either at the front ('reverse' type) or the back ('reach-through' type) of the diode. One such configuration of the 'reach-through' type is shown in Figure 7.3.

Primary electrons and holes generated in the low field region will contribute to the signal in the normal way, but the signal will be predominately attributable to those electrons that reach the avalanche region. All these electrons will experience the same gain once they reach the region. Because of the relatively deep low field region, the capacitance is comparable to a conventional p–i–n diode. Both the field and depth over the whole avalanche region must be carefully controlled so that good gain uniformity can be obtained.

APDs were pioneered many years before the advent of lithographic techniques by researchers such as Gerald Huth (GE Space Sciences Centre) [13]. The high fields required can obviously lead to high surface current, and Huth was instrumental in introducing contouring in the form of a 'bevelled edge'. However, one characteristic due to the high field region is that the charge collection times are extremely fast (~2 ns), and they can therefore be operated at very short processing times well away from current noise and, if required, at very high rates.

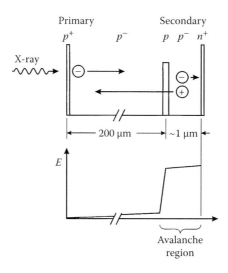

Figure 7.3 Diagram of an Avalanche Photodiode showing the Avalanche Region. (From Spieler, H., *Semiconductor Detector Systems*, Oxford University Press, 2005. With permission.)

The higher signal levels and lower energy thresholds can be advantageous, but the excess noise factor [14], which varies roughly as $G^{0.28}$ and is due to the statistics of the multiplication process, gives no resolution advantage. Despite continued activity, APDs do not feature to a great extent in present-day energy dispersive x-ray spectroscopy (EDXRS). We will therefore not go into any more details of the variants but refer to other discussions in, for example, Spieler's book [7] and other references quoted below. Commercially, APDs have been developed by Advanced Photonix, Radiation Monitoring Devices, and Hamamatsu and have been evaluated as x-rays detectors respectively by Moszynski et al. [15], Ludhova et al. [16], and Kataoka et al. [17]. A reach-through APD with an active area of 28 mm² and a depletion depth of 130 μm (manufactured by Hamamatsu) had a detection threshold energy of ~500 eV and a resolution of 380 eV at 5.9 keV (see Figure 7.4) when cooled to –20°C.

Although the depletion depth and resolution is not as good as that achieved by a p–i–n diode, APDs do have the advantage of better high rate performance. Their main importance lies in their use as a replacement for conventional photomultiplier tubes when optically coupled to a suitable scintillator. This extends their energy range and has the added advantage of being compact and relatively insensitive to magnetic fields.

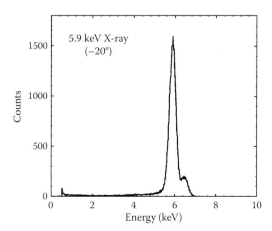

Figure 7.4 Results obtained using an avalanche photodiode. (From Kataoka, J. et al., *NIM A*, 541, 398, 2005. With permission.)

7.3 Pixellated X-Ray Detectors

Pixel detectors are sometimes called PADs (pixel array detectors). However, the term 'pad detector' sometimes applies to just the opposite, that is, a single element detector with a 'pad' contact. To avoid this confusion, we prefer the term 'pixellated x-ray detector' or PXD. We will, however, extend this definition to gamma-ray detectors that use the same concept. The PXDs overcome the disadvantage of the high capacitance/unit active area of single element planar detectors such as p–i–n diodes and APDs by arranging them into an array of smaller area devices, usually square, on a single wafer of silicon (see Figure 1.29c). Large area arrays can also have the potential to reconstruct an *image* of the x-ray source both in two dimensions and time. For this reason, the elements are referred to as 'pixels'. A great deal of research and development is being carried out on PXDs for medical imaging, time-resolved diffraction patterns (e.g., in protein crystallography), extended x-ray absorption fine structure (EXAFS), particle tracking (CERN, Medipix-based, and CPPM detector) and x-ray astronomy. Because of the imaging potential (shared with CCDs), publications are often in the *Proceedings of SPIE* (Society of Photographic Instrumentation Engineers; founded in 1955) Symposia. The state of the art of PXDs for general applications, but particle tracking in particular, has been summarised in a recent monograph by Rossi et al. [18].

In this chapter, we are not concerned with imaging but with the narrower field of energy resolving medium-area silicon detectors where we take advantage of the low current (operation at or near room temperature) and low capacitance of the individual pixels of which it is composed and of the overall high rate capability. In this, they of course share a commonality with CCDs that are discussed in Chapter 8. The cost of the pixellation is in the complexity of the readout of the charge deposited by x-ray events in each individual pixel. The most obvious method of achieving this is by bonding the PXD to a separate readout chip [application-specific IC (ASIC)] behind the PXD as shown in Figure 7.5.

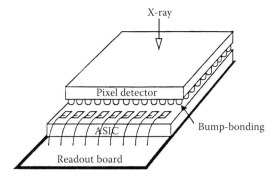

Figure 7.5 Pixel detector bump bonded to a separate readout chip.

Such an arrangement is called a 'hybrid detector', because the two silicon chips require very different starting material and processing and are manufactured independently of each other. This approach has the advantage in that PXDs of different characteristics (resistivity, thickness, and even elemental material) can, in principle, be bonded to the same ASIC. We will discuss the use of different materials (CdZnTe, GaAs, etc.) for PXDs in Chapter 10. Before looking at the construction of PXDs in detail, let us look at some of their general problems and characteristics.

7.3.1 Pileup in PXDs

These low-capacitance detectors can potentially analyze at very high rate because the probability of pileup (see Section 1.14.2) in any given pixel is small. To illustrate the low probability of pileup, consider a PXD composed of an area divided into N individual pixels. The probability of two randomly distributed x-rays (emitted within the integration time τ of the signal recognition circuitry) falling on any one pixel will be $1/N$. With a conventional detector of the same area this, would of course give a definite pileup ($N = 1$). With N x-rays falling randomly onto the pixellated area within the integration time τ, about one-third ($1/e$) will have no pileup. Pileup is determined by the rate/pixel and is the same for all values of N in the above example; then, as N (and as a consequence, the overall input rate) is increased, the extent of pileup remains the same. In practice, the input count rate is kept much lower (number of x-rays within the integration time $\tau < N/100$) so that the pileup is <1%. This is known as 'sparse counting'. The actual throughput rate depends on how fast all N pixels can be read out (without unduly affecting the noise), but with a low capacitance/pixel the integration time τ can be chosen to be fairly short (say, ~2 µs) and very high rates can be achieved by small pixels even under sparse counting conditions. There are, of course, practical limits on how small the pixels can be made, and 'charge sharing' between adjacent pixels (see below) effectively limits their number and the effective pileup. This potential performance of PXDs (and, of course, other pixel arrays such as CCDs) means that they can take advantage of the increasingly available high-intensity monochromatic synchrotron x-ray beams as they become available.

7.3.2 Charge Sharing

In Section 1.7.1, we discussed the diffusion radius R of the charge cloud associated with an x-ray photoelectric absorption. The spatial spread of the charge generated by a low energy x-ray is dominated by diffusion during drift rather than by the range of the photoelectrons [19]. In practice, the field will not be uniform, but in order to obtain an estimate of the importance of charge diffusion, let us assume we apply a bias V_B to achieve a high overdepletion and an approximately uniform field E. We can then assume $E \sim V_B/d$, where d is the thickness of the detector. The drift velocity of carriers (μE) will be uniform, and the transit time of a carrier across the whole thickness d is therefore

$$t = d^2/\mu V_B.$$

At this time, the carrier diffusion causes an outward expansion to a radius approximating to the rms diffusion length $L = \sqrt{(2Dt)}$, where D is the diffusion constant. If the carrier mobilities are high, as they are in HPSi and HPGe detectors, the drift time is short and they have less time to diffuse. Thus, the diffusion radius R is not a strong function of material or temperature. Equating R to L,

$$R = d\sqrt{(2D/\mu V_B)}.$$

The diffusion radius R increases linearly with detector depth d and the term $\sqrt{(2D/\mu V_B)}$ is a dimensionless constant that can vary typically between 1/60 to 1/10 for moderate V_B so that even for silicon wafer-based thin detectors it leads to considerable diffusion distances (5–30 µm) and potential charge sharing between pixels. The equation also shows the need to use as high a bias voltage V_B as the leakage current and avalanche onset will allow.

Charge sharing in PXDs has been investigated in detail [19]. The effect can be viewed simply as follows. If we consider a single square pixel of side p and uniform x-ray irradiation, charge will be lost from a peripheral area of roughly $4pR$, and a fraction of events $4R/p$ (proportional to d/p) will go into to the low-energy background unless it is recovered from the adjacent pixels. If $p = d = 300$ µm, even for a high field (>2 kV/cm) $R \sim 5$ µm and on this simple model ~6.7% of charge will be shared with adjacent pixels. Charge will not necessarily be lost from the region *between* adjacent pixels if the oxide passivation there is neutral, because the field lines will terminate only on the metallised pixels. The capacitance of a pixel due to the adjacent pixels is significant [20] and can be reduced by increasing the separation between them. However, this is limited by the positive charge in the passivation, which may cause interpixel trapping. The pixel array is usually surrounded by one or more guard rings. By summing the charge from adjacent pixels, many of the background events can be returned to the peak (or at least to the low energy tail). It is clear, however, that a compromise pixel size must be reached based on the performance in terms of resolution, rate, and spectral peak/background ratio. For most PXDs, the pixel size in silicon chosen is $p \sim 100–300$ µm. Another consideration is the area required by the readout IC on the ASIC, but pixel sizes down to 30 µm can normally be accommodated.

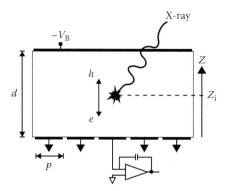

Figure 7.6 Induced charge experienced by an individual pixel. (From Eskin, J.D. et al., *J. Appl. Phys.*, 85, 647, 1999. With permission.)

7.3.3 Small Pixel Effect

The small pixel effect, sometimes called the 'near-field effect', concerns the signal induced on any one particular pixel by the motion of the charge carriers in the field applied between the pixels and the window contact as depicted in Figure 7.6.

One approach to analyzing this problem is to apply the Shockley–Ramo theorem (see Section 1.7.2). To find the signal induced on any one pixel, we have to calculate the 'weighting potential' for that pixel. We calculate this by setting the potential of the measuring electrode (the chosen pixel) to unity and all the others to ground as the charges move across the detector. Such calculations have been carried out, for example, by the group at the University of Arizona [21,22]. Figure 7.7 shows the results of such calculations.

These plots represent the weighting that needs to be applied to the induced signal on the pixel centered at 0 as the charges

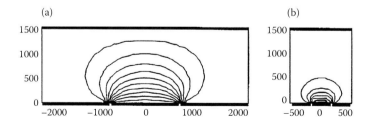

Figure 7.7 Weighting potential for two different size pixels (a and b). (From Eskin, J.D. et al., *J. Appl. Phys.*, 85, 647, 1999. With permission.)

move in the field in the detector. Referring to Figures 7.6 and 7.7, we can see that the holes from an interaction near the center of the detector, in moving away from the pixels, will induce much less signal than the electrons moving toward the pixels through the higher weighting potential. We also see that the effect is enhanced for smaller pixels, as in Figure 7.7b. Most of the signal is generated from electrons as they get close to the pixel. In this way, the small pixel effect approximates a PXD to a 'charge delivery' type of detector (see Section 1.15). To a first approximation, the signal is only generated when one polarity of carrier (electrons in the example above) are delivered to the pixel. This behaviour is in contrast to that observed in the simple planar detector (such as the p–i–n diode), which is a 'charge induction detector'. In this case, we get equal contributions from holes and electrons throughout their drift between the electrodes. It is also clear that the signal induced in a pixel is no longer independent of where the interaction took place. Movement of charge outside the region where the weighting potential for a pixel is strong induces little signal in that pixel. Referring again to Figures 7.6 and 7.8, we see how the weighting function varies with the depth of interaction z.

The small pixel effect has practical implications for cases where we have significant hole-trapping. Because the hole signal can be suppressed, as we have seen above, then any hole trapping is largely irrelevant. The PXD then operates as a 'unipolar' or single-carrier (in this case, electron) device.

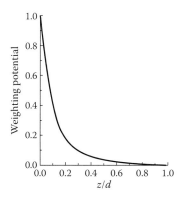

Figure 7.8 Variation in weighting potential with depth of interaction. (From Eskin, J.D. et al., *J. Appl. Phys.*, 85, 647, 1999. With permission.)

This is most effective when the size of the pixel is significantly less (~1/10th) than the thickness of the detector [23]. Most compound semiconductors have a hole-trapping issue, and we shall return to PXD detectors and the relevance of the small pixel effect for such materials in Chapter 10.

7.3.4 Hybrid PXDs and ASIC Readout

In designing the ASIC, it is a challenge to fabricate low-noise circuits with limited elements and negligible cross talk under each pixel area. The mix of digital triggers and analogue readout can give rise to cross talk through the detector element into the ASIC. We will not be concerned here with the details of the various ASIC readout chips and related electronics and refer the reader to details published elsewhere [3,23,24]. Bump bonding ('flip chip') technology for PXDs has also been reviewed elsewhere [25]. The conventional solder or indium bump bonds that have been used at wafer level are deemed unsuitable for PXD x-ray spectrometers, because of the XRF of the materials. Furthermore, the high pressures required during bonding may damage the wafer (particularly in the case of CZT). At the detector chip level, gold bumps or 'studs' (Matsushita process) have been used successfully [24,26,27]. Gold is more acceptable as a spectral contamination in the XRF community. The bumps are made from 17-μm gold wire ultrasonically bonded to the aluminium pixel bond pad. The wires are cut and planarised by applying pressure and screen-printed with a conductive adhesive for bonding to the ASIC chip. Usually, an insulating underfill is used to increase mechanical strength as shown in Figure 7.9.

We will point out some aspects that have bearing on the ultimate performance of PXDs. Because each pixel has its own discriminator and amplifier channel, it is necessary to equalise each individually to allow for variations in gain and noise. It may also be necessary to 'mask' off individual noisy pixels. The readout of the signal can be 'event-driven', meaning that only the pixels that have a 'hit' are readout (unlike CCDs where pixel contents are read out regardless of whether there was a hit). This is termed 'multiplexing' and leads to reduced dead time. Furthermore, because the pixels are small, the chance of pileup events in any pixel is low for 'sparse counting' as

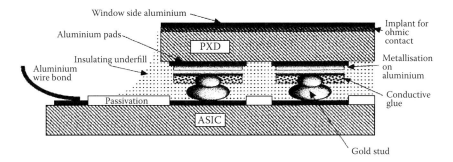

Figure 7.9 Bump bonding and insulating underfill.

described previously. As currents and capacitance are usually low the processing times can be short, and each pixel is capable of counting at a relatively high rate (~1 MHz). When all the pixels are accounted for, the rate performance of PXDs is unsurpassed and this, with their room temperature operation, potentially makes them an ideal choice where high intensity beams are available such as at synchrotrons and applications such as EXAFS and time-resolved work.

7.3.5 Performance of Hybrid PXDs

A number of groups have made significant progress with energy-resolving PXDs as described above. These include researchers from Berkeley [24,28], CERN [29], University of Arizona [30], Rutherford-Appleton Lab (RAL), UK [31], and Brookhaven [32]. In the case of silicon PXD detectors, useful energy resolutions of 480 eV at 300 K (>1 MHz pixel rate) [24], 330 eV at 300 K [33], 265 eV at 300 K [19], 205 eV (at −30°C) and 300 eV at a pixel rate of 100 kHz [32], were found. However, in all these cases, the spectral background has been high largely because of charge sharing and trapping in the interpixel oxide passivation [28]. Direct detection of x-rays by silicon PXDs is not widespread at present, but the prospect for a useful very high rate large area detector with reasonable efficiency (50% at 20 keV) operating at room temperature remains. Because of the impressive performance of SDDs and depleted p-channel FETs (DEPFETs; see below), little work has been carried out recently on silicon hybrid PXDs. The trend has been moving toward CZT hybrid

PXD detectors, making use of their greater stopping power and the small pixel effect to suppress the effects of hole trapping. These topics will be discussed in Chapter 10.

7.3.6 Monolithic PXDs

An alternative to bump bonding is to wire-bond pixels to an ASIC out of plane of the PXD [32]. Besides being able to use single-side processing and avoiding the bump bonding, which does not have a 100% yield, this allows more flexibility for the PXD and ASIC and largely avoids pickup problems from the ASIC [27]. However, the pixel stray capacitance is higher (~0.3 pF, depending on the wire length) and the detector is obviously not as compact. Yet another approach is to combine the readout on the same chip as the PXD (as in the CCD) to make a *monolithic* PXD. In 1990, Holland and Spieler [34] developed an IC-compatible detector process that allowed the monolithic integration of the ASIC and fully depleted silicon. This was later extended to full CMOS (complementary metal–oxide–semiconductor) processing. This showed that the processing was not a fundamental limitation, but the manufacturing yield was very low because of the multiplicity of channels. A simpler monolithic detector was described by a Stanford University–Hawaii University [35] collaboration. This used HPSi with wells of n-type silicon between the p+ collection pixels as shown in Figure 7.10. The pixel electronics was formed in these n-Si wells.

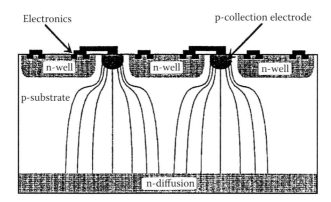

Figure 7.10 CMOS structure with integrated electronics. (From Snoeys, W. et al., *NIM A* 326, 144, 1992. With permission.)

This was shown to give excellent imaging resolution but has not been exploited as a spectrometer. These techniques are used widely for optical imaging where a few microns of depletion is adequate for good sensitivity.

7.3.7 APD and DEPFET PXDs

Two other monolithic PXDs have been recently developed: APD and DEPFET arrays. Monolithic 8×4 pixel arrays of APDs are now commercially available (e.g., Hamamatsu) and have been evaluated as an x-ray spectrometer [17]. The full active area is $128 \ mm^2$, and the resolution at room temperature for individual pixels is ~400 eV with a threshold energy of ~800 eV. The gain variation is $<\pm 3\%$ for $G \sim 50$. Clearly, there is some potential for such devices for a high rate soft x-ray spectrometer in some applications.

Commercial junction FETs (JFETs) have long been known to be sensitive to x-rays. X-rays interact in the depletion region of the FET that offers an ideal low internal capacitive match. Resolutions of ~210 eV can be obtained at room temperature [36]. The sensitive volume, however, is miniscule. Kemmer and Lutz [37] were the first to suggest that the depletion region can effectively be increased by integrating a p-channel JFET onto a high-resistivity n-type wafer that can be fully depleted as shown in Figures 7.11 and 7.12.

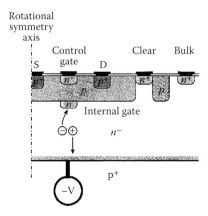

Figure 7.11 Integrated FET cross section.

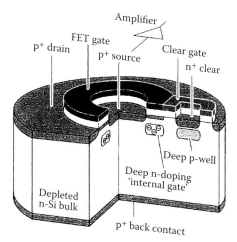

Figure 7.12 Integrated FET on a fully depleted n-type bulk substrate. (From Treis, J. et al., *SPIE*, 6276, 627610, 2006. With permission.)

This structure is called a 'depletion mode' or 'depleted p-channel FET' (DEPFET). The charge collection electrode is a deep n⁺ implant called the internal gate (rather like the tetrode JFET where the substrate itself is the internal gate). Like the SDD, charge generated by radiation moves to the potential valley and in the case of the DEPFET modulates the drain current much as an internal gate of a MOSFET. In an array [38], all the DEPFETs will store the collected charge independent of whether they are switched on or off. For readout, all the FETs in a row can be switched on to connect to a charge-sensitive amplifier. In this respect, sparse counting mode is applicable. Accumulated charge at the internal gate is cleared by applying a forward biasing positive pulse to the *clear* contact (Figure 7.11) that is shielded from the bulk by a deep p⁺ implant. The charge from each event is measured by comparing the voltage levels at the source before and after the clear process (as in CCDs). The intrinsic low capacitance at the internal gate gives very low noise. Being a series readout, the rates are comparable to those of CCDs. Also, like CCDs, the current noise only depends on the exposure time between readouts, which is independent of the shaping time of the amplifier. The development of DEPFETs has largely taken place in Germany, spearheaded by groups working with the MPI semiconductor laboratory in Munich (MPI Garching, and Universities of Bonn, Dortmund, Mannheim, and Milan). This development started in the 1990s [39–42], making such devices a practical reality. This activity has increased beyond the millennium with impetus for the need of Vertex tracking detectors, particularly for the proposed International Linear Collider. DEPFETs have now also been based on a p-channel MOSFET (DEPMOSFET) structure and incorporated as the readout FET for SSDs by PNSensor GmbH [43]. Results from recent DEPFET PXDs are shown below. Figure 7.13 shows the Fe-55 spectrum from a MOSFET structure [44] at room temperature.

Particularly impressive is the noise measured as 19 eV. The spectrum is taken from a single pixel, and so there is no selection to reduce charge loss at the periphery, which accounts for the peak broadening (131 eV) and high background. Figure 7.14 shows the Fe-55 spectra taken at −60°C [45] (although the authors state that it does not change much below −20°C) for a DEPMOSFET device but on a logarithmic scale.

The pixels were 75 μm^2 and covered an area of 4.8 mm^2. The leakage current at room temperature was ~0.3 nA/cm^2 and the measured resolution was ~130 eV. The x-ray illumination

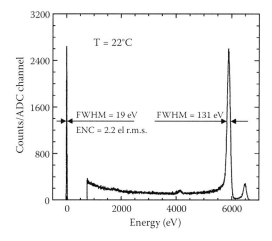

Figure 7.13 [44] Results of a MOSFET structure using Fe-55 at room temperature. (From Fischer, P. et al., *NIM A*, 582, 3, 843, 2007. With permission.)

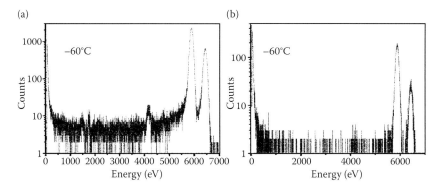

(a) −60°C

(b) −60°C

Counts / Energy (eV)

Figure 7.14 (a) Pixel illumination. (From Treis, J. et al., *SPIE*, 6276, 627610, 2006. With permission.) (b) p+ face illumination.

is on the pixellated side in Figure 7.14a and on the continuous p⁺ contact in Figure 7.14b. The tailing and background are higher in the former because of the charge loss due to dead layers on the pixel side, but they are sufficiently low to show both the Kα and Kβ silicon escape peaks. Figure 7.15 shows a spectrum taken with a DEPMOSFET array at −40°C with better statistics [43].

These results indicate the potential for DEPFET arrays as large-area room temperature spectrometers. However, CCDs and SDDs (discussed in Chapters 8, 9) do have better spectral characteristics and have the advantage of a single readout

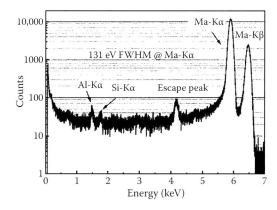

Figure 7.15 [43] DEPMOS FET at −40°C showing tailing and a P/B at 1 keV of ~1000:1. (From Porro, M. et al., *IEEE* NS-53, 3, 401, 2006. With permission.)

node and, consequently, higher production yields. This is particularly true of SDDs in which the FET is discrete and not monolithically integrated.

References

1. Keil G. and Lindner E., *NIM* 104, 209, 1972.
2. Gaalema S., *IEEE* NS-32, 417, 1985.
3. Heijne E.H.M. et al., *NIM A* 349, 138, 1994.
4. Pullia A. and Kraner H.W., *IEEE* NS-42, 585, 1995.
5. Catura R.G. and Smithson R.C., *Rev. Sci. Instrum.* 50, 219, 1979.
6. Gatti E. and Rehak P., LBL Workshop on collider detectors, LBL 15973 Feb/Mar 1983.
7. Spieler H., *Semiconductor Detector Systems*, Oxford University Press, 2005, ISBN0-19-852784-5.
8. Nowotny R. et al., *NIM* 147, 477, 1977.
9. Nowotny R. et al., *NIM* 153, 597, 1978.
10. Desi S., *NIM* 164, 201, 1979.
11. Evensen L. et al., *NIM A* 326, 136, 1993.
12. Rossington Tull C. et al., *IEEE* NS-45, 3, 421, 1998.
13. Huth G.C. et al., *Rev. Sci. Instrum.* 35, 1220, 1964.
14. Allier C.P., *IEEE* NS-45, 3, 576, 1998.
15. Moszynski M. et al., *NIM A* 485, 504, 2002.
16. Ludhova L. et al., *NIM A* 540, 1, 169, 2005.
17. Kataoka J. et al., *NIM A* 541, 398, 2005.
18. Rossi L. et al., *Pixel Detectors*, Springer, 2006, ISBN-10-3-540-28332-3.
19. Mathieson K. et al., *NIM A* 487, 113, 2002.
20. Kavadias S. et al., *NIM A* 335, 266, 1993.
21. Barrett H.H. et al., *Phys. Rev. Lett.* 75, 1, 156, 1995.
22. Eskin J.D. et al., *J. Appl. Phys.* 85, 647, 1999.
23. Raymond M. et al., *NIM A* 348, 673, 1994.
24. Beuville E. et al., *IEEE* NS-43, 3, 1243, 1996.
25. Hallewell G.D., *NIM A* 383, 44, 1996.
26. Iles G. et al., *NIM A* 381, 103, 1996.
27. Mathieson K. et al., *NIM* 460, 191, 2001.
28. Datte P. et al., *NIM A* 391, 471, 1997.
29. Weilhammer P. et al., CERN, 2nd Int. Symp. on Dev. and Appl. of Semiconductor Tracking Detectors, Hiroshima, 1995.
30. Marks D.G. et al., *IEEE* NS-43, 1253, 1995.

31. Seller P. et al., *SPIE* 3774, 30, 1999.
32. De Geronimo G. et al., *IEEE* NS-50, 885, 2003.
33. Pullia A. and Kraner H.W., *NIM A* 395, 452, 1997.
34. Holland S. and Spieler H., *IEEE* NS-37, 463, 1990.
35. Snoeys W. et al., *NIM A* 326, 144, 1992.
36. Lund J.C. et al., *IEEE* NS-40, 47, 1993.
37. Kemmer J. and Lutz G., *NIM A* 253, 365, 1987.
38. Cesura G. et al., *NIM A* 377, 521, 1996.
39. Kemmer J. and Lutz G., *NIM A* 288, 92, 1990.
40. Klein P. et al., *NIM A* 392, 254, 1997.
41. Fischer P. et al., *NIM A* 451, 651, 2000.
42. Neeser W. et al., *IEEE* NS-47, 3, 1246, 2000.
43. Porro M. et al., *IEEE* NS-53, 3, 401, 2006.
44. Fischer P. et al., *NIM A* 582, 3, 843, 2007.
45. Treis J. et al., *SPIE* 6276, 627610, 2006.

8

CCD-Based X-Ray Detectors

We briefly touched on the use of charge-coupled devices (CCDs) as x-ray detectors in Section 1.15, where we gave them as an example of a 'charge delivery detector'—that is to say, that one polarity of charge (usually electrons) generated by an x-ray event is effectively stored until a predetermined time when it is transferred and deposited onto a 'readout node'. Then, and only then, is it measured. In CCDs, the charge packet is transferred by a 'clocking' procedure from one pixel to the next, as described below, until it is eventually deposited on the readout node. There is no signal at the readout node from this charge packet while it is being transferred between pixels. CCDs were developed in the 1970s as imaging devices that could replace the light-sensitive film in cameras. Indeed, the digital camera is still by far their largest application and they are mass produced cheaply for this purpose. Their first use as a direct sensor for x-rays was reported in 1979 [1], and since then their main application has been in x-ray astronomy, for which purpose they have been reviewed [2,3]. Here, we will describe the basic structure and leave the reader to find more details in the aforementioned references. We will concentrate on their use as x-ray spectrometers and not as imaging devices, although it should be noted that a CCD device was recently chosen for the Mars rover *Curiosity* chemical analyzer [4] because of its capability of combined x-ray fluorescence (XRF) and x-ray diffraction (XRD) (imaging diffraction patterns).

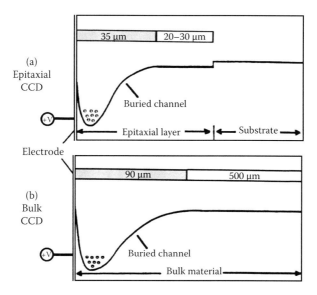

Figure 8.1 (a) Buried channels in an epitaxial and (b) a bulk silicon charge-couple device (CCD).

8.1 Introduction

The basic pixel structure of the CCD is manufactured on a p-type silicon wafer (or a p-type epitaxial silicon layer grown on a wafer) by an n-type implant to form a channel buried under the surface. The buried channel of an epitaxial silicon (epi-Si) and a bulk silicon structure are shown in Figure 8.1a and b.*

When the surface electrode is positively biased, the potential energy profile for electrons is as shown. The three significant regions in the wafer consist of a potential energy minimum for electrons, a depletion region, and a field-free region. In the case of epi-Si, the interface with the bulk silicon also acts as a reflective field boundary, which helps to contain diffusing charge in the field-free region. The benefit of using an

---

* Note that Figures 8.1 to 8.11 in this introduction to CCDs have been adapted from presentations given by personnel from the University of Leicester (Andrew Holland, George Fraser et al.) and e2v Ltd. (Peter Pool et al.) and are hereby gratefully acknowledged.

epitaxial layer of silicon (usually grown from the vapour phase using the substrate as the 'seed' crystal) is that the purity and doping can be well controlled, and oxygen in the substrate getters out impurities from the epi-Si layer. The yield of good pixels, that is, pixels that do not have high leakage current due to defects, is also higher for epi-Si. In the bulk silicon case, the depletion depth may be greater but charge diffusion from the bulk results in charge sharing over many pixels. To complete the structure, a layer of oxide (which also acts as a reflective barrier to electrons) and a metal that defines electrodes (or 'gates') are deposited on the surface as shown in Figure 8.2, giving a basic metal–oxide–silicon (MOS) structure.

A p-type diffusion in the form of a strip acts as a 'channel stop' that isolates one row of electrodes from the adjacent rows in the pixel matrix. Referring to Figure 8.2, the electrons are 'stored' under the positively biased electrode in the depletion region in the buried channel as long as the potentials shown in the figure are held. This keeps electrons away from oxide interface traps and increases the charge lifetime. The electrons are held in this potential well with virtually no loss, whereas the holes are lost to the silicon substrate. The electrons can be transported by first being shared with the next electrode and then being driven entirely into the next in line in a three-stage process shown schematically in Figure 8.3.

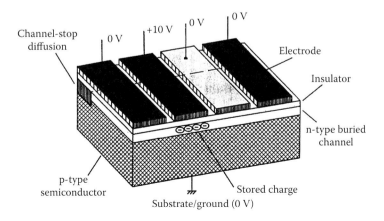

Figure 8.2 Metal–oxide–silicon (MOS) CCD.

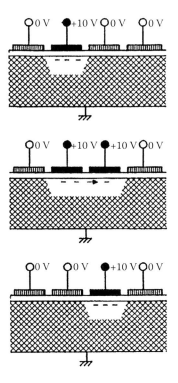

Figure 8.3 Charge transfer in a CCD.

The clever part of this process is that all the electrodes requiring the same bias phase, that is, every fourth electrode in the line, can be ganged together so that the clocking transports all the charge stored under the electrodes toward a readout node at the same time, as shown in Figure 8.4.

Six cycles of clocking takes the charge from one set of three electrodes to the next set, and each set of three electrodes (phases φ_1, φ_2, φ_3) comprises a *single pixel*. Note that the whole pixel area is sensitive to x-rays or is 'active' as the charge is drawn to the biased electrode of each pixel. However, diffusion in the weak field regions may cause some charge sharing between pixels. In practice, the oxide gap between electrodes may lead to a potential barrier, which makes the transfer of charge inefficient. To overcome this, the electrodes overlap slightly with the oxide continued between the layers. The charge transport efficiency (CTE, the equivalent of charge

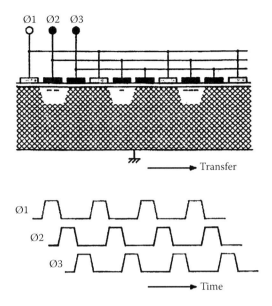

Figure 8.4 Clocking sequence in a CCD.

collection efficiency in conventional detectors) between pixels turns out to be very high in good-quality silicon, and large areas (many cm²) can be covered by quite small (~10 μm) pixels. However, the speed of transfer is limited, ultimately, by the RC time constants and the drift velocity of electrons. The series readout time and increased noise above a clocking rate ~300 kHz put limits on the throughput rate.

The conventional imaging CCD and most 'scientific CCDs' require that the charge accumulated in pixels in each row is transferred to a separate readout register, which itself is a row of pixels (shielded from light and x-rays) running normal to the image rows as shown in Figure 8.5.

Often, the whole image is transferred rapidly to a frame store that is a shielded pixel matrix and is then read out during the 'integration time' for the next image (frame transfer mode). This speeds up the readout process. A problem with CCDs when used for imaging is their continuous sensitivity that includes the transfer time to the readout. We do not impose any 'dead time' to stop the process, and we therefore want the integration time to be long compared to the transfer

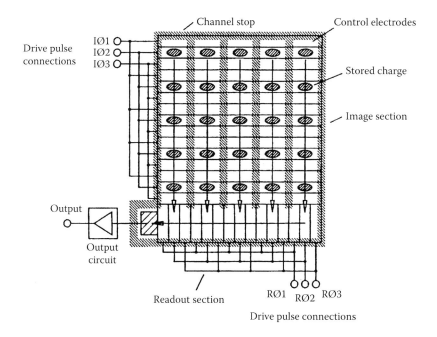

Figure 8.5 Transfer of charge to the output circuit in a CCD.

time. The simplest form of CCD x-ray detector (without imaging capability) is a linear device (i.e., a single row of pixels) read out on a continuous basis. In this case, the pixels could be large enough to compensate for the lack of adjacent rows (effectively p–i–n photodiodes) as shown in Figure 8.6.

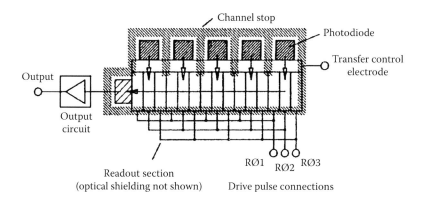

Figure 8.6 Single-row CCD.

Another way of achieving the functional equivalent of a linear array with continuous readout and yet covering a large active area is the swept charge device (SCD) [5], shown schematically in Figure 8.7.

In nonimaging devices, the concept of a pixel ('picture element') is not appropriate, and we will refer to the area defined by the three electrodes as merely 'elements'. The elements in the SCD run diagonally across the square area, but the isolation channels are arranged parallel to the sides. The latter also include a central diagonal isolation channel ending in the readout node at one corner, as shown in Figure 8.7. To show the similarity to the linear array, we will describe the basic concept. When we start clocking, the charge under the elements is transferred along the rows defined by the channel stops (i.e., parallel to the sides as shown by the dark arrows in the figure) toward the central channel. At the central channel, they combine with charge packets that have arrived from other elements that have moved through the *same number* of clocking cycles. They, then move along the diagonal channel and are presented to the readout node consecutively. Thus, the charge readout is an accumulation of all the charges that have travelled through the same number of transfer clocks to get to the readout node. A little consideration will confirm that this is the total charge that was deposited during the time in which elements were

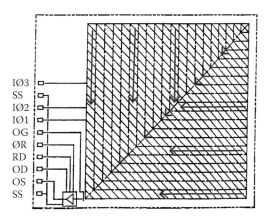

Figure 8.7 Swept charge device (SCD). (From Lowe, B.G. et al., *NIM A*, 458, 568, 2001. With permission.)

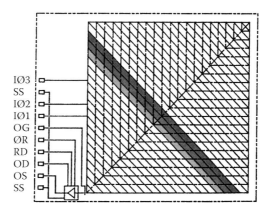

Figure 8.8 (**See colour insert.**) Movement of charge in diagonal areas to readout node.

active in diagonals across the device. Figure 8.8 shows three such diagonal regions as shaded, (highlighted green, red, and blue in colour insert) which will be read out consecutively. We will refer to these charges and areas as 'samples' and 'sample regions' from now on.

In a sense, the device 'sweeps' the electrons deposited in these diagonal sample regions across the area to the readout node like a linear CCD, hence the term 'swept charge device'. Note that if charge is shared between adjacent elements, it arrives at the readout either simultaneously (in same sample region) or consecutively (in the preceding or following sample regions) and can, if required, therefore be binned together. The pileup and effective leakage current associated with a sample will vary with the area (or length) of the diagonal sample region over the device as it is read out.

8.2 Techniques of Scientific CCDs

CCD foundries use specialised lithography processes and scientific CCD manufacturers are even more specialised. The latter companies work closely with universities and customers (which are often government laboratories or space agencies). Examples are Loral Fairchild Imaging Sensors: working with Penn State University, the Jet Propulsion Laboratory (JPL), NASA; Scientific Imaging Technologies (SITe, formerly Tektronix): working with Berkeley; MIT (Massachusetts Institute of Technology) Lincoln Laboratory: working with NASA and Niton Corp. in the United States; e2v (formerly EEV Ltd): working with Leicester University and Brunel University, European Space Agency (ESA), Rutherford-Appleton Lab (RAL); and Oxford Instruments in the UK and the Extraterrestrial Physics group at the Max Planck Institute (MPE): working with ESA, University of Milan, CERN, Ketek, PNSensors, etc., in continental Europe. These centres have developed similar techniques, and we will now look at some of them.

8.2.1 Noise Reduction

The overarching advantage of CCDs, as in all pixel detectors, is that they have very low output capacitances and their integrated readouts are designed to match these. In the simplest case, the pixel charge is transferred to a reverse biased output gate in parallel with the capacitance C_n of the output node and the gate of a MOSFET (MOS field effect transistor) amplifier. All these components are integrated onto the CCD chip to avoid stray capacitance. Figure 8.9 shows the arrangement schematically.

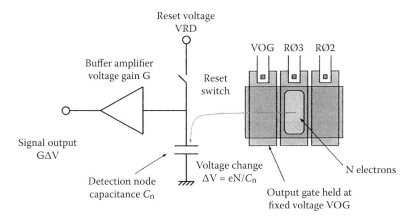

Figure 8.9 Transfer of charge to a low-capacity output node.

It was once thought that the $1/f$ noise in MOSFETs excluded them as a low noise output sensor but this was questioned [6], and the present technology, as we shall see, is capable of very low noise equally as good as any junction FET (JFET). After the signal step has been sent to an external high-speed digital signal processor at the load resistor of the MOSFET, a reset transistor (also integrated on the chip) discharges the capacitance ready for the next signal. The principal noise sources are from the integrated leakage current (during the active time of the pixel) that can be reduced by cooling and from the readout noise. The readout noise has a component caused by the channel resistance of the reset transistor that causes noise fluctuations, which can be shown to be of the form $\sqrt{(kTC_n)}$ e-rms but can be reduced by correlating the signal level with the reset level. This is called correlated double sampling [7,8]. This samples the baseline during reset before the signal and a cycle later, during the signal plus baseline. It then takes the difference, leaving just the signal. This removes any DC offsets and is effective in reducing noise because the readout noise pattern is largely repeated for each sample. As would be expected, the readout MOS transistor also introduces noise that has both white and $1/f$ components. For absolute minimum noise (<2 e-rms), the readout node capacitance is reduced to ~10 fF. Clock-induced pickup can also be suppressed by using an adjacent on-chip 'dummy' readout node (having no signal charge). The outputs of both are taken on the same readout cycle and fed to a differential amplifier.

The main source of leakage current during integration is the surface component that is similar in magnitude to that of p–i–n diodes and pixellated x-ray detectors (1–2 nA/cm^2 at room temperature). These surface currents may be reduced by using an IMO (inverted mode operation) mode [9], also referred to as multipinned phase [10]. In this mode, the CCD silicon substrate is biased at a positive level with respect to the clock lower level in order to induce holes to accumulate at the silicon surface under the 'off' pixel electrodes. This effectively inverts the surface from n-type to p-type, thereby suppressing the surface current (see Section 1.11.3). In effect, the surface-generated electrons recombine with the holes to give a flat-band neutral condition. This depresses the surface component of leakage current, and currents <10 pA/cm^2 are routinely achieved without cooling. Because the surface states responsible for leakage current are relatively long-lived, they do not have time to generate leakage current after the inversion, providing the clocking rate is sufficiently high [11].

For low rate operation, non-destructive read-out techniques can be used to measure the same charge signal several times. If this is carried out n times, the noise might be expected to reduce by $1/\sqrt{n}$. This can be achieved by replacing the conventional floating diffusion amplifier, which effectively destroys the signal on reading, with a floating gate amplifier [12], where the signal charge is only capacitively coupled to the output node. The signal charge can be sent to a series of floating gates (distributed or 'skipper' amplifiers) or presented several times to the same floating gate ('dithered'). If the dithering is rapid enough, the effect is that the whole surface area is inverted in IMO so reducing the surface leakage current. In practice, all phases over the whole area must spend at least ~5 μs in the 'low' condition, and this puts a further limit on the readout speed. The first practical demonstration of nondestructive readout to reduce noise in CCDs was presented by Janesick [13] in 1988, and sub-1 e-rms noise has been achieved, for example, by Penn State University [14].

8.2.2 Low Energy Sensitivity

Conventional CCDs, with their overlapping metal electrodes and thick dielectric insulation and passivation as described above, present a significant barrier to low-energy x-rays. A major step forward to improving low-energy sensitivity was the employment of thin polysilicon electrodes [15] and dielectric thinning over the third electrode phase [16] that was expanded in area relative to the other phase electrodes. Most is to be gained by thinning the phase-3 electrode, because it is not overlapped by other electrodes. With the removal of the polysilicon altogether they can even operate as an 'open window' using a shallow implant as shown in Figure 8.10.

In the 'open window' version shown below, the oxide thickness can be about 40 nm and the shallow surface pinning implant of boron maintains the surface potential. The open window area is 30%–40% of the total pixel area, but when

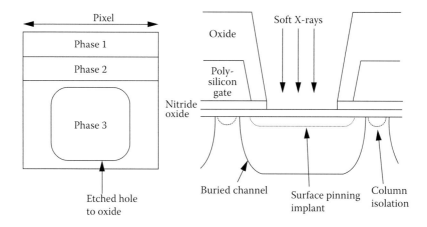

Figure 8.10 'Open window' CCDs for good low energy performance.

thinned (40 nm polysilicon) it is 60%–70% [17]. The efficiency [usually the term 'quantum efficiency' (QE) is used in this context] at C Kα (277 eV) is usually ~15%.

A different approach is to 'back-thin' the CCD substrate by etching and using the backside as the entrance window for x-rays. In the case of epi-Si, the thinning can be right back to the epitaxial interface. This has the effect of removing most of the low field region and allows the possibility—with a suitable continuous contact such as a shallow boron implant—to deplete from the nonpixel side as shown in Figure 8.11 that should be compared to Figure 8.1.

By doing this, the QE at low energies is substantially improved (~80% at C Kα). However, diffusion causes increased charge sharing among pixels, and adding charge from adjacent pixels always results in poorer spectroscopy. If the fields are not

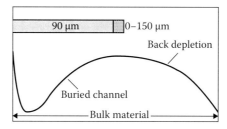

Figure 8.11 Back-thinning the CCD substrate.

very high, there may be charge loss to the entrance window contact itself. As a consequence, spectral quality is not as good as that from the thin polysilicon or open windows described above (Figure 8.10). As we might expect, back-thinned CCDs are fragile and yields are low. As a consequence, although they offer the best low-energy efficiency, they are an expensive option. Vibration of the structure may also give rise to microphony.

8.2.3 High Energy Sensitivity

The depletion depths in MOS CCDs are very limited. The attenuation, and therefore the efficiency for 20 keV x-rays—for example, in the depletion depths mentioned below—can be estimated from Figure 1.1. Standard CCDs are manufactured on silicon wafers which, in x-ray detector terms, are low resistivity (~20 Ω cm epi-Si) as the depletion depths are perfectly adequate for light or infrared sensitivity. High-resistivity epi-Si (~1.5 kΩ cm) can be used, and was used, for example, in the SCD [5], where the depletion depth (30–40 μm) is limited by the thickness of the epi-Si layer. Fully depleted back-illuminated CCDs have been manufactured on high-resistivity n-type silicon for UV and light imaging [18] for earth-bound telescopes, and recently [19] CCDs have been manufactured on high resistivity (8 kΩ cm) bulk p-type silicon for x-ray detectors. With no bias on the substrate the depletion depth is 93 μm, but when raised to −110 V it is ~300 μm, so that it is fully depleted and the QE is similar to that for most p–i–n diodes. Another approach to full depletion that has been developed over the past 20 years is the pn-CCD; it relies on p–n junction technology rather than an MOS and will be discussed below (Section 8.4). But before we do that, let us summarise the state and performance of all MOS CCDs discussed above.

8.3 Performance of MOS CCD X-Ray Detectors

The known limitations of MOS CCDs with respect to QE at high energies have been discussed above. The 30–40 μm depletion in high resistivity epi-Si results in QE of 20%–25% at 10 keV and 3%–4% at 20 keV. This does not, of course, mean that they cannot be used in the higher energy range, only that the sensitivity will not be as good as a p–i–n diode and certainly will not compare with a Si(Li) detector 95%–98% at 20 keV. In a study using Loral Fairchild thinned poly-Si electrodes, the group at Penn State University and Janesick at JPL [14] on a NASA grant, demonstrated state-of-the-art low noise and good low energy QE under conditions where count rate is not an issue, in 1995. They used a 1024×1024 18-μm pixel CCD manufactured on 30 μm epi-Si of 50 Ω cm resistivity. They cooled to –110°C to reduce dark current and used a readout rate of 63 kHz. They thinned the poly-Si contact to 40 nm on 100 nm dielectric on one electrode, which represented two-thirds the total pixel area. This gave ~17% QE at C Kα (277 eV). The noise was dominated by the output amplifier, but by using 16 distributed floating gates per pixel this was reduced to 0.9 e-rms (7.8 eV), giving a resolution of 120 ± 8 eV (the statistics were poor in the report). The low noise allowed them to accurately select single pixel events and set a low threshold (see Figure 8.12).

This was the first time the Al L line at 66 eV had been detected by any semiconductor x-ray detector. Being below

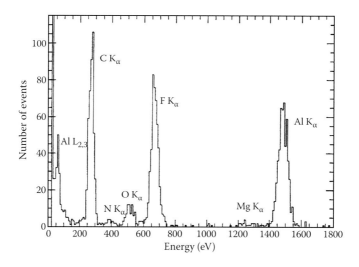

Figure 8.12 Spectrum obtained using a thinned poly-Si MOS CCD. (From Kraft, R.P. et al., *NIM A*, 361, 372, 1995. With permission.)

the Si L absorption edge (~100 eV), the QE was relatively high (~11%) at this energy. Again, although the statistics are poor, the peak widths are impressive and consistent with $F = 0.115$ and $\omega = 3.68$ eV down to very low energies (see Section 2.4.2). This performance is, of course, limited to applications where count rates are very low.

As previously noted, the SCD was designed to overcome such count-rate limitations. The device was developed by e2v Ltd, University of Leicester, and OI in the late 1990s to be used in XRF applications where count rate and noncryogen cooling were primary considerations. The prototypes were manufactured on 1.5 kΩ cm epi-Si giving a depletion depth of ~36 μm. They were illuminated on the pixel side (pixels ~25 μm), and there was no attempt at this stage to improve low energy efficiency through thinning, although detection of C Kα and O Kα x-rays was demonstrated on a scanning electron microscope [5]. The noise was ~8 e-rms at –15°C and the resolution was 450 eV at room temperature, 265 eV at 0°C and 140 eV at –20°C. Figure 8.13 shows the Fe-55 spectra at various temperatures clocking at 100 kHz and demonstrates a peak/background ratio (2 keV) of ~3000:1.

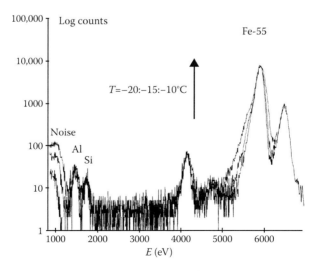

Figure 8.13 Results obtained with a SCD. (From Lowe, B.G. et al., *NIM A*, 458, 568, 2001. With permission.)

By resetting periodically on demand (rather than sample by sample), the maximum output was ~23 kHz and the resolution was 228 eV. With further progress and improvements in the low-energy efficiency (QE >30% at C Kα), 100 of these devices flew successfully in the x-ray spectrometer on ESA's SMART-1 mission to the moon launched in 2003 (impacted the moon in 2006) [20] and on the Chandrayaan Indian moon mission in 2008. Figure 8.14 shows a Fe-55 spectrum from one of the Chandrayaan SCDs at −30°C for all events (black) and single pixel events (light) before proton irradiation tests [21]. The resolution is 133 eV.

Three further variations of SCD have been manufactured at e2v Ltd. One had the same active area (~1 cm²), one was larger (~4 cm²), and one was smaller (5.2 mm²). All of them use two-phase clocking for increased speed and larger elements (100 µm) to obtain a higher proportion of single pixel events. The larger device is basically made up of four of the 1-cm² devices, where charge is swept to the center of the single chip. It is intended that it would fly on the first dedicated astronomy satellite, the Chinese Lunar Mission Hard X-ray Modulation Telescope. The other two are intended for XRF applications at or near room temperature. Preliminary measurements of the

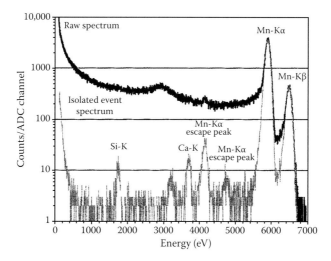

Figure 8.14 An SCD operating at −30°C. Black, all events; gray single, pixel events. (From Smith, D.R. et al., *NIM A*, 583, 270, 2007. With permission.)

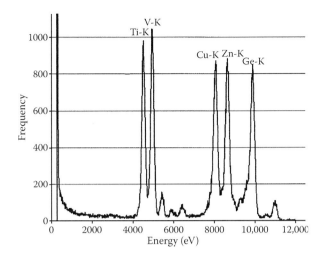

Figure 8.15 An SCD used in an XRF spectrometer and operating at +27°C. (From Xcam Ltd. UK, www.xcam.co.uk. With permission.)

smaller device at a readout rate of 100 kHz, in dither mode, give a noise of ~5 e-rms (43 eV) at 0°C and increasing to ~11 e-rms (98 eV) at +19°C and ~14 e-rms (121 eV) at +27°C. Figure 8.15 shows a spectrum from a composite XRF target taken in air at +27°C for single pixel hits.

The measured resolution on the Cu Kα peak (8.05 keV) was 180 eV at room temperature. In all these measurements, the CTE was shown to be excellent and not affected by the larger element size or two-phase clocking. Preliminary measurements of count rate capability suggest that pileup of events starts to decrease the throughput at around 2.1 kHz, but a faster clocking rate would increase this proportionately. These results have been improved upon and devices are available commercially [22].

8.4 pn-CCD

A review of pn-CCDs (and other imaging x-ray detectors) is given by Struder et al., in a contribution to the book *XRS—Recent Technical Advances* [23]. The p–n junction CCD (pn-CCD) was first postulated in the same seminal paper as the SDD by Emilio Gatti and Pavel Rehak [24] (see also Chapter 9), and indeed it has more affinity to the SDD than it does to the CCD. Recall (Section 1.17) that in the SDD an n-type silicon wafer is depleted from both p+ faces to achieve full depletion, forming a channel in the center. One of the faces has strip electrodes, as in a CCD, but unlike the CCD the depletion regions spread transversely to join up and, by stepping up the bias one to the next, electrons are transported continuously in the channel running under one electrode to the next. A similar concept is used in the pn-CCD, except that the transport of electrons is achieved by *clocking* the electrodes (p+ implants with no MOS structure) much as in a conventional CCD. Figure 8.16 shows such an arrangement for a pn-CCD constructed on n-type epi-Si on HPSi.

The device is fully depleted and irradiated from the continuous entrance window side as shown, and the resulting electrons are clocked in a regimented way transversely to the readout node rather than continuously down a sloping potential gradient as in the SDD. Note that, unlike MOS CCDs, p–n CCDs are majority carrier devices like SDDs.

Strips of deep n-type implant are put in at right angles to the electrodes forming 'pixels' as shown in Figure 8.17.

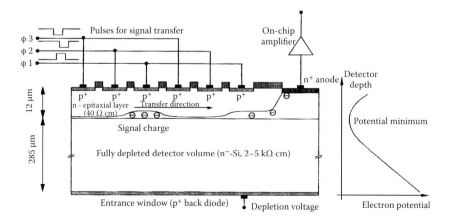

Figure 8.16 A pn-CCD constructed on HPSi. (From Struder, L. et al., *Rev. Sci. Instrum.*, 68, 4271, 1997. With permission.)

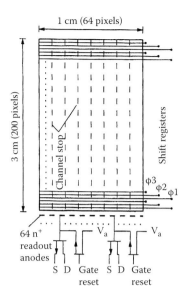

Figure 8.17 pn-CCD layout. (Struder, L. et al., in Tsuji, K. et al. (eds.): *X-ray Spectroscopy—Recent Technological Advances*. 2004. Copyright Wiley-VCH Verlag GmbH & Co. KGaA. Reproduced with permission.)

These act not only as channel stops but also increase the depth of the potential minimum profile below the surface to allow greater modulation of the potentials between the electrodes without increasing the hole current. As shown in the example above, three-phase clocking is usually used to transfer charge in each row of pixels to the row's own dedicated JFET readout node.

8.4.1 Performance of pn-CCD X-Ray Detectors

Since their common conception in the early 1980s, the pn-CCD and the SDD have been developed in parallel at about the same rate and by the same groups (MPI Garching and Milan University). Much experience has been gained, and very large area devices (2.9 cm^2) have been manufactured [25]. Twelve have been flown successfully on the ESA XMM (X-ray Multi-Mirror) satellite launched in 1999 and are still in operation, although there has been some radiation damage due to protons [26]. They have been used in the CERN solar telescope [27], and further improvements have been implemented [28,29] for future missions (DUO, ROSITA). Figure 8.18 shows the Fe-55 spectrum from a 4-cm², 450-μm fully depleted frame store pn-CCD cooled to −75°C.

The quality of the spectrum is apparent with a good tail and peak/background ratio of ~3000:1. The image is transferred to the frame store in <100 μs, and 'out of time' events (those occurring during transfer and which will therefore affect the image quality) have been reduced to <0.1%. This has been achieved with an x-ray photon integration time 50 ms and 75 μm pixels. The CTE is corrected for but is <0.5% over the whole area for a Fe-55 source. The device has 256 × 265 pixels with 250 JFETs integrated on chip. The entrance window implant gives >90% QE at 300 eV but is rendered light-tight by 70 nm dielectric plus 100 nm aluminium, which reduces the QE. Figure 8.19 shows C Kα photopeak with FWHM (full-width half-maximum) of 48 eV (single pixels) and 50 eV (4-fold charge sharing pixels). The noise is quoted as 2.3 e-rms (20 eV).

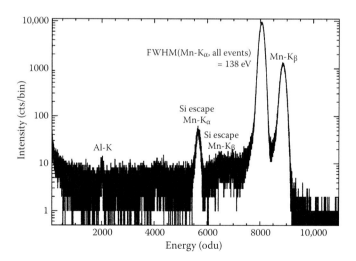

Figure 8.18 Fe-55 spectrum from a 4-cm², 450-μm fully depleted frame store pn-CCD cooled to −75°C. (From Meidinger, N. et al., *SPIE*, 5898, 58980 W 1-9, 2005. With permission.)

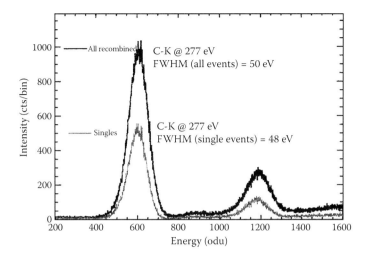

Figure 8.19 Detection of carbon K using a pn-CCD. (From Meidinger, N. et al., *NIM A*, 568, 141, 2006. With permission.)

8.5 CCDs—Summary

Pn-CCDs have several advantages over most MOS CCDs. They have deeper depletion and therefore have reasonably good QE at moderate energies (90% at 10 keV). In principle at least, the continuous entrance junction can be made very shallow to give a good QE at low energy as in the SDD [30]. Because the charge transfer is deep below the surface, they can be operated at fairly high clocking rates, and because the impurity concentration is low and the electrons are far from any surface states the CTE is good. MOS CCDs have limited pixel size because the transverse field, electrode to electrode, is limited by the pixel size. This is not true for the pn-CCD, where the pixel size is typically 75–150 μm. Because they are based on nonoxide p–n junctions rather than MOS, they are also more radiation tolerant. This, of course, will not apply to protons, which will give rise to lattice displacement.

In comparison with the SDD and SCD, the pn-CCD is far more complex (and potentially expensive) but combines spectrometry and imaging capability in the same device. pn-CCDs also do not have the speed of the SDD and SCD and even with multiple readout nodes the readout time is of the order of a few milliseconds. pn-CCDs have not been commercially exploited as x-ray spectrometers, but this could change if an application integrating the excellent energy dispersive characteristics can be combined with the imaging capability (e.g., XRF with XRD) under low flux conditions.

References

1. Catura R.C. and Smithson R.C., *Rev. Sci. Instrum*. 50, 219, 1979.
2. Fraser G.W., *X-ray Detectors in Astronomy*, Cambridge Astrophysics, 1989, ISBN-10: 0521106036.
3. Janesick J.R., Scientific CCDs, *SPIE Monograph* PM83, 2001.
4. http:/msl-scicorner.jpl.nasa.gov/Instruments/CheMin/Chemistry and Mineralogy, David Blake, 2007 (access. 2013).
5. Lowe B.G. et al., *NIM A* 458, 568, 2001.
6. Kandiah K. and Whiting F.B., *NIM A* 326, 49, 1993.
7. White M.H. et al., *IEEE J. Solid State Circuits* SC-9, 1, 1974.
8. Spieler H., *Semiconductor Detector Systems*, Oxford University Press, 2005, ISBN0-19-852784-5.
9. Burt D.J., EEV Technical Note 11, 1990.
10. Janesick J. et al., in *Astronomical CCD Observing and Reduction Techniques*, Howell S.B. (Ed.), p. 1, 1992.
11. Burke B.E. et al., *IEEE Trans Elect Dev* 38, 285, 1991.
12. Wen D.D., *IEEE J. Solid State Circuits* SC-9, 6, 1974.
13. Janesick J. et al., *SPIE X-ray Instrumentation in Astronomy II*, 982, 70, 1988.
14. Kraft R.P. et al., *NIM A* 361, 372, 1995.
15. Bertram W.J. et al., *IEEE* ED-21 No. 12, 758, 1974.
16. Castelli C. et al., *NIM A* 310, 240, 1991.
17. Skinner M.A. et al., *SPIE* 215, 163, 1995.
18. Holland S.E., *NIM A* 579, 653, 2007.
19. Murray N.J. et al., *NIM A* 604, 180, 2009.
20. Grande M. et al., *Planet. Space Sci*. 55, 4, 494, 2007.
21. Smith D.R. et al., *NIM A* 583, 270, 2007.
22. Xcam Ltd. UK, www.xcam.co.uk.
23. Struder L. et al., *X-ray Spectroscopy—Recent Technological Advances* (Chapter 4), Tsuji K., Injuk J., Van Grieken R. (Eds.), Wiley, 2004, ISBN-10 047148640X.

24. Gatti E. and Rehak P., *NIM* 225, 608, 1984.
25. Struder L. et al., *Rev. Sci. Instrum.* 68, 4271, 1997.
26. Struder L. et al., *Astronomy Astrophys.* 365, L18, 2001.
27. Kuster M. et al., *SPIE* 5500, 139, 2004.
28. Meidinger N. et al., *SPIE* 5898, 58980 W 1-9, 2005.
29. Meidinger N. et al., *NIM A* 568, 141, 2006.
30. Hartmann R. et al., *NIM A* 387, 250, 1997.

9

Silicon Drift X-Ray Detectors

9.1 Introduction

A brief introduction to the SDD (silicon drift detector) x-ray detector was given in Section 1.17. Struder et al. give a detailed review in the book *X-Ray Spectroscopy: Recent Technical Advances* (see Table 1.1 [8]). There is no better introduction to the principles of the SDD than to refer back to the original description by the inventors, Emilio Gatti and Pavel Rehak [1,2], in their paper presented at the 2nd Pisa Conference on Advanced Detectors in 1983 (see also Section 11.26). To demonstrate full depletion of a silicon wafer for the detection of ionising radiation, they constructed a test device using an n-type Si wafer with p^+ contacts (Au) on both faces and several small n^+ contacts (Al) around the edges (see Figure 9.1a).

When this p–n structure was reverse-biased on both sides as shown, the depletion grew into the wafer uniformly from both wafer faces, at first forming a 'quasi-n^+ contact' in the center of the wafer (see Figure 9.1b). The capacitance at this point between all the p^+ and all the n^+ contacts connected together is effectively that of two thin p–i–n diodes in parallel and therefore relatively high. However, at full depletion (Figure 9.1c) the intermediate n^+ quasi-contact layer pinches off, and the capacitance drops dramatically as the inventors demonstrated in Figure 9.2.

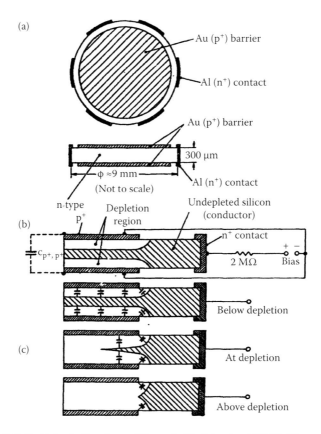

Figure 9.1 Full depletion in a silicon drift detector (SDD)–like structure.
(a) Test device to demonstrate full depletion, (b) partial depletion showing
'quasi-n contact' in the centre of the section, and (c) full depletion. (From
Gatti, E. and Rehak, P., *NIM*, 225, 608, 1984. With permission.)

This behaviour is quite different to that of a conventional
p–i–n diode, in which the capacitance C varies with the bias
V_b as $C \sim V_b^{-1/2}$ (see Section 1.6.1). Besides the relatively low
capacitance of such a structure at full depletion, two other
properties should be noted. First, when fully depleted, a minimum potential exists at the central plane of the wafer, and
this acts as a collection channel for electrons, for example,
those electrons produced by ionising radiation. This is much
like the buried channel of a charge-coupled device (CCD) element. Thus, like the CCD, the SDD is a *single carrier* detector. Second, in the 'back to back diodes' depicted before full

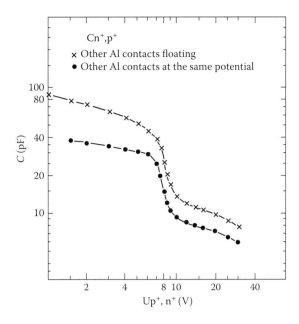

Figure 9.2 Drop in capacity of the structure at full depletion. (From Gatti, E. and Rehak, P., *NIM*, 225, 608, 1984. With permission.)

depletion in Figure 9.1b, the depletion depth W is $\sim V_b^{-1/2}$ for each diode. Thus, depleting the silicon to a depth equal to the sum of the thicknesses of the two diodes (i.e., $2W = d$, the full wafer thickness) requires one-quarter of the voltage that would be required for a simple p–i–n structure of thickness d. For example, using $W \sim \sqrt{(3\rho V)}/100$, a p–i–n structure on 5 kΩ cm n-type silicon 500 μm thick will fully deplete at 200 V, but in the above example at only 50 V. The lower bias requirement for full depletion is realised when the anode is on the side of the device as in Figure 9.1. However, in the case of an SDD with a central anode located on the back face, the volume above this anode will require the full depletion voltage.

9.1.1 Depletion

For an explanation of the depletion of the SDD, again it is useful to include the reasoning from the seminal paper of 1983 (see Figure 9.3).

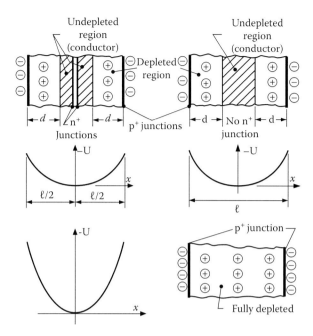

Figure 9.3 Depletion in an SDD. (From Gatti, E. and Rehak, P., *NIM*, 225, 608, 1984. With permission.)

The top left-hand schematic shows two partially depleted *discrete* p–i–n diodes back to back. The situation on the right shows the SDD where the two diodes are joined together in the same wafer and the n⁺ contact junctions are effectively replaced by a conducting layer (the undepleted n-type silicon) that is connected to the small n⁺ contact on the edge of the wafer (not shown). Below these schematics (Figure 9.3), are the corresponding potential profiles for electrons. The bottom profile shows the situation at full depletion with the potential minimum that acts as the electron channel. Figure 9.4 shows the potentials for electrons as the bias is raised in overdepletion together with the field (which is the negative gradient of the potentials shown and does not vary with bias) shown on the right-hand scale.

This is unlike the p–i–n diode, where the field at the front surface increases with bias in overdepletion (see Figure 1.23 in Section 1.6.1). This field can be increased in SDDs by using lower resistivity silicon (with corresponding higher bias needed to deplete), but there is a compromise between this and

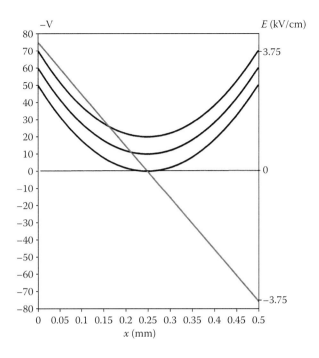

Figure 9.4 Potential and field for electrons in an overdepleted SDD.

keeping the leakage currents low. In practice, the field at the wafer surface, and the charge collection at the x-ray entrance widow, can be improved using ion implantation techniques (see Section 4.5.1 and discussion below).

9.1.2 Transverse Drift Field

Figure 9.4 also indicates how the electrons that collect in the central channel can be drifted transversely in the wafer toward an anode at one edge (or any other position). With overdepletion, there is a potential difference between the central channel and the anode, usually held near ground potential, and current will flow. If one or both p⁺ contact planes are divided into separate strip elements with the bias on each increasing away from the anode (in the *y*-direction referring to Figures 9.3, 9.4, and 9.5), the transverse field can be made linear and the electron potential profiles will form a tilted channel, as indicated schematically in Figure 9.5.

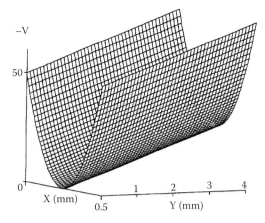

Figure 9.5 Tilted channel in an SDD.

In the center of the channel, the electrons experience the transverse drift field (in the y-direction) and drift down the channel toward an anode and a field effect transistor (FET) amplifier. It therefore effectively forms a *charge delivery detector* (see Section 1.15), rather like a continuous charge transfer CCD. Because this drift field is relatively weak (0.4–0.6 kV/cm), the drift times for delivery to the anode can be of the order of µs and more than one charge packet from an ionisation event may be drifting in the channel at any one time. The above has assumed that the electrode elements are parallel strips much as in a silicon strip detector, and the technology of planar processing and lithography developed for strip detectors can be carried over to manufacture the elements of the SDD. In practice, of course, other effects such as charge diffusion and electrostatic repulsion will also act on the electrons.

9.2 Concentric Ring SDD X-Ray Detectors

The SDD was first conceived for particle tracking as a position-sensitive detector (PSD; see Section 11.27) but for energy dispersive x-ray spectrometry (EDXRS), solid angle, count rate, and energy resolution are the prime requirements. The position sensitivity was therefore sacrificed in much the same way as it was in the SCD (see Section 8.1) that evolved from the CCD. The transverse drift field no longer needed to be linear or stable and a concentric ring structure on one side with a plain implanted side on the other, evolved (see Figure 9.6). The concentric ring structure [3] avoids the problem of high fields at the strip ends, and the single-sided structure [4] avoids, to some extent, the problems associated with the alignment of the mask patterns on the opposing faces during manufacture. Combining these features led to the design shown in Figure 9.6 (see also Figure 1.64).

The drift field is now radial taking the charge to a centrally located anode on the bottom face of the structure. This anode can be relatively small with the consequent advantage of low capacity as seen by an amplifier (FET) connected to it. This has major benefits for low noise, and most importantly at short measuring times, for high rate spectrometry and element mapping. The top face (x-ray entrance window) of the detector is now a continuous surface that can be engineered for the virtually uninterrupted

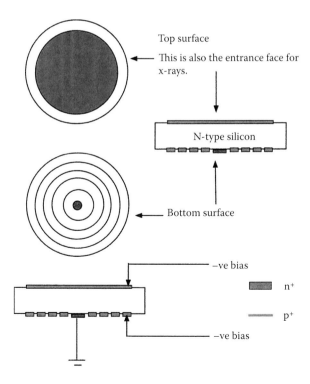

Figure 9.6 Design of an SDD.

detection of the incident x-rays and for optimum electron collection. The original inventors also noted that the ring structure could be self-biased by the chain or 'punch-through' effect.

9.2.1 Biasing

N-type silicon wafers are usually used because this gives a greater depletion depth for the same bias value (see Section 1.6.1). Referring to Figure 9.6, a negative bias is applied to the top p-type face formed by a boron implant, and the diode starts to deplete toward the n-type anode region that is held at or close to ground potential on the bottom face. The anode is surrounded by a number of concentric p-type contact rings again formed by implanting boron. These rings are also negatively biased with respect to the anode, and each ring will attempt to deplete part of the bulk silicon toward the anode.

The onset of depletion starts as shown in Figure 9.7, where one ring only is shown for clarity. The bulk starts to be depleted via the bias applied to the top face and from the ring toward the anode. Any electrons liberated in the bulk silicon will move to the minimum potential and travel toward the anode. Any holes will move to the negative surfaces and be lost as far as signal is concerned.

The rings can be biased to form a transverse drift field in a number of different ways. As noted previously, they can be self-biased by the depletion region between them reaching through from one to another in a chain or punch-through effect. A relatively large negative voltage is applied to the outer ring, and the inner ring is tied close to ground potential by a large resistor, whereas the intermediate p$^+$ rings are left 'floating'.

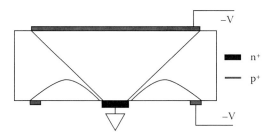

Figure 9.7 Depletion in an SDD.

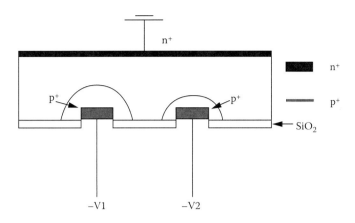

Figure 9.8 Self-biasing the rings in a silicon strip detector.

Self-biasing by punch-through is commonly used for silicon strip detectors. In order to explain this concept, consider Figure 9.8, which shows two adjacent p^+ implant rings on a single-sided n-type silicon *strip detector*.

If these strips are biased negatively, with V1 more negative than V2, reverse-biased junctions are formed that inhibit the flow of holes (the majority carriers in the p^+ regions) from one to the other. However, as we increase the bias the depletion regions punch through to each other, and each ring will be connected to the next ring by a depletion layer, except at the surface. There will then be a leakage current of holes under the oxide layer that, if controlled, can be used to bias the strips automatically to form the drift field for the collection of electrons in the bulk. In SDDs, this can be achieved by applying the bias between only the outermost ring and the anode on the same surface. Once the space between all the rings has been depleted, a current consisting of holes moving toward the outermost ring is set up, forming a potential divider chain between them. This current can be relatively high (>15 μA) but, as it flows from the inner most ring to the outermost ring, does not enter the amplifier circuit and does not add noise. Even though this current is not seen in the signal path, it will have implications on the amount of power the device consumes and internal heating. The top n^+ plane in Figure 9.8 can also be biased by punch-through in this way if it is left floating without being held by an applied bias. This also applies to the top p$^+$ plane of an SDD detector (as in Figure 9.10). It was shown [4] that the 'floating rings' need to be capacitively coupled to the ground plane in order to avoid hole accumulation and 'ghost peaks' (reduced magnitude signals from the rings).

Other designs take each ring to one element in a resistor chain, and a negative voltage is applied to the outer ring end of this chain, the inner ring again being close to ground potential. Another design uses resistors [5] formed in the space between the rings, and a further design [6] replaces the concentric rings with a continuous resistive spiral. Rawlings [7] proposed to maintain the drift field by replacing the ring structure with a continuous 'conducting layer oxide semiconductor' on the outer surface, that is, on top of the normal existing oxide layer. By controlling the resistivity of this layer, the current between the inner and outer rings could maintain the strength of the drift field. The choice of biasing method can in part be related to reducing the complexity of the process recipe and in part as an attempt to obtain technical advantages. Usually, however, the rings are used to give a self-bias drift field by the punch-through effect and the top p$^+$ plane is biased separately.

An ingenious method of imposing a lateral drift field was used by Luke [8] in the case of a HPGe drift detector. He pointed out that you cannot process semiconductors other than silicon in a silicon-based foundry because it introduces impurities, and there is also no equivalent for the MOS processes for these other materials. He used a tapered structure as shown schematically in Figure 9.9, using continuous contacts on both

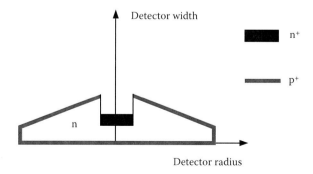

Figure 9.9 Achievement of a lateral field gradient in an HPGe detector.

sides. The field strength increases toward the edges of the detector, thus supplying a transverse field gradient. The same effect could also be achieved by fabricating the semiconductor with an impurity concentration gradient. He obtained excellent results on a 27-mm-diameter germanium detector with a thickness starting at 3 mm in the center and tapering to 1 mm at the rim.

In the immediate neighbourhood of the anode, the field in an SDD approximates to what one would expect for a conventional planar detector. Here, the field falls linearly (see Section 1.6) from a maximum at the junction, and there is no minimum potential for the collection of electrons: they are simply swept toward the anode. In this localised region, it acts as a *charge induction detector* and, because of the charge diffusion, will suffer from charge division between the anode and concentric rings (and FET, if it is integrated) much as in a PXD. This is a disadvantage for a central anode as this gives rise to incomplete charge collection (ICC) (see Section 9.3).

With asymmetric biasing of the top and bottom surfaces using rings, the potential minimum for electrons is no longer at the center of the wafer. As we make the rings more negative, moving outward, the potential minimum in the silicon moves toward the opposite continuous face. Thus, the maximum transverse drift field achievable in a device is restricted by the potential on the outermost ring that must be less than that which drives the minimum to the surface. In other words, the difference in potential between the outermost ring and that of the opposite continuous face must be less than the full depletion value for the wafer. Conversely, as we make the rings less negative, moving inward, the potential minimum in the silicon moves toward the rings, eventually delivering the electrons to the anode. This is illustrated in Figure 9.10.

Thus, the SDD under these conditions requires the same full depletion bias as a conventional planar detector of the same thickness. Overdepletion, as used in the latter to improve carrier transport, cannot be used in the same way for an SDD. Practical drift lengths are also limited and the active area for premium performance is <30 mm^2 but large solid angles can be achieved, for example, in TEM applications, by using multiple SDDs [9] surrounding the EM beam.

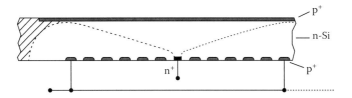

Figure 9.10 Asymmetric biasing of an SDD with rings.

9.2.2 Leakage Current in SDDs

In the quiescent state of a fully depleted SDD, thermal carrier pairs are continuously being generated in the bulk. The electrons in the bulk will be delivered to the drift channel and thence to the anode, and holes will go to the surfaces where they will need to be cleared for stability. The electrons, along with any free surface electrons moving to the anode, will contribute to the noise of the device. The bulk leakage current in SDDs is low because of their small overall volume, and the surface currents are low because of the relatively small exposed oxide passivated surfaces and the low tangential fields across them. They can be operated with moderate cooling (single-stage thermoelectric). However, as for Si(Li) detectors, the most common cause of failure during manufacture is high leakage current. In SDDs, the leakage currents can be reduced by improved planar processing [10] and by using good-quality p–n junctions (implants), good-quality surface oxides, and the use of guard rings. An n+ guard ring [5,11] can be used between the outermost p+ ring and the undepleted wafer edge as a sink for electrons generated in this region. On the entrance window side, multiple floating guard rings (see Figure 9.11) can be used to avoid high fields near the contact termination in the same way as for p–i–n diodes (see Section 7.1).

Leakage currents can be reduced to <100 pA/cm^2 at room temperature and much lower when moderately cooled (–10°C). It must be noted that leakage currents much higher than those of Si(Li) detectors can be tolerated as the low capacitance allows operation at much shorter processing times (see Figure 1.43).

Figure 9.11 Multiple floating guard rings designed to reduce leakage currents.

9.2.3 Capacitance

The dynamic capacitance per unit of the anode area of a fully depleted device is—as a first approximation—given by $C \sim \varepsilon\varepsilon_0/d$, where d is the wafer thickness and $\varepsilon\varepsilon_0$ is the dielectric permittivity of the wafer. As the anode area can be very small, the capacitance can be extremely low but would be expected to be much larger than this in an SDD, because of the additional capacitance between the anode and the first ring and the effects of fringe fields. To a second approximation, we can consider a disk with radius a suspended a distance d above a ground plane [12].

$$C = \frac{a^2 \pi \varepsilon \varepsilon_0}{d}\left\{1 + \frac{2d}{\pi \varepsilon a}\left[\ln\left(\frac{a}{2d}\right) + (1.41\varepsilon + 1.77) + \frac{d}{a}(0.268\varepsilon + 1.65)\right]\right\}$$

In Gatti and Rehak's original paper [3], anode capacities of ~60 fF were quoted for 200-μm-diameter anodes on a 380-μm-thick silicon wafer, which agrees reasonably well with this formula. Compare these capacitance values with those of many pF for p–i–n diodes of the equivalent x-ray sensitivity. It should also be noted that there is another capacity in parallel with the main capacity. This is the capacity between the rim of the anode (C_{rim}) and the first ring. This capacity is expressed as

$$C_{\text{rim}} = C = \frac{2\pi \varepsilon \varepsilon_0}{\ln\left(\dfrac{r}{a}\right)}$$

per unit of implant depth. This calculates as ~1 fF for $r = 1$ μm thick ion-implanted depth and for $a = 100$ μm separation between the anode and ring. Both the fringe capacity and the rim capacity become important, for example, in FET design when gate lengths approach 1 μm.

9.2.4 Charge Diffusion

With the restrictions on the magnitude of the radial drift field discussed in Section 9.2.1, the drift velocities are relatively low and the drift time t from the outer rings may be high (~μs). As the diffusion length of charge varies with time t as $\sqrt{t}$ (see Section 1.7.1), the electron charge packet diffuses along the channel, spreading it out by as much as a few millimetres in the case of large-area SDDs [13] and is often larger than the anode dimensions. This has the effect of spreading the signal over time as originally shown by the inventors [14]. Figure 9.12 shows their oscilloscope traces for a collimated beam of particles as it is moved 2.5 mm away from the anode.

This shows the drift time as well as the spreading due to diffusion (the horizontal scaling is 200 ns/cm). This spreading is obviously important for timing measurements, as in a PSD, and it was shown [2] that the position resolution is proportional to the inverse square of the drift field. Note that charge diffusion of electrons during the drift time to the anode in charge induction detectors (e.g., Si(Li) and p–i–n diodes) has little effect on the signal as this is developed during the whole of the carrier trajectory and the drift times are smaller (50–100 ns) because of the high fields in overdepletion. In this case, diffusion is only significant as it affects ICC at the entrance cathode contact (see Section 2.7.4). The rings of the SDD structure screen the anode until the moving charge passes the final ring. Then it is strongly coupled via the depletion field to the other face of the structure, and in fact only in this small volume behind the anode does the detector act as a conventional charge induction detector. A theoretical study on the effect of charge diffusion on energy resolution in a double-sided radial detector was conducted by Rawlings [13]. The diffusion increases the signal rise time and can lead to a signal loss if the peaking time of the amplifier is not sufficiently long. The

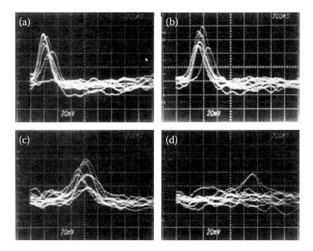

Figure 9.12 Oscilloscope traces for the charge signal at various radial positions. (a) Signal when radiation enters close to the anode, (b) signal from radiation entering a short distance away from the anode, (c) signal roughly half way between anode and outer ring showing spreading, and (d) signal generated when radiation enters close to the outer ring. (From Gatti, E. et al., *NIM*, 226, 129, 1984. With permission.)

signal charge develops a pseudo-Gaussian shaped signal at the output of the amplifier as the charge moves the distance δr, that is, the distance between the final ring and the anode. If the velocity of the charge is the radial drift velocity v_r, the standard deviation on the signal time is $t_{sg} = \dfrac{\delta r}{2v_r}$, that is, half the width of the output signal. The characteristic time t_c is the diffusion length $r_L = \sqrt{D_e t}$ divided by the drift velocity where D_e is the electron diffusion coefficient and t is the drift time from the outermost regions to the central anode. In other words, it is the duration of the charge packet entering the anode space or the characteristic time length of the signal t_c. Taking as an example a 10-mm² ring SDD with −100 V bias on the outer ring and the anode close to ground potential, the drift time is ~240 ns and the diffusion length (r_L) ~80 μm. This gives a characteristic time t_c ~10 ns. Because the optimum resolutions are achieved at integration times of ~1 μs, this is short enough not to be a problem. Diffusion of electrons out of the channel might be expected to cause charge loss to the inner rings, but in practice the fields are sufficiently strong here to prevent this from happening.

9.2.5 Integrated FET

Above we have considered an anode connected to a discrete FET gate by a bond pad in the same way as, for example, a p–i–n diode or a Si(Li) detector. This allows the anode to be very small, but the bond pad itself and the flying wire introduce significant added capacitance. The FETs used in this way are typically junction FET (JFET) tetrodes, which are commercially available. The integration of a JFET onto depleted HPSi by implantation was first described by Radeka et al. [15]. Although the fully depleted n-type HPSi of an SDD and the JFET substrate characteristics are incompatible, the former normally being a few kΩ cm p-type silicon substrate, they showed that it was technically feasible, and the German company Ketek soon went into production with a product based on this technology. Details of an integrated JFET structure have been given by Pinotti et al. [16]. The cylindrical geometry JFET (see Figure 1.48) was implanted into the center of an annular anode as shown in Figure 9.13 and the micrograph in Figure 9.14.

These figures show only the central region near the anode. A negatively biased inner guard ring, as shown, can be used to ensure that electrons move to the anode and shield the JFET

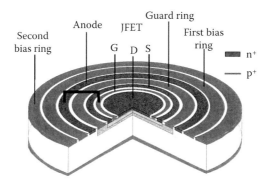

Figure 9.13 Integrating a field effect transistor (FET) into anode of an SDD.

Figure 9.14 Micrograph of an integrated FET structure. (From PNSensors website, www.pnsensors.de. With permission.)

from the effect of the signal electrons other than through the gate. This deep p⁺ implant extends below the FET structure as shown in Figure 9.13 and acts to isolate it. However, the presence of the FET in the center of the sensitive region does lead to some ICC, and this is discussed further below. Typical biasing values are –15 V on the guard ring, –15 V on the first ring, and –90 V on the outer ring, whereas the anode is close to ground. The entrance window side is typically held at about –100 V. Although these integrated JFETs were not necessarily the best in terms of the low noise achievable with discrete devices, the advantages of the low capacitance were quickly apparent. The gate capacitance of such an integrated FET is ~200 fF, which is well matched to the SDD capacitance. The integration of the JFET onto the SDD also makes it less sensitive to microphony and electrical pickup. With moderate cooling, the fact that they are at the same temperature as the SDD is not a problem. The charge reset of the JFET can be electronically implemented as for the discrete JFET. Despite the yield for a good SDD and integrated JFET combination being naturally reduced with respect to discrete combinations, most manufacturers now incorporate a JFET integrated onto the device. The sensitive area of the SDD is limited by the biasing arrangement as described but can be manufactured up to at least 50 mm². For larger areas, monolithic arrays can be used. For example, an annular array of twelve 5-mm² elements, each with an integrated FET, has been described [18].

9.2.6 Sensitive Depth

Our interest as far as EDXRS is concerned is in the application of the SDDs to high-resolution measurements of x-rays mainly in the 0.1 to 25 keV energy portion of the spectrum. The gold standard for such x-ray measurements is the Si(Li) detector, as discussed in Chapter 5. The Si(Li) detector evolved into a deep structure (often >3 mm), giving very high efficiency in this entire energy range. Its partner, the 3-mm HPGe x-ray detector (see Section 6.2), extended this efficiency range up to ~100 keV. In a parallel development, large-volume coaxial HPGe detectors having high efficiencies in the hundreds of keV to several MeV range for the measurement of gamma rays are available. The technology drive has been to obtain deep structures to give both high efficiency and low electrical capacity.

The original SDD was designed to replace the silicon strip detectors used to locate a particle's position in high-energy physics experiments. For this purpose, the detector thickness, typically ~300 μm, is adequate, but in any case was limited by wafer availability and handling during the planar processing. In increasing the depth, there is a balance between the use of low-resistivity material to form a sufficiently steep potential profile and the high voltage bias needed to deplete it. However, SDDs are now available with a depth of 450 μm giving >90% efficiency for x-rays of energy up to 10 keV [19]. Figure 9.15 shows a comparison between the efficiency of a 450-μm-deep SDD and a typical 10-mm^2, 3-mm-deep Si(Li) detector.

The depth of a Si(Li) crystal is usually designed to increase in proportion to the active area but is normally <5 mm. This present limitation of SDD efficiency due to the exponential falloff in x-ray attenuation (Beer's law) results in an efficiency of only ~20% at 25 keV, but for most applications this is usually *practicable*. Increasing the thickness further presents difficulties in production because of increased leakage current. Stacking two SDDs together has been evaluated [20]. For higher energies and gamma rays, SDDs can be used as photodetectors coupled to scintillators, replacing the more cumbersome photomultipliers [20]. This is at considerable cost in

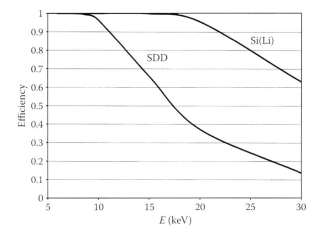

Figure 9.15 Efficiency of an SDD compared with a Si(Li) as a function of energy.

terms of complexity and performance compared to the direct conversion SDDs.

9.3 Droplet SDD X-Ray Detectors

As mentioned above, the central anode (and integrated JFET if used) in the ring structure of an SDD gives rise to ICC. As a result, the peak-to-background (P/B) measurements, for example, on a Fe-55 source (see Section 1.14.3), were rather poor (~1000:1). To overcome this, further symmetry was sacrificed by moving the anode and JFET to the edge of the device (see Figure 9.16) and so, when appropriately collimated, away from x-ray exposure [21]. Figure 9.17 shows a mounted example.

The effect of moving the anode to one edge is to increase the drift time and diffusion spread for a given active area. The example depicted here is a 10-mm² sensitive area SDD, but the droplet geometry also lends itself to arrays that increase the sensitivity and even allows annular designs such as shown in Figure 9.18.

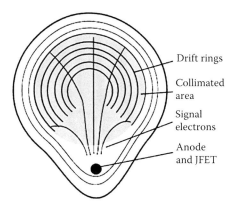

Figure 9.16 Droplet structure. (From Fiorini, C. et al., *NIM B*, 266, 2173, 2008. With permission.)

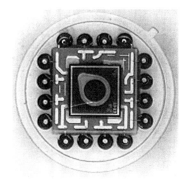

Figure 9.17 Chip arrangement. (From PNSensors website, www.pnsen sors.de. With permission.)

Figure 9.18 Annular SDD array of 'droplet' structures. (From Fiorini, C. et al., *NIM B*, 266, 2173, 2008. With permission.)

An added advantage of the droplet configuration is that because the electrons are approaching the anode from one side only, the integrated FET does not need to be completely surrounded by the anode and can be reduced in size and capacitance. The total capacitance at the amplifier input can be reduced to ~120 fF [19].

9.4 SDD Performance

The performance of x-ray SDDs has improved beyond all expectations since their conception in 1983, and it continues to improve. It is possible that the figures quoted below may be improved upon by the time of publication. The quality of low-energy spectra is now comparable or superior to that of Si(Li) detectors, but it has taken more than 25 years (see Section 11.26). The original workers [3] published a Cd-109 radioactive source spectrum (see Figure 11.78) in 1985 that was already remarkable because it was taken at room temperature. With the proviso of poor high energy efficiency (Section 9.2.6), we can compare the performance of the SDD to that of a Si(Li) detector using the parameters provided by Fe-55 spectra discussed in Section 1.14.3—specifically resolution [full width at half-maximum (FWHM) at 5.9 keV], tail (deviation of photopeak from Gaussian), and P/B (proportion of counts falling into the flattish region between noise and photopeak usually taken about 1 keV)—although we must apply caution that these measurements are defined in the same or even in a similar manner. Added to this, count rate and low energy performance may be important for many applications. Finally, convenience and cost will always be considerations.

The optimum noise and resolution for Si(Li) detectors (~40 and ~127 eV, respectively) are achieved at processing (integrating) times $\tau \sim 40$ μs. We will see below that although results for SDDs are similar (in some cases, better), they are achieved at much shorter τ and hence can have potentially higher input/output rates. The performance of an SDD at high rate (>100 kHz) for EDXRS and imaging in an SEM has been reported [22]. The optimum resolution of a 50-mm^2 SDD was 134 eV at $\tau = 8$ μs and 188 eV at $\tau = 0.5$ μs (rate ~280 kHz). This enabled far superior performance for elemental mapping and comparable performance on resolution although in this case, at least, spectral performance at low energies was inferior to that of Si(Li). As in the Si(Li), the JFET was a discrete element. There has been no serious attempt to integrate a JFET onto a Si(Li) crystal, the main problem being the differential in temperature required (see Section 1.16.4). The improvement in P/B by moving the anode/JFET combination into a screened area at the edge of the SDD in the droplet SDD (SD³) is shown in Figure 9.19 [21] using a Fe-55 source.

This improvement is such that the gross features are now dominated by the physics of the device, as described in Chapter 2, rather than the manufacturing technology and can be compared to a Si(Li) detector spectrum, as shown in Figure 9.20.

The performance of the droplet SDD has been reported in detail [21,23]. The resolution of a 300-μm-deep, 5-mm² droplet SDD is 128 eV at −10°C and 123 eV at −20°C, which is superior to that achieved with Si(Li) detectors although for the equivalent area (10 mm²) it is 147 eV. The low value of the 'dispersion' (see Section 1.10.1) for SDDs is partly attributable to the semiconductor bandgap decrease (see Section 2.4.3) at higher

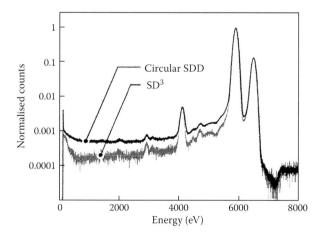

Figure 9.19 Improved background achieved by moving the FET to the side of structure. (From Lechner, P. et al., *Adv. X-ray Anal.*, 47, 53, 2004. With permission.)

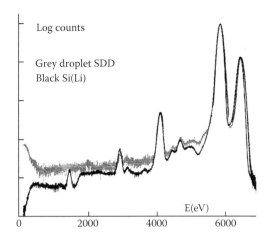

Figure 9.20 Backgrounds in an SDD (grey line) and a Si(Li) (black line).

temperatures. The value of the P/B when it is collimated to ~3 mm^2 is about 6500:1 but more typically 3000:1. This compares to the Si(Li), where it is generally >10,000:1 (Figure 9.20). The high rate performance of the droplet SDD degrades the resolution from ~140 eV at low rate to ~180 eV at 100 kHz input rate using a processing time of 210 ns [17]. By reducing the processing time to 70 ns, the SDD is still able to cope with 1 MHz input rate (400 kHz output rate). Typical process times for a Si(Li) are about an order of magnitude longer.

The excellent low-energy performance of some SDDs has been achieved using implant technology (discussed in Section 4.5). The efficiency above 300 eV is >90%, and the resolutions are equally impressive. For example, the FWHM on CKα (277 eV) can be <48 eV [23–25]. The apparent low value of the dispersion for SDDs suggests that Si(Li) detectors may suffer from ICC at low energies, but the superior tail on MnKα at 5.9 keV (Figure 9.20) in general does not support this assertion. A decrease in electron mobility (see Section 1.7.1) at the higher temperature might explain the lower ICC at low energies. Ongoing improvements in the entrance window technology of SDD are illustrated in Figure 9.21.

This shows a standard p–n window compared to an optimised one, giving a P/B of ~16,000:1. The latter used a tight

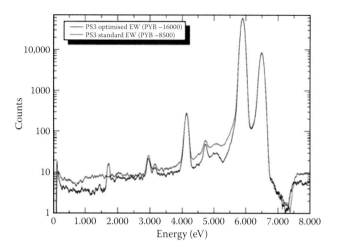

Figure 9.21 Improvements in SDD spectral shapes due to entrance window (EW) technology. (From Niculae, A. et al., *Microsc. Microanal.*, 13, supp. 2, 1430, 2007. With permission.)

collimator integrated directly onto the SDD chip and is compared to a Si(Li) spectrum in Figure 9.22.

The Si(Li) spectrum was taken with a standard collimated 10-mm^2 detector using technology that is now more than 20 years old. The collimator is not in contact with the silicon (0.1–0.2 mm separates them; see Figure 1.62). It is remarkable that such spectra are converging despite the totally different

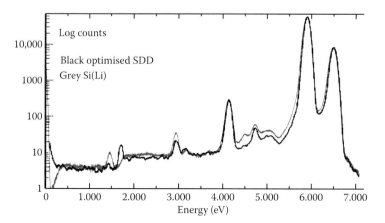

Figure 9.22 Collimated SDD compared to a Si(Li).

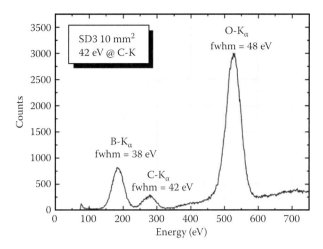

Figure 9.23 Very low energy performance of an SDD. (From PNSensors GmbH, Figure from Guazzoni C., 11th European Symposium on Semiconductor Detectors, June 2009. With permission.)

approaches to the entrance window contact technologies (surface barrier and ion implantation). The tail and resolution for the optimised SDD is superior to that of the Si(Li), although there is a contribution in the latter from photoelectrons from the 10-nm Ni contact at around 4.5 keV (see Figure 2.21). Figure 9.23, shows the excellent performance [20,26] of SDDs for ultralow energy XRS.

9.5 SDD Manufacture

The design of an SDD detector and its associated manufacturing processes are closely intertwined with, and rely on, an exchange of knowledge between the sensor designers and the silicon foundry engineering staff. The detector scientists need to be familiar with the methodology of working with foundries. Questions such as the ownership of processes, masks, and intellectual property need to be clarified early in the project. This is a departure from the traditional in-house processes that have been used for many years to make Si(Li) and HPGe detectors. The 'black magic' used in the manufacture of the latter is largely replaced by the investment and complexity of the many steps used in a planar lithographic process. The processing is long and complicated and involves the accurate alignment of double-sided masks. Moreover, the experts in the silicon foundries need to be persuaded to cooperate with those in the detector industries in what may be a heuristic approach to development. This is very similar to the situation existing among CCD detector manufacturers. Present-day successful commercial SDD manufacturers (see Section 11.28) have relied to date on major polytechnic facilities in Germany (MPI) and Italy (Milan) and in the United States (Stanford University) and government laboratories (Brookhaven National Laboratory).

By the mid 1980s, SDDs had been made at a number of different laboratories. Brookhaven made some small 5-mm^2 cylindrical ones [5] with nonintegrated FETs. SINTEF (Norway) have made SDDs for CERN (CERES experiment). From the first designs at Brookhaven, new designs appeared in Germany at the MPI and one of its industrial partners, a small company called Ketek. Experimental designs were made in the UK at Micron and SemiFab, in France at Enertec, in Finland at Metorex, and in the United States at Berkeley. Eventually, the x-ray market settled on the designs offered by Ketek and PNSensors in Germany, Radiant (now owned by Seiko), and Amptek in the United States. As we have shown, some of these designs have been integrated with the FET used as the first stage of amplification. The mask design and processing is elaborate, and it is interesting to observe that the really successful designs result from companies either embedded in or in close proximity and partnership arrangements with the silicon foundry. At the time of writing, PNSensors and Ketek appear to be the leaders, but useful contributions have come from other companies.

9.5.1 Silicon Wafers

The SDDs used for x-ray spectroscopy are currently made on 500-μm-thick silicon wafers (the original designs were on 280- and 350-μm-thick wafers), and it is likely that future trends will be toward 1000 μm. There are several silicon suppliers who can make silicon of the required quality and specification. Wacker Chemie (Germany), Topsil (Denmark), and Komatsu (Japan) are the leaders. The authors purchased silicon wafers from Wacker Chemie in Germany with the following specification:

N-type silicon wafers 100 mm diameter
$\langle 1:1:1 \rangle$ orientation
500 µm thick
Lifetime greater than 1 ms
Resistivity 6 k Ω cm
Polished both sides

One of the advantages of using a foundry-based manufacturing process is that a large number of devices can be made on one wafer, and devices can also be appropriately scaled to give different sized sensors without the need for changing any parts of the process. The scaling is incorporated into the mask set used at the lithography stage. It is important at this stage to consider the footprint and shape of the die, because individual devices have to be cut from the wafer. For simplicity, we used a 10 × 10 mm die size and left 120-µm spaces between die to allow easy access for sawing at the final stage. This meant that the largest designs filled the individual die footprint, and the smaller designs were located in the center of the die. It is important to stress from the beginning that because the process will be double-sided, the foundry must have good mask alignment technology and methods of protecting sensitive surfaces. This is very important particularly at the sawing stage. Figure 9.24 shows a 100-mm wafer with the various designs laid out, giving 100 die per wafer.

We chose to have two types of sensors on the wafer. One sensor was an experimental p–i–n diode and the other family of sensors were all SDDs but with different areas and numbers of rings. Examples of these individual designs are discussed below.

Figure 9.25 shows a design of an SDD where the side that includes the n⁺ anode in the center is surrounded by 11 concentric p⁺ rings and an outer guard ring. Each ring has an associated track to take the ring out to a resistor chain. This way,

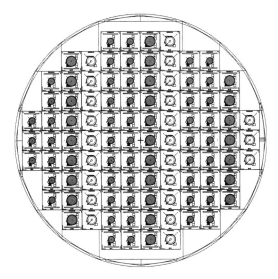

Figure 9.24 Arrangement of SDDs on a wafer.

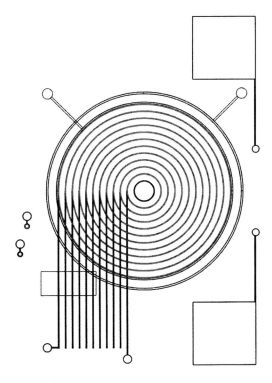

Figure 9.25 SDD architecture.

the bias can be applied to the outer ring and each successive ring can be appropriately biased. The inner ring is grounded through a resistor. The two large square pads are mounting pads for either one or two FETs. Also shown to the left of the lines connecting the rings to the resistor chain are two small contacts to a diode used as a temperature sensor. We chose to investigate several variations of this design:

   7 mm^2 design with 11 rings
   7 mm^2 design with 18 rings
   12 mm^2 design with 11 rings
   12 mm^2 design with 18 rings
   12 mm^2 p–i–n diode

   Part of the design goal was to achieve a collimated area of either 5 or 10 mm^2 for the two devices and to test the concept of depleting the volume from both surfaces to a small central anode. Using a discrete FET allowed us to simplify the processing and to investigate basic parameters such as leakage current, required voltage for depletion, ring bias and its relationship with punch-through currents. Having the two mounting pads allowed us to compare results using different FETs, where the change was made by a new anode to gate connection.

9.5.2 Ring Designs

With 6 kΩ cm n-type silicon, we expect the volume to be fully depleted when 60 V is applied simultaneously to the top face and the outer ring. The resistor chain should ensure that on the 11-ring detector, we get a voltage drop of 5.45 V per ring and a field of 0.54 kV/cm. This is much less than the upper limit of 3.6 kV/cm suggested by Rehak et al. [6]. Figure 9.26 shows the layout and ring dimensions.

   We chose quite a wide ring spacing (100 µm) for a low value for the field (less than 54 mV/µm) in the drift channel. Carrier velocity saturation at room temperature in silicon is ~10^7 cm/s, and this occurs at fields in the range of 10–100 kV/cm (see Figure 9.27 [27]).

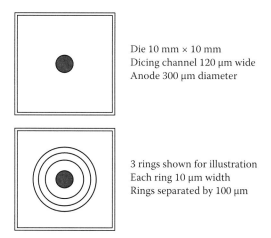

Die 10 mm × 10 mm
Dicing channel 120 μm wide
Anode 300 μm diameter

3 rings shown for illustration
Each ring 10 μm width
Rings separated by 100 μm

Figure 9.26 SDD ring designs.

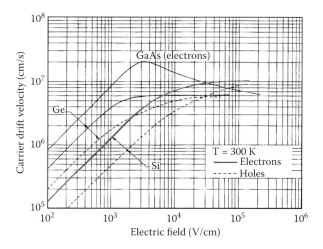

Figure 9.27 Carrier drift velocity in various semiconductors at room temperature. (Sze, S.M.: *Physics of Semiconductor Devices*. 1981. Copyright Wiley-VCH Verlag GmbH & Co. KGaA. Reproduced with permission.)

This gave the flexibility of increasing the ring voltage at the expense of the front face voltage if we so chose. Even in the reduced field of 0.54 kV/cm, the signal rise-time would be significantly less than the charge integration times in the following amplifier.

This experimental SDD was part of a feasibility study looking at the design and manufacturing trade-offs needed when working with a semiconductor foundry. In some ways, the program was similar to the familiar steps used in making a JFET. The results obtained were quite encouraging with a resolution on Fe-55 of better than 150 eV and a P/B of 3000:1. However, the yield on the wafer of good working devices was quite low, and we abandoned the program in favour of the very cost-effective solutions being offered by the companies that had established themselves in close proximity to a foundry (PNSensors and Ketek at MPI). Again, this gave us an indication of how the industry has changed—moving away from the 'cottage industry' concept of making Si(Li) and HPGe detectors in semiclean room conditions and moving into properly controlled semiconductor foundries having a range of processing tools including ion implantation, double-sided processing, simulation software, and intricate mask design.

9.6 SDDs—Summary

The SDD is an extremely ingenious device making use of the modern technologies developed in the extensive silicon microelectronics foundries. It moves detector manufacture away from the traditional wet-chemistry small batch process toward mass production. As we have seen, the SDD performance is impressive. Furthermore, they do not require liquid nitrogen and can be easily incorporated into portable and online industrial instruments. Much credit must go to Technical University Munich and the German companies Ketek and PNSensors for their investment and perseverance with the technology (see Chapter 11). Further work is needed to improve high energy efficiency, and deeper devices are being progressed. Alternative materials to silicon are also being considered (see Chapter 10). Background fluorescence peaks from heavy materials behind the SDD at high energies, are a particular problem for EDXRF. The SDD has replaced the Si(Li) detector for many applications, but it remains to be seen whether it will replace them completely.

References

1. Gatti E. and Rehak P., 2nd Pisa Meeting on Advanced Detectors, 1983.
2. Gatti E. and Rehak P., *NIM* 225, 608, 1984.
3. Rehak P. et al., *NIM* 235, 224, 1985.
4. Kemmer J. et al., *NIM A* 253, 378, 1987.
5. Jalas P. et al., *IEEE* NS-41, 4, 1048, 1995.
6. Rehak P. et al., IEEE NS-36, 203, 1989.
7. Rawlings K.J., *NIM A* 256, 297, 1987.
8. Luke P.N. 1988, *NIM A* 271, 567.
9. von Harrach H.S. et al., *Microsc. Microanal.* 15, suppl. 2, 208, 2009.
10. Kemmer J., *NIM* 226, 89, 1984.
11. Kemmer J. et al., *NIM A* 253, 365, 1987.
12. Chew et al., *IEEE Trans. Microwave Theory Tech.* MTT-28, 2, 1980.
13. Rawlings K.J., *NIM A* 253, 85, 1986.
14. Gatti E. et al., *NIM* 226, 129, 1984.
15. Radeka V. et al., *IEEE Electron Devices lett.* ED-10, 91, 1989.
16. Pinotti E. et al., *NIM A* 326, 85, 1993.
17. PNSensor website, www.pnsensors.de.
18. Longoni A. et al., *IEEE* NS-49, 3, 1001, 2002.
19. Fiorini C. et al., *NIM B* 266, 2173, 2008.
20. Schlosser D.M. et al., *NIM A* 624, 270, 2010.
21. Lechner P. et al., *Adv. X-ray Anal.* 47, 53, 2004.
22. Newbury D.E., *Scanning* 27, 227, 2005.
23. Niculae A. et al., *Microsc. Microanal.* 13, suppl. 2, 1430, 2007.
24. Oxford Instruments Plc. website, www.oxford-instruments.com.
25. Ketek GmbH website, www.ketek.net.
26. PNSensors GmbH, Figure from Guazzoni C. 11th European Symposium on Semiconductor Detectors, June 2009.
27. Sze S.M., *Physics of Semiconductor Devices*, John Wiley & Sons, 1981.

10

Wide Bandgap Semiconductors

We define wide bandgap semiconductors (WBGSs) as those semiconductors having a wider bandgap than Si at room temperature (1.12 eV). In all cases except elemental diamond © they are compounds. We briefly touched on the properties of the more commonly used compounds and WBGS in Chapter 1 (see Tables 1.1 through 1.4). There we compared the mobility–lifetime product $\mu\tau$ and dispersion (or 'Fano noise') of some materials, rather unfavourably with those of Si and Ge. Recall that the carrier drift length L ($\mu\tau$ product/unit applied field) appears in the Hecht equation to describe trapping and recombination. The higher bandgap results in higher average energy per electron–hole pair values, ω, and therefore potentially poorer energy resolutions. They are also mostly indirect bandgap semiconductors (see Section 1.7.3), so that direct recombination can take place and charge lifetime τ is short. So what are the attractions of the WBGSs? First, being wide bandgap, the thermal leakage current, being dominated by the term $\exp(-E_g/2kT)$ (see Section 1.11.1), is much reduced compared to Si at room temperature. This presents the possibility of x-ray spectroscopy without the inconvenience and cost of cooling to low temperatures. It has obvious attractions for portable radiation survey equipment and industrial energy dispersive x-ray fluorescence (EDXRF) applications. Other opportunities open up, for example, medical, security, and nonproliferation applications. Added to this, the stopping power, being largely determined by the highest atomic number (Z) element in the material, is—with the exception of diamond and silicon carbide (SiC)—as good as, or better than Ge so that very thin detectors can be used for x-rays, thus reducing the bulk leakage current and partly compensating for a short carrier drift length. Higher Z not only increases the stopping power as mentioned, but gives a higher Peak/Compton scattering ratio for high energy x-rays and gamma-rays, which is where the main applications currently lie for most of them. However, the large escape peaks and fluorescence peaks from the high Z materials present can be problematic for spectroscopic analysis (see Section 1.14.1). Figure 10.1 [1] shows the detection efficiency of x-rays as a function of energy for 500-µm-thick detectors for a number of WBGSs (with Si for comparison).

The intrinsic resistivities are naturally high at room temperature, being usually in the range of 10^7–10^{14} Ω cm (cf. Si ~ 10^4 Ω cm), and enable high fields to be applied with low current, partially compensating for the low values of µτ. Radiation tolerance is enhanced because the energy of defect formation in WBGS is larger than that in Si and Ge. This is particularly true in the case of thin detectors (having reduced

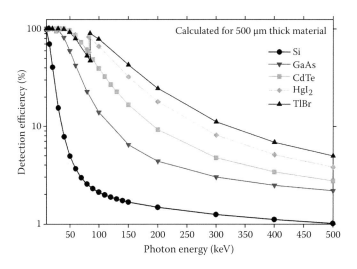

Figure 10.1 **(See colour insert.)** Detection efficiency of a range of WBGS compared to Si. (From Sellin, P.J., Development of Semiconductor Gamma-ray and Neutron Detectors for Security Applications (presentation pdf: Dept. of Physics, Univ. of Surrey, UK), 2009. With permission.)

volume) and makes them attractive for high-energy physics (HEP), military, and space applications [2].

Because of the reasons given above, there is a great deal of activity in the development of WBGS detectors. The progress is documented in the proceedings of the series workshops on 'Room Temperature Semiconductor X-ray and Gamma-Ray Detectors', held every 4 years or so and reported in the journal *Nuclear Physics and Methods in Physics Research* (North Holland). A recent book, *Compound Semiconductor Radiation Detectors* [3], gives more detail and reviews the present status. Compound semiconductor radiation detectors have a very long history, in fact predating that of the elemental semiconductors, Si and Ge. 'Crystal detectors' (see Section 11.4) were studied in 1945, moving on to Si and CdTe around 1957 (for early references, see, e.g., the *Asheville Conference Proceedings, 1960*, p. 194) [4]. There have been a number of other excellent reviews on the status of WBGS detectors over the years (e.g., [5–12]), and these and other publications show how the interest in various materials has risen and fallen (and sometimes risen again as with HgI_2 and CdTe). Whereas Si (and to some extent, Ge) detectors rode on the back of the advancements in the materials for the microelectronics industry, the progress has been slow and incremental for the other materials—although with interest in applications for light-emitting devices and solar energy conversion, this trend is changing. At present, CdTe, CdZnTe (CZT), GaAs, and HgI_2 still show the most promise for WBGS for x-ray spectrometry, a situation that has lasted more or less for the past 30 years [6]. CdZnTe dominates room temperature energy dispersive x-ray spectrometry (EDXRS), at least commercially, and there is a major driving force for better material for security applications. The near-field (or small pixel) effect in coplanar grid detectors [13] and pixellated x-ray detectors (PXDs) [14] has given them a new lease of life.

10.1 Candidates

In Section 1.3.2, we showed how semiconductors are formed from the group 4 elements (C, Si, Ge), or binary compounds of them (SiC, Si:Ge), as well as **3–5** compounds (such as GaAs) and **2–6** compounds (such as CdTe). They can also be formed from group 7B binary compounds (such as the heavy metal halides HgI_2, PbI_2, and TlBr). Ternary compounds, such as HgCdTe and InGaAs (which are not WBGS), have been investigated but, being more complex, intrinsic defects in the materials are problematical. Incidentally, CdZnTe (mentioned above) is not a ternary compound but an alloy of CdTe and ZnTe, where the proportion of Zn replacing Cd can be varied continuously along, to a certain extent, with the bandgap energy.

For the elemental semiconductors in group 4, the melting point and bandgap decrease with atomic number Z, and the only elemental WBGS candidate is C in the single (or, more likely, polycrystalline) form, diamond. One recent allotrope of carbon, graphene, a material much in the news, is formed as an atomic layer of carbon atoms arranged in a honeycomb structure and does not have a bandgap. Carbon, in its diamond form, immediately signals problems in terms of cost, volume, and stopping power for x-rays (although its density exceeds that of Si). But, as the energy gap is very large (5.4 eV), it is a serious contender for radiation detectors operating at high temperatures in high radiation environments such as found in HEP, the military, and the nuclear industry. It is also very stable, hard, and resistant to corrosion.

Table 10.1 shows the physical properties at room temperature of most of the WBGSs that have been investigated to date. The order in Table 10.1 represents the degree of success they have achieved in terms of practical stable media for x-ray detection and will be the order in which they will be discussed in the following sections.

Although Table 10.1 presents the physical properties, it is often the chemistry or physical chemistry rather than the physics that dominates the degree of success, both with regard to the crystal growth and to the detector fabrication. Handling problems—for example, due to brittleness, softness, or layering—and stability of operation (so-called 'polarisation') are important factors. The toxicity of materials (Cd, Hg, Tl, As) is also a factor. The stability (time-dependent effects) in terms of resolution and gain has been a long-term problem yet to be fully understood and solved.

TABLE 10.1
Properties of Wide Bandgap Semiconductors (WBGSs)

Semiconductor	Formula	Average Atomic Number, Z	Density, ρ (g/cm³)	Bandgap (eV)	Energy per Electron–Hole Pair (eV)	$\mu\tau$ Product Electrons (cm²/V×1M)	$\mu\tau$ Product Holes (cm²/V×1M)
Cadmium zinc telluride	$Cd_{0.9}Zn_{0.1}Te$	49.1	5.78	1.57	4.64	4000	120
Cadmium telluride	CdTe	50	5.85	1.44	4.43	3000	200
Gallium arsenide	GaAs	31.5	5.32	1.43	4.2	80	4
Mercuric iodide	HgI_2	62	6.4	2.15	4.2	300	40
Lead iodide	PbI_2	63	6.2	2.32	4.9	10	0.3
Silicon carbide	SiC	10	3.21	3.26	7.8	400	80
Silicon germanium alloy	Si:Ge			0.67–1.1			
Diamond	C	6	3.5	5.4	13.25	20	15
Thallium bromide	TlBr	58	7.56	2.68	6.5	500	2

10.2 General Comments on WBGSs

We now consider some of the commonalities of various WBGSs.

10.2.1 Ultimate Resolution

The dispersion D appears in the resolution R formula (Equation 1.35)

$$R(E)^2 = D(E)^2 + N^2$$

and is given, as we saw, by Equation 1.34

$$D(E)^2 = 2.355^2(F\omega E)$$

The relationship between the pair creation energy ω and the bandgap E_g is not the same for all semiconductors. Si, Ge, diamond, and SiC are indirect bandgap transition semiconductors (see Section 1.7.3), but the other WBGSs considered are direct bandgap transition semiconductors. However, as

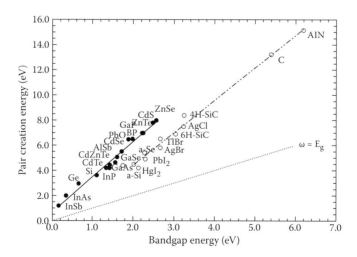

Figure 10.2 Electron–hole creation energy vs. bandgap energy for a range of semiconductors. (From Owens, A., Peacock, A., *NIM A*, 531, 18, 2004. With permission.)

Figure 10.2 shows, for all the semiconductors roughly only ~1/3 of the x-ray energy is converted to charge generation, with the remainder going into phonon generation. In detail two distinct branches are found [11] as shown in Figure 10.2.

Also shown is the limiting line, where $\omega = E_g$. We see that diamond, SiC, and the heavy metal halides HgI_2, PbI_2, and TlBr have high value bandgaps combined with relatively low pair creation energies. This gives them the advantage of low leakage current and potentially good resolution. The degree to which cooling below room temperature is required for spectrometry decreases with bandgap energy and only those above ~1.5 eV can be regarded as 'room temperature' detector materials. Estimates of the Fano factor have only been made for a few WBGSs, but these are not expected to deviate far from the range 0.10 to 0.14, and an average of 0.12 might be used to predict the dispersion D of any of them. Figure 10.3 shows this for some WBGSs at 5.9 keV with Si and Ge added for comparison.

Recall (Section 1.7.4) that the drift length L of a carrier in a field E is given by

$$L = \mu\tau E$$

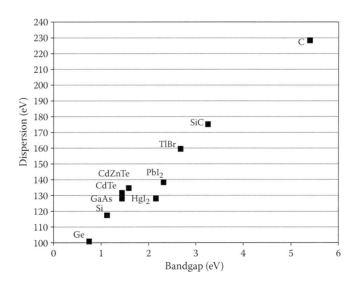

Figure 10.3 Dispersion for a range of WBGS compared with Si and Ge together with their bandgaps.

Unlike Si and Ge detectors, incomplete charge collection (ICC) usually dominates the resolution R in WBGS detectors except in the case of very thin detectors. For a planar detector the latter will, however, have increased capacitance and increased noise N.

The Hecht equation predicts the peak broadening and tailing for a given field E and $\mu\tau$ product (see Table 10.1), which depends on the material quality (impurities, poor stoichiometry, plastic deformation, dislocations, etc.). It does not include the detrapping of charge which, if not occurring within the charge integration time, will give a 'slow charge component' to the signal. Thus, the product $\mu\tau$ can be regarded as a 'quality factor' for the WBGS. To take into account broadening due to ICC in WBGS detectors, an extra important component can be introduced [11]

$$R(E)^2 = D(E)^2 + N^2 + aE^{b2}$$

The values of a and b are determined by the density of trapping centres in the material, and the last term is usually very small in Si and Ge. In order to maximise the field E, the resistivity ρ and electrical breakdown potential should be high. The electrical properties of the WBGS at room temperature are tabulated in Table 10.2 (where Si and Ge are included

TABLE 10.2
Electrical Properties of WBGSs

Semiconductor	Formula	Dielectric Constant	Resistivity (Ω cm)
Silicon	Si	11.9	1.00×10^4
Germanium	Ge	16	50
Cadmium zinc telluride	$Cd_{0.9}Zn_{0.1}Te$	10	1.00×10^{11}
Cadmium telluride	CdTe	10.9	1.00×10^9
Gallium arsenide	GaAs	12.5	1.00×10^7
Mercuric iodide	HgI_2	8.8	1.00×10^{13}
Lead iodide	PbI_2	21	1.00×10^{14}
Silicon carbide	SiC	9.2	1.00×10^{16}
Diamond	C	5.7	1.00×10^{14}
Thallium bromide	TlBr	29.8	1.00×10^{12}

for comparison). For thin detectors, the capacitance must also be considered because it may give rise to significant added noise. In this respect, the dielectric constant is a factor to be considered.

10.2.2 Spectral Enhancement Techniques

In the absence of crystal growth techniques to improve the hole $\mu\tau$ product (see Table 10.1) of the material, the only alternative has been to discard the hole signal. This results in a 'single carrier' or 'unipolar' detector, and there are a number of techniques by which this can be approached. For thick detectors and low-energy x-rays, the difference in $\mu\tau$ for holes and electrons renders the detectors to some extent single carrier devices by default. Better results are obtained when the negative bias electrode is also the x-ray entrance window, that is, any contribution from holes is minimised. For more penetrating x-rays or gamma rays, other techniques can be used. The signals from the holes will in general have a longer rise time, and this can be detected and the signal rejected by the amplifier electronics (rise time discrimination) [15–18]. The tailing and resolution are improved by rise-time discrimination, but this is at the expense of some loss in efficiency. Pulse rise-time correction [19–21] has also been used. This relies on a relation between the rise time and reduced amplitude due to ICC and has obvious limitations. The geometry of the crystal or field can also be chosen to enhance the electron signal, for example, by choosing a hemispherical or 'quasi-hemispherical' geometry [19,22,23], coaxial [24], or trapezoidal geometry [25,26]. With the outer surface entrance contact as cathode, the hemispherical geometry increases the volume from which electrons are collected and the field increases quadratically toward the end of their trajectory. The coplanar grid structure [13] has been described (see Figure 1.29b). Here, the anodes are arranged into interdigital grids, effectively making two planar detectors in one crystal or a 'coplanar grid' structure. A bias is applied between the grids so that the electrons move preferentially to one set rather than the other. The signal induced on one set of grids is subtracted from the signal on the other set. As the component of the signal induced during the motion of the holes and electrons, by virtue of the Shockley–Ramo theorem, in the bulk of the crystal will be the same for both sets of grids, the difference will essentially be due to the *electrons* as they approach the anode. This results in signals that are attributable to electrons only and are insensitive to where the charge was originally generated in the volume. The coplanar grid is analogous to the Frisch grid used in gas proportional detectors. An example of how effective the technique can be for gamma rays that penetrate to an appreciable depth is given in Figure 10.4. Here, spectra from a simple planar structure are compared to that from a coplanar grid structure in a 10-mm-deep CdZnTe detector.

The concept is very similar to the 'small pixel' effect [14] (Section 7.3.3) that occurs in PXDs. In this case, the signals are largely generated by the electrons as they approach the small pixel. Their effectiveness as a single-carrier detector has been demonstrated for CdZnTe (see below). PXD detectors, with small pixels, suppress the background due to hole trapping and have low noise, but one of the penalties is that they introduce artefacts (fixed photopeaks) resulting from fluorescence x-rays generated in surrounding pixels. In the case of

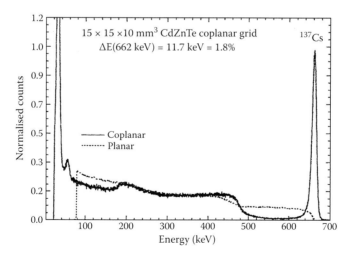

Figure 10.4 Planar detector Cs-137 spectrum compared to that from a coplanar detector. (From Owens, A. et al., *J. Synch. Rad.*, 13, 143, 2006. With permission.)

CdZnTe, the lines of Te (27.5 KeV, 31 KeV), Cd (23 KeV, 26 KeV), and Zn (8.6 KeV, 9.6 KeV) will be prominent if the x-ray energy is high enough to excite them. This is, of course, in addition to the usual escape peaks that move in relation to the parent photopeak energy. We have already seen in previous chapters that CCDs and SDDs are 'single carrier' detectors by their nature. There has been some success in making such devices from WBGSs with a GaAs CCD [28,29], a HgI_2 drift detector [30], and a CdZnTe drift detector [31]. These will be discussed further under their material headings.

10.2.3 Polarisation

The so-called polarisation or the degradation of spectra with time, radiation intensity, and intensity–energy product, has plagued WBGSs from the start. It can have two sources. First, the trapping of holes in the bulk causes a build up of space charge and a collapse of the electric field, and second, the accumulation of charge at the contact interfaces that can give rise to field collapse and instabilities as we have seen even in the cases of Si and Ge (see Section 4.6.4). The susceptibility of the various materials to polarisation will be discussed under their individual headings.

10.2.4 Crystal Growth

The success of semiconductor detector manufacture hinges on the technology and availability of the starting material. In Chapter 11, we will see that, despite the huge investments made in the microelectronics industry, detector-grade Si, Si(Li), and HPGe detectors did not come easy. Ideally, materials need to be chemically pure, have perfect crystallinity, and have exact stoichiometry. Strain caused by any lattice mismatching should be minimal (<15%). In some cases, suitable doping techniques need to be developed. Whereas Si and Ge are grown by Czochralski methods (see Section 1.16), this cannot be used directly for WBGSs. In fact, there are nearly as many growth techniques as there are WBGSs, and we will discuss briefly those preferred for each one under their material headings.

With modern epitaxial growth techniques such as chemical vapour deposition (CVD) and molecular beam epitaxy, it is literally possible to grow the materials one atomic layer at a time and technically feasible to engineer the properties for any application. The bandgap and melting point increase with the *ionicity* (see Section 5.1.9) of the compound. These are lowest for the group 4 elemental semiconductors (essentially covalent), increasing in the **3**–**5** compounds, and highest in the **2**–**6** compounds. The bandgap also increases as the lattice constant decreases (see Figure 1.9). This can be achieved by changing the composition of an alloy, for example, in $Cd_{(1-x)}$ $Zn_x Te$, where x is continuously variable. The resolution is optimal at 300 K for $x \sim 0.7$ and for moderate cooling (–30°C) for $x = 0.1$. This is called 'bandgap engineering'. The addition of Zn is also found to lower the defect density.

10.2.5 Contacts

For WBGSs with very high resistivity (such as HgI_2), we might expect to be able to treat the material as 'semi-insulating' and to be able to form a simple photoconducting type detector with the application of ohmic contacts, allowing the free flow of electrons and holes from the contacts into the bulk. The current–bias (I/V) characteristic will be linear, and a high field can be applied. Even though the holes may be quickly trapped, the electron signal should result in good spectroscopy. Indeed, it has been argued [32] that such a model would result in better signal rise time than in the absence of hole trapping. The faster rise time allows shorter integration times and reduction of current noise. In such a model with no rectifying junctions, the polarity of the bias in respect of leakage current will be immaterial. A further advantage would be that ohmic contacts provide a mechanism for the freeing of charge trapped in deep levels in the energy gap, thus avoiding polarisation problems. In practice, truly ohmic contacts are very difficult to achieve. The interfaces generally pin the Fermi level so that the barrier heights are independent of the metal work function. The situation may be improved by greater control in modern CVD techniques.

The lower resistivity materials (such as CdTe and GaAs) can be grown or doped to produce p-type or n-type material. The leakage current can be reduced considerably by using a rectifying (blocking) contact such as that achieved with a Schottky, implanted, or diffused junction contact. Recall that in Si, for example, where the resistivity is relatively low, it is always necessary to use a blocking contact. Then, with the correct polarity bias, a depletion layer is formed which, because of the high field across it, forms most of the 'active volume' for x-ray detection. However, in many WBGS detectors, it is difficult to achieve a depletion depth of more than a few 100 μm, which is adequate for low-energy x-rays but not for the penetrating x-rays (and gamma-rays) that will interact in the undepleted region. Even the lower resistivity WBGS will deplete with just the 'built-in' contact potential. When bias is applied, they can act as 'quasi-ohmic' or 'resistive' [33]. Whatever type of contact is used, it is essential that it be noninjecting. The contacts that have been used in practice will be discussed for the individual WBGS headings below.

10.2.6 Passivation

The passivation of surfaces of WBGSs is not the issue that it is for Si and Ge. Leakage currents are usually dominated by bulk rather than surface conductivity. As long as the surfaces are clean, the dangling bonds (see Section 5.1.9) redistribute the charge at the surface by distortion and accompanying surface strain. As a result, the surfaces are found to be virtually flat band with few surface states. Oxide growth is also much slower. For example, for GaAs, the initial sticking coefficient for oxygen is ~2×10^{-5} compared with 5×10^{-2} for $\langle 111 \rangle$ silicon. Particular passivations, when used, will be mentioned under the WBGS headings below.

10.3 Present Status of WBGS X-Ray Detectors

In the following subsections, we review the present status of WBGS for x-ray detectors individually. Although we have attempted to include the most recent results, it is inevitable—because of the rapid advances being made in certain areas (in particular, PXDs, drift detectors, molecular engineering of contacts, etc.)—that there will be some omissions. The reader is encouraged to follow this progress by consulting some of the more recent work by the authors referenced.

10.3.1 CdTe, CdZnTe, and CdMnTe

CdTe was first investigated for x-ray and gamma-ray detectors around 1957 by Russian researchers, but it was over a decade later before any energy spectra were reported [34,35] (see Section 11.8). The main problem was the high melting point (1041°C) and the high vapour pressure of Cd that causes poor stoichiometry. One method used to overcome this problem is to dope with Cl. This drives the material toward n-type although it usually still remains p-type. The processing normally involves etching in 2%–5% Br–methanol solution (damage etch) followed by 2% Br/20% lactic acid (polish etch) and quenching in methanol. Edge passivation (if used) involves the application of a dielectric because there are no stable native **3–5** oxides.

Improvements were incremental [36–38], giving resolutions ~7.5 keV at 60 keV (Am-241 is a convenient 60 keV gamma-ray source), until the alloying with ZnTe to form CdZnTe at Aurora Technologies in 1992 [39], which gave a resolution of ~3.6 keV at 60 keV. The material and Peltier cooled detectors are now commercially available [40–45]. Using rise-time discrimination detectors give resolutions of 930 eV at 60 keV and sub-300 eV resolution at 5.9 keV. Amptek literature, for example, shows resolutions of 600 eV at 60 keV. A CdZnTe detector was incorporated into ROVER, the mobile x-ray detector on the Mars Pathfinder Mission in 1997. Figure 10.5 shows both Am-241 and Fe-55 spectra taken from a 3×3 mm^2 2.5-mm-thick CdZnTe detector.

Resolutions from 2-mm-thick CdZnTe detectors at −40°C used with low noise electronics of better than 250 eV have been reported (240 eV [18], 249 eV [47]). Figure 10.6 [48] shows the Fe-55 spectrum from a $5 \times 5 \times 2$ mm detector using Digirad material at −40°C. The Peak/Background exceeds 200:1.

Peltier cooled crystals for hard x-rays are commercially available [49]. CdZnTe is normally grown with a high pressure

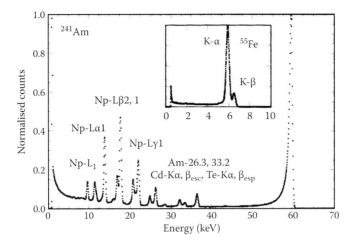

Figure 10.5 Response of a 2.5-mm-thick CdZnTe detector to radiation from Am-241 and Fe-55. (From Owens, A., *Compound Semiconductor Radiation Detectors*, Taylor & Francis, CRC Press, 2013. With permission.)

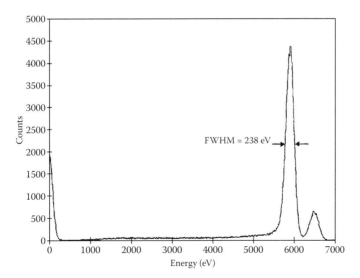

Figure 10.6 Response of a 2-mm-deep CdZnTe detector cooled to −40°C made from material supplied by Digirad. (From Bale, G., Ph.D. thesis, Univ. of Leicester, UK, 2001. With permission.)

Bridgman (HPB) technique [50]. Typically, a pressure of 100 atm is used to grow ingots of up to ~10 kg. The addition of Zn increases the bandgap by more than 100 meV and also the resistivity (by 1 to 2 orders of magnitude), such that it does not necessarily require doping [32]. As an added bonus, the mechanical strength is also improved and the number of traps reduced so that it suffers less compared to CdTe from polarisation effects, although this latter fact has been contested for the more recently grown CdTe (see below).

Various contacts to CdZnTe have been discussed [51], but the general consensus is that Au Schottky contacts are preferred. It does not appear to matter how the Au is deposited (electroless plating or evaporation). They give reproducible noninjecting contacts and low leakage currents with negligible generation–recombination noise. Detectors with an insulator passivation are also found to have lower $1/f$ noise.

A 4 × 4 array CdZnTe PXD developed at the Arizona University [52] gave a resolution of 1.2 keV at 60 keV. A hybrid PXD has been manufactured at Caltech [53] using the Arizona University ASIC (application-specific integrated circuit). Indium bump bonding was used on the $8 \times 7 \times 2$ mm^3 CdZnTe array, which also gave 1.2 keV resolution at 60 keV at room temperature and 670 eV when cooled to −10°C. The In photopeak would, of course, be a spectral contaminant in EDXRS applications. The CdZnTe PXDs developed by NASA have been reviewed [54]. A CdZnTe drift detector, albeit a very simple one, has recently been described [55]. The crystal has dimensions of $5 \times 5 \times 1$ mm^3 and has just two rings surrounding an 80-μm-diameter central anode. The preliminary results are encouraging, giving a resolution of ~660 eV at 10 keV and ~840 eV at 60 keV.

Other alloys of CdTe have been investigated, for example, at Brookhaven National Laboratory, the most interesting one being CdMnTe [56,57]. It is produced by a low-pressure Bridgman technique (LPB) rather than HPB as in the case of CdZnTe and is thus less expensive. By varying the Mn proportion, a much larger range of bandgap (up to 160 meV taking it to ~1.60 eV) is available. X-ray diffraction shows that the lattice constant decreases linearly, and the energy gap (and ω) increases linearly with the addition of Mn. However, the energy gap changes by more than twice that for Zn and, as a consequence, less Mn than Zn is required. The resistivity and $\mu\tau$ product is comparable to CdZnTe, but the homogeneity is better [58]. Although preliminary results have been poor (~21 keV resolution at 60 keV), this material still holds great promise. One problem is the availability of the material. Longxia Li's company [59], which was founded in 1997 is the only source. He is now a consultant on the work being carried out at Brookhaven National Laboratory.

With improved crystal growing techniques, such as the travelling heater method (THM), CdTe has recently made a comeback. The relative merits are still controversial [60], but a number of companies [61–65] now specialize in its growth and use as x-ray detectors [43,63]. It is claimed [63] that the homogeneity and stability are superior to CdZnTe although the latter has been disputed [10]. Planar detectors are available as semi-insulator (Pt contacts on both faces) or as Schottky diodes (Pt and In contacts). The latter have produced excellent spectral performance. Figure 10.7 shows the results from an Amptek [43] 1-mm-thick CdTe detector. The resolution is 290 eV at 5.9 keV and 800 eV at 60 keV (600 eV with rise-time corrector).

A hybrid CdTe PXD has recently been manufactured [66] using an Acrorad 12 × 12 array of 200 µm pixels. The authors experienced problems with digital-to-analogue interference, but with an external trigger they achieved a resolution of 870 eV

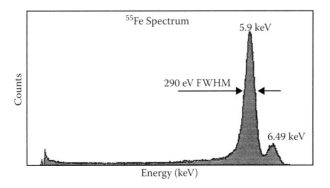

Figure 10.7 From Amptek website. Specification Data Sheet. (From Amptek, Bedford, MA, USA. With permission.)

at 60 keV at −20°C. Interference shields are planned for the next version.

10.3.2 GaAs

GaAs has a similar stopping power to Ge and was first used for gamma-ray detectors using material from TI and Merck in 1960 [67]. GaAs ingots were first grown by liquid encapsulated Czochralski, but better results were achieved with epi-GaAs [8,68,69]. For example, 60-μm-thick detectors with Au contacts were manufactured in 1970 [68] and demonstrated a resolution of just less than 1 keV at 60 keV at a temperature of 130 K. In recent times, the technology of growth, processing, and contacting has been more advanced than for other WBGSs because of parallel developments in the mainstream electronics industry. Here, wafers are used as substrates and devices commonly used in solar cells and high-speed electronics. GaAs is attractive because of the exceptional mobility of electrons μ (8500 cm^2/V s compared to the value for Si of 1500 cm^2/V s). The mobility of the holes is similar to that in Si. With GaAs, there is also the future possibility of on-chip processing. The energy gap is similar to that of CdTe, but unfortunately the charge lifetime τ is only of the order of ~10 ns, so the μτ product is rather poor (see Table 10.1). The resistivity is also rather low (see Table 10.2), and the very high value of electron mobility does not compensate for this.

Recently, thin detectors of high-quality epi-GaAs have been grown by CVD in a collaboration involving the European Space Agency (ESA) [12]. These are grown on an n$^+$ substrate of up to 40–400 μm thickness with a p$^+$ layer (patterned to make a guard ring or pixel structure) deposited on the top face to complete a p–i–n diode. The leakage current at room temperature, with a bias of 100 V, was <4 nA/cm^2. Figure 10.8 shows spectra from a 1-mm-diameter, 40-μm-thick detector at −20°C with 160 V bias.

The resolution (~400 eV) is just sufficient to resolve the Mn Kα and Kβ lines and at 60 keV is 670 eV. PXDs that demonstrate even better resolution have also been manufactured [70–72]. For example, a 5 × 5 mm^2 prototype of 40-μm-thick epi-GaAs also developed for ESA gave a resolution of 266 eV at

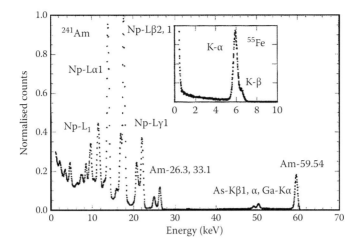

Figure 10.8 Spectrum of both Am-241 and Fe-55 from a GaAs detector 1 mm diameter, 40 μm thick, at −20°C with 160 V bias. (From Lumb, D.H. et al., *NIM A*, 568, 427, 2006. With permission.)

room temperature and 219 eV at −31°C. The room temperature resolution at 60 keV was 487 eV, superior to that of CdZnTe, and the photopeaks are Gaussian up to ~60 keV. A Fano factor of 0.12 [70] to 0.14 [71–72] was estimated from this work.

Schottky-type contacts are made to epi-GaAs detectors by metal deposition (Au/Pt/Ti), with the ohmic contact (to the n+ substrate) made noninjecting with a diffused layer of Ge [73]. It is possible that contacts to GaAs (and to many WBGS) may be improved by engineering the barrier height using an 'interface control layer' [74]. The radiation hardness of GaAs detectors has been discussed [69], but polarisation under normal operating conditions for x-rays has not been reported to be a problem.

10.3.3 HgI$_2$

Mercuric iodide was first used as the medium for gamma-ray detection in 1971 [75] and as a gamma-ray spectrometer in 1974 [76]. Vapour phase methods of crystal growth were shown by the EG&G Energy Measurements group at Santa Barbara, CA, USA, to be superior to solution (KI) growth methods [77]. The method was further refined at Purdue

University, USA [78]. Thin x-ray detectors are still most commonly made from small 'platelets' formed by time-consuming multiple sublimations of the red crystals in a sealed evacuated ampoule (vertical ampoule method) [79]. These were the room temperature detectors of choice during the 1980s and 1990s that were taken up commercially by a number of companies such as Kevex and Tracor (Spectrace) and, despite several technical problems (see below), are still supplied commercially, for example, by TN-Technology (Thermo Scientific)—The Metallurgist, a handheld XRF alloy metal analyzer; Spectrace9000 (also part of Thermo Scientific), a field portable XRF system for the analysis of soils, thin films, and lead in paint; and Mercury Module (Constellation Technology Corporation) [80]. The latter company has recently grown crystals up to 9 mm thick epitaxially, again using vapour transport, with promising spectroscopic results [81]. Most of the instruments mentioned are used for x-ray energies above ~2 keV with resolution of 300–500 eV. In Europe, ETH (Zurich, Switzerland) and Eurorad (France) produce material for research organisations. There are many problems associated with the growth, storage, and application of HgI_2. First, the growth of large crystals is complicated by the low melting point (~260°C) and a low-temperature solid phase transition at 127°C that prohibits it from being grown from the melt. It has high vapour pressure and is prone to evaporation (especially under vacuum) and sensitive to relative humidity [82]. A passivation (e.g., paralene; Union Carbide), or a-C and good encapsulation is essential. All heavy metal iodides have a layered structure and are soft with low mechanical strength and poor chemical stability. Hydrocarbon contamination, for example from microcracks in the glass ampoule, is responsible for reducing the charge lifetime τ. Contacting is complicated by the fact that it reacts with some metals (e.g., Ag, Al) because of the presence of mercury and iodine. Contacts can be painted with aquadag (carbon) or evaporated Pd or Au. Because the hole drift length is so short (see $\mu\tau$ in Table 10.1), the best results come from using a negatively biased entrance window so that the contribution from holes is minimised. The resistivity is very high (Table 10.2) so that a p–n junction is not necessary, and it can be treated as a semi-insulator. But the major problem in the past has been polarisation. The performance can degrade [83] or (sometimes) improve [84] with time. These problems have largely been overcome by improvements in surface protection and encapsulation methods [83,85]. Despite such problems, there are some very impressive results from x-ray spectroscopy, particularly at low energy. Resolutions of sub-200 eV are reported, but usually for this the field effect transistor (FET) has to be cooled. Figure 10.9 shows the Fe-55 spectra from a collaboration of NASA, University of Southern California, and Tracor X-ray Inc. in 1983 [86].

Figure 10.10 shows composite spectra taken on an SEM illustrating the low energy performance. The same work also detected an O Kα peak at 523 eV [87].

The peak/background ratio at 5.9 keV is ~200:1 [88], and the resolutions at 60 keV are below 1 keV. A number of new electrode structure prototypes have been investigated. For example, PXDs [87,89,90], some of which exhibited good count rate performance, and a coplanar grid and drift detector [30],

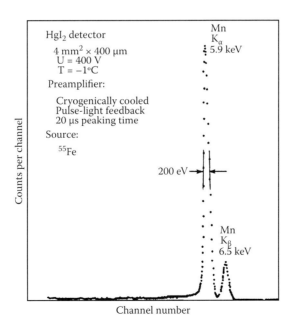

Figure 10.9 Spectrum of Fe-55 using an HgI_2 detector. (From Dabrowski, A.J. et al., *NIM*, 213, 89, 1983. With permission.)

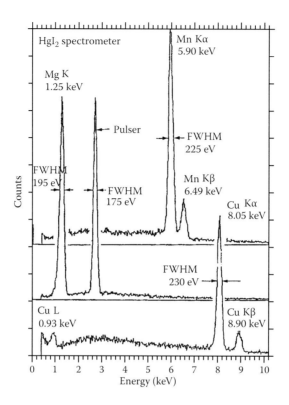

Figure 10.10 Spectrum taken on an SEM using a HgI_2 detector. (From Iwanczyk, J.S., *NIM A*, 283, 208, 1989. With permission.)

showed reduced tailing at 60 keV. There is renewed interest in thick HgI_2 for gamma-ray detectors using PXDs [91] at the University of Michigan and in polycrystalline form for room temperature x-ray imaging [92].

10.3.4 PbI_2

PbI_2 has similar density and stopping power to HgI_2 but its electron $\mu\tau$ product is down by a factor of ~1/30. However, it does have several advantages. Its bandgap is slightly higher, it has a lower vapour pressure at room temperature, and it does not have the destructive solid phase change between melting point and room temperature that HgI_2 has. The latter property enables PbI_2 to be grown from the melt by the Bridgman

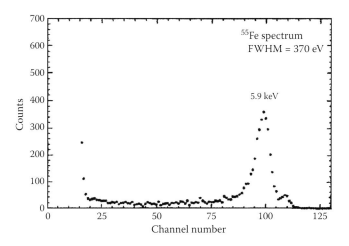

Figure 10.11 Fe-55 spectrum taken using a PbI_2 detector. (From Shah, K.S. et al., *IEEE* NS-44, 3, 448, 1997. With permission.)

method and zone refined. It is also very soft and difficult to handle but is not as soft as HgI_2.

Although particle detectors had been manufactured [93], their use as x-ray detectors was only taken up by the group at Radiation Monitoring Devices (RMD) [94] about 1987 [95,96]. They used thin platelets (100–200 μm thick) cleaved from melt-grown ingots. Both carbon dag and Au or Pd contacts were used. This group has gradually improved the performance, resulting in a resolution of 370 eV [96] as shown in Figure 10.11.

This was taken with a 1-mm², 100-μm-thick detector at room temperature. Polarisation has been studied in PbI_2 detectors [98]. The authors concluded that the instabilities were due to the accumulation of the charge in trap levels in the bandgap.

10.3.5 Diamond and SiC

Diamond has the highest atomic density of any material. This makes it hard and strong, with a high thermal conductivity and bandgap energy. To act as a semiconductor it needs to be doped, but this has proven difficult. From a radiation detector point of view, it is interesting in that it has the potential to operate at very high temperatures (~200°C) under harsh corrosive environments and is extremely radiation hard. To date, however, it is not 'spectroscopic' in terms of x-rays. Diamond has been used extensively as a rad-hard room-temperature particle detector in HEP [99], featuring in the ATLAS detector at the Large Hadron Collider at CERN. Strip detectors [100] and PXDs [101] have been manufactured for HEP.

SiC HV Schottky diodes have been commercially available since 2001. Like diamond, it is grown as thin films by CVD and is normally grown as 4H-SiC, that is, the form with the largest bandgap in the polytype hexagonal structures. Although diamond has a 50% larger bandgap, SiC (carborundum) has many of the features of diamond with high thermal conductivity, very high resistivity, and breakdown field (>2 MV/cm cf. 0.3 MV/cm for Si). In this respect, SiC can be regarded as a 'high energy resolution' version of diamond or a high temperature version of Si. The main interest is in its performance at high temperatures, and detectors have been operated at 200°C. The leakage currents at 127°C are similar to those achieved by Si at room temperature (~1 nA/cm^2). It can usefully be compared to Si. It has a stopping power that is similar, but it has eight times the breakdown field, which helps ICC and partly compensates for the poor $\mu\tau$ product (which is, however, better than HgI_2). It also has twice the thermal conductivity of Si, a feature that aids thermal stability. In the thin CVD SiC detectors, the depletion depths are a modest 10–20 μm and so efficiency is comparable to conventional Si CCDs.

SiC sensors were reviewed by Wright and Horsfall [102]. The first x-ray spectra were obtained in 2001 from small planar 70-μm-thick epi-SiC manufactured by Alenia Marconi for the Milan group [103] with a Au Schottky contact. The 60 keV peak of Am-241 was clear at a noise level of ~466 eV. The Milan group further improved on this [104,105], demonstrating a noise of 315 eV at room temperature and 797 eV at 100°C as shown in Figure 10.12.

This was achieved with a 200-μm-diameter, 70-μm-deep n-type epi-SiC planar detector with a Au Schottky contact and 200 V bias. The 60-keV photopeak shows little tailing. Recently, thin (18 nm) Ni/Ti Schottky contacts have been developed

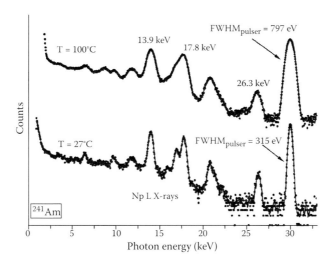

Figure 10.12 Spectrum of Am-241 using a SiC detector at 27°C and 100°C. (From Bertuccio, G. et al., *Sensor Electron. Lett.*, 40, 3, 173, 2004. With permission.)

[106,107] that allow low-energy x-rays to be detected. Figure 10.13 shows a Fe-55 spectrum acquired at room temperature.

The depletion depth at 60 V bias was estimated as ~15 μm and the resolution as ~1.5 keV (it was about the same at 22 keV) and was again limited by the electronics. The spectrum was

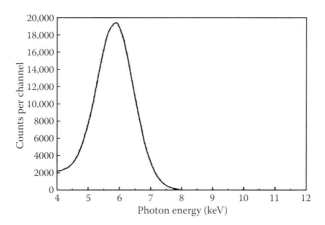

Figure 10.13 Spectrum of Fe-55 taken with a SiC detector. (From Wright, N.G., Horsfall, A., *J. Phys. D Appl. Phys.*, 40, 6345, 2007. With permission.)

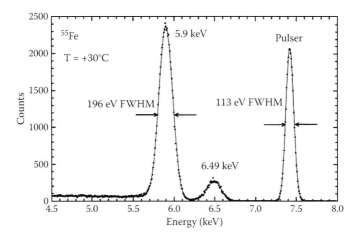

Figure 10.14 Fe-55 spectrum acquired with single pixel at room temperature. (From Bertuccio, G. et al., *NIM A*, 652, 193, 2011. With permission.)

accumulated over many hours. Also recently [108,109], some results have been achieved by the Milan group with pixel detectors manufactured on 70 μm epi-SiC, which are encouraging in that they indicate that it is only the electronic noise (mainly 1/f) that is limiting the resolution. Figure 10.14 shows a Fe-55 spectrum taken with a single pixel at 30°C with their CMOSFET (complementary metal–oxide–silicon FET) preamplifier. The pixel was 200 μm in diameter with a Au Schottky contact. The depletion depth was estimated to be ~25 μm and the Fano factor to be 0.10.

To summarise, SiC shows promise as a high-temperature semiconductor for spectroscopy. Substantial efforts are being expended to improve the quality of SiC for photocells, and lifetime τ has been much improved recently.

10.3.6 TlBr

Recently, there has been renewed interest and activity in the use of the material TlBr, which has the highest bandgap of the three heavy metal halides we have considered and the greatest stopping power of all the WBGSs. Although it is the hardest of the three heavy metal halides, it is still rather soft and easily damaged, for example, by cutting or mounting. Like

PbI_2, it does not exhibit any solid phase transition and can be grown directly from the melt using a Bridgman method. Because TlBr is an ionic semiconductor with a high dielectric constant, the application of bias causes the Tl^+ and Br^- ions to accumulate under the electrodes resulting in a degradation of the field and causing polarisation, which is normally irreversible. This has dogged TlBr detectors. A reduction in temperature is thought to suppress this electrolytic current [110]. The RMD group [111–113] used highly purified material to investigate the potential of TlBr x-ray detectors in the 1990s. Figure 10.15 shows the Fe-55 spectrum from this work with a resolution of ~1.5 keV at room temperature.

The Am-241 peak at 60 keV was only just resolved at high bias (>100 V on 100-μm-thick crystal) and showed hole-trap tailing. More recently, better material has been produced by THM [114] and the Bridgman–Stockbarger [115,116] method. The former method has improved high-energy performance, giving a resolution of 6.4 keV at 60 keV [117] with much reduced low energy tailing. Moreover, the polarisation problem seems largely to have been solved in this work by an intermediate layer of Tl deposited by evaporation under the Au electrodes [118]. The resolution at low energies has not improved much on the earlier work but by cooling a 2.8-mm², 0.8-mm-thick

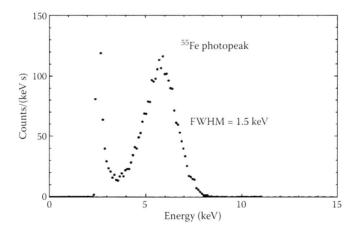

Figure 10.15 Spectrum of Fe-55 using a TlBr detector. (From Shah, K.S. et al., *NIM A*, 229, 57, 1990. With permission.)

crystal to −30°C, a resolution of ~800 eV was achieved at 5.9 keV and 2.6 keV at 60 keV [115]. By strict collimation (50 μm), the resolution at 5.9 keV could be reduced to 664 eV, showing that edge effects (exacerbated by the large dielectric constant) dominate charge collection efficiency. In this work, hole trapping was given as the likely cause of polarisation.

A prototype 3×3 array of 2.7 mm pixels, 1 mm deep, TlBr detector has been manufactured by ESTEC [119] (under an ESA contract) [10,120]. This gave a resolution of 20 keV at 60 keV at room temperature and 4 keV at −30°C. However, it suffered from severe polarisation problems.

10.4 Summary

We have discussed the properties and status of the principal WBGS detectors more or less in the order of their present success or promise of success in replacing Si and/or HPGe for x-ray spectroscopy without the need for a cryogen. CdTe and CdZnTe have already done so for some less demanding applications. We have omitted many more materials that have been studied or might be promising, most notably InP, GaN, GaSe, GaP, InN, AlN, ZnSe, CdSe, AlSb, and BiI_3, not to mention some ternary semiconductors such as CdSeTe and HgBrI. During the 1980s and 1990s HgI_2 was a popular choice, but we have seen this largely replaced by commercially available Peltier cooled Si (p–i–n and SDD) and CdTe (or CdZnTe). The situation for WBGSs in general is now rapidly changing, and new methods of purification and crystal growth are appearing. Developments in other areas, such as fiber optics, could be a driving force for better materials and techniques, and it may still be a case of 'watch this space'. Front end electronics to match some of the WBGSs needs further development to exploit their real potential but may still involve cooling the FET. Also, geometrical methods to suppress the effects of hole trapping have not yet been implemented to full effect. As has been pointed out [22], if the latter route is to be pursued, we do not really want to improve the hole trapping properties of the material.

None of the materials discussed come near to the performance of Si, Si(Li), and HPGe detectors, particularly for low-energy EDXRS. For quality analytical work, as well as industrial, medical, and security operations, the whole perspective would be changed with the invention of a quiet, high-power, cheap, and reliable cooling engine for Si(Li) and HPGe x-ray detectors. Cooling engines are discussed further in Section 11.25.2.

References

1. Sellin P.J., Development of Semiconductor Gamma-ray and Neutron Detectors for Security Applications (presentation pdf: Dept. of Physics, Univ. of Surrey, UK), 2009.
2. Lumb D.H., *Astrophys. Space Sci. Lib.* 300, 2004.
3. Owens A., *Compound Semiconductor Radiation Detectors*, Taylor & Francis, CRC Press, 2013, ISBN 978-1-4398-0.
4. Dabbs J.W.T. and Walter F.J. (Eds.), Semiconductor nuclear particle detectors (Asheville Conf. 1960), NAS-871, 1960.
5. Prince M.B. and Polishuk P., *IEEE* NS-14, 1, 537, 1967.
6. Sakai E., *NIM* 196, 121, 1982.
7. Schlesinger T.E. and James R.B. (Eds.), *Semiconductors for Room Temperature Radiation Detector Applications*, Academic Press, 1995, ISBN 10-0127521437.
8. Schieber M. et al., *NIM A* 377, 492, 1996.
9. McGregor D.S. and Hermon H., *NIM A* 395, 101, 1997.
10. Sellin P.J., *NIM A* 513, 332, 2003.
11. Owens A. and Peacock A., *NIM A* 531, 18, 2004.
12. Lumb D.H. et al., *NIM A* 568, 427, 2006.
13. Luke P.N., *IEEE* NS-42, 207, 1995.
14. Barrett H.H. et al., *Phys. Rev. Lett.*, 75, 156, 1995.
15. Jones L.T. and Woollam P.B., *NIM* 124, 591, 1975.
16. Jordanov V.T. et al., *NIM A* 380, 353, 1996.
17. Niemela A. et al., *NIM A* 377, 484, 1996.
18. Niemela A. et al., *IEEE* NS-43, 1476, 1996.
19. Richter M. and Siffert P., *NIM A* 322, 529, 1992.
20. Arlt R. et al., *NIM A* 322, 575, 1992.
21. Lund J.C. et al., *IEEE* NS-43, 1411, 1996.
22. Luke P.N., *NIM A* 380, 232, 1996.
23. Bale D.S. and Szeles C., *SPIE* 6319, 1, 2006.

24. Lund J.C. et al., *NIM A* 377, 479, 1996.
25. McGregor D.S. et al., *NIM A* 422, 164, 1999.
26. McGregor D.S. et al., *IEEE* NS-46, 250, 1999.
27. Owens A. et al., *J. Synch. Rad.* 13, 143, 2006.
28. Passmore S. et al., *NIM A* 434, 30, 1999.
29. Ludwig J. et al., *NIM A* 460, 72, 2001.
30. Patt B.E. et al., *NIM A* 380, 276, 1996.
31. Owens A. et al., *J. Appl. Phys.* 102, 545, 2007.
32. Lachish U., *NIM A* 403, 417, 1998.
33. Nemirovski Y. et al., *J. Electron. Mater.*, 26, 756, 1997.
34. Arkadeva E.N. et al., *IEEE* NS-15, 3, 258, 1968.
35. Mayer J.W., *NAS Publ.* 1592, 377, 1969.
36. Zanio K. et al., *IEEE* NS-21, 1, 315, 1974.
37. Dabrowski A.J. et al., *Rev. Phys. Appl.* 12, 297, 1977.
38. Ryan F.J. et al., *Appl. Phys. Lett.* 46, 274, 1985.
39. Butler J.F. et al., *IEEE* NS-39, 605, 1992.
40. Endicott Interconnect (EI), Saxonburg, USA (formerly eV Products).
41. Digirad (formerly Aurora, San Diego), Poway, CA, USA.
42. Redlen, Redlen Technology, Sidney, BC, Canada.
43. Amptek, Bedford, MA, USA.
44. Technion, Israel Institute of Technology.
45. Ritec, Ritec Ltd., Riga, and Baltic Scientific, Latvia.
46. Niemela A., *J. Phys. III* 6, C4-271, 1996.
47. Bale G. et al., *NIM A* 436, 150, 1999.
48. Bale G., Ph.D. thesis, Univ. of Leicester, UK, 2001.
49. Redus R.H. et al., *NIM A* 458, 214, 2001.
50. Raiskin E. and Butler J.F., *IEEE* NS-35, 1, 81, 1988.
51. Nemirovski Y. et al., *J. Electron. Mater.* 26, 756, 1997.
52. Ramsey B. et al., *NIM A* 458, 55, 2001.
53. Bolotnikov A.E. et al., *NIM A* 458, 585, 2001.
54. Stahle C.M. et al., *NIM A* 436, 138, 1999.
55. Owens A. et al., *J. Appl. Phys.* 102, 545, 2007.
56. Burger A. et al., *J. Cryst. Growth* 198–199, 872, 1999.
57. Mycielski A. et al., BNL-93721-2010-CP, 2010.
58. Cui Y. et al., BNL-81493-2008-CP, 2008.
59. Yinnel Tech., South Bend, IN, US.
60. Fougeres P. et al., *NIM A* 428, 38, 1999.
61. Imarad, Tel Aviv, Israel.
62. Eurorad S.A., Strasbourg, France.
63. Acrorad and Acrotec, Okinawa, Japan.
64. Kromek, Durham, UK.
65. New Semiconductors, Kiev, Ukraine.

66. Sato G. et al., *IEEE Nuclear Science Symp., Orlando, Conf. Record*, NS-41-6, 2009.
67. Harding W.R. et al., *Nature* 187, 405, 1960.
68. Eberhardt J.E. et al., *Appl. Phys. Lett*. 17, 427, 1970.
69. Buttar C.M., *NIM A* 395, 1, 1997.
70. Bertuccio G. et al., *IEEE* NS-44, 1, 1, 1997.
71. Owens A. et al., *J. Appl. Phys*. 90, 5367, 2001.
72. Owens A. et al., *NIM A* 466, 168, 2001.
73. Grovenor C.R.M., *Solid State Electron*. 24, 8, 792, 1981.
74. Hsu J.W.P. et al., *J. Vac. Sci. B* 21, 4, 1928, 2003.
75. Willig W.R., *NIM* 96, 615, 1971.
76. Swierkowski S.P. et al., *IEEE* NS-21, 1, 302, 1974.
77. Schieber M. et al., *Proc. Inter. Conf. on Crystal Growth*, Tokyo, EGG 565-119, 1974.
78. Faile S.P., Patent US 4282057, Texas Nuclear, 1981.
79. Van den Berg L., *NIM A* 458, 148, 2001.
80. Constellation Technology Corp., Largo, FL, USA.
81. Saleno M.R. et al., *IEEE Nuclear Science Symp. Orlando, Conf. Proc*. NS-41-7, 2211, 2009.
82. Ponpon J.P. et al., *NIM A* 412, 104, 1998.
83. Gerrish V., *NIM A* 322, 402, 1992.
84. Iwanczyk J.S., *IEEE* NS-36, 841, 1989.
85. Squillante M.R. et al., *NIM A* 288, 79, 1990.
86. Dabrowski A.J. et al., *NIM A* 213, 89, 1983.
87. Iwanczyk J.S., *NIM A* 283, 208, 1989.
88. Bao X.J. et al., *NIM A* 380, 58, 1996.
89. Iwanczyk J.S. et al., *IEEE* NS-39, 1275, 1992.
90. Saleno M.R. et al., *NIM A* 652, 197, 2011.
91. Baciak J.E. and He Z., *IEEE* NS-50, 4, 1220, 2003.
92. Schieber M. et al., *IEEE* NS-53, 4, 2385, 2006.
93. Manfredotti C. et al., *IEEE* NS-24, 1, 158, 1977.
94. RMD, Radiation Monitoring Devices, Watertown, MA, USA.
95. Lund J.C. et al., *IEEE* NS-35, 1, 89, 1988.
96. Lund J.C. et al., *NIM A* 283, 299, 1989.
97. Shah K.S. et al., *IEEE* NS-44, 3, 448, 1997.
98. Ponpon J.P., *NIM A* 526, 447, 2004.
99. Krammer M. et al., *NIM A* 436, 326, 1999.
100. Borchelt F. et al., *NIM A* 354, 506, 1991.
101. Adam W. et al., *NIM A* 436, 326, 1999.
102. Wright N.G. and Horsfall A., *J. Phys. D Appl. Phys*. 40, 6345, 2007.
103. Bertuccio G. et al., *IEEE* NS-48, 2, 232, 2001.
104. Bertuccio G. et al., *Sensor Electron. Lett*. 40, 3, 173, 2004.

105. Bertuccio G. et al., *NIM A* 522, 413, 2004.
106. Lees J.E. et al., *NIM A* 578, 226, 2007.
107. Lees J.E., *NIM A* 613, 98, 2010.
108. Bertuccio G. et al., *IEEE* NS-53, 4, 2421, 2006.
109. Bertuccio G. et al., *NIM A* 652, 193, 2011.
110. Kozlov V., Dissertation, University of Helsinki, 2010.
111. Shah K.S. et al., *IEEE* NS-36, 199, 1989.
112. Shah K.S. et al., *NIM A* 229, 57, 1990.
113. Olschner F. et al., *NIM A* 322, 504, 1992.
114. Hitomi K., *NIM A* 428, 372, 1999.
115. Owens A. et al., *NIM A* 497, 370, 2003.
116. Kozlov V. et al., *NIM A* 607, 126, 2009.
117. Hitomi K. et al., *NIM A* 579, 153, 2007.
118. Hitomi K. et al., *NIM A* 585, 102, 2008.
119. ESTEC, The Research & Technical Centre for ESA, Noordwijk, Netherlands.
120. Owens A. et al., *NIM A* 497, 359, 2003.

11

History of Semiconductor X-Ray Detectors and Their Applications

11.1 Introduction

There are good reasons for the inclusion of a history of semiconductor detectors. It is an established educational approach and thus augments the foregoing chapters. We can see why certain development routes were taken and why others failed. Furthermore, although there have been many previous reviews at different periods (which will be referenced below), these have not generally detailed the significant work that took place within the industrial companies involved, particularly from around 1966 onward. Much of this development took place in competitive circumstances, in different countries, and details were not published. The authors, working not only in x-ray detector manufacturing [Oak Ridge Technical Enterprises Company (ORTEC), Link, OI, e2v] but also on repairs and modifications to most other manufacturer's detectors, were in the privileged position of seeing firsthand the progress unfolding over a period of 40 years.

It is never possible to know precisely how scientific history is made and who exactly was involved. Publications and patents give only one side of the story. The situation was perhaps best described by E.O. Wilson (the American biologist accredited as the father of sociobiology): 'The best of science doesn't consist of mathematical models and experiments, as textbooks make it seem. Those come later. It springs fresh from a more primitive mode of thought wherein the hunter's mind weaves ideas from old facts and fresh metaphors and the scrambled crazy images of things recently seen'. He carried on this thought and stated, 'to move forward is to concoct new patterns of thought, which in turn dictate the design of the models and experiments. Easy to say, difficult to achieve.... Genius is the summed production of the many with the names of the few attached for easy recall, unfairly so to other scientists'. This chapter is dedicated to all those stubborn engineers who, following a heuristic approach, strove to control the vagaries of semiconductors, their bulk properties and defects, and their surface states and contacts, to create workable devices, and particularly to those who further attempted to make a living by manufacturing commercial devices for the scientific community. Occasionally, they were rewarded by transforming a dull gray uninspiring slab of material into an ultraresponsive x-ray sensor within a precision analytical instrument: truly crystal divination! Often, it was the meticulous, usually unsung, work of technicians, which played an inordinate part in the success stories. Usually, these highly skilled workers combined production and R&D. As the French scientist Claude Bernard said, 'Science is a superb and dazzlingly lighted hall which may be reached only by passing through a long and ghastly kitchen'. Below, we will visit some of these 'ghastly kitchens', where two of the major ingredients were curiosity and luck, while striving to give a balanced account and the hope that the cooks we mention will forgive any errors, and the ones we have not mentioned will accept our apologies.

11.2 Beginnings

The design of semiconductor x-ray detectors has evolved slowly and empirically from nuclear particle detectors. Their history is inextricably tied to the history of the semiconductor materials themselves and to the early history of charged particle detectors in general. After all, the x-ray detector is nothing more than a photoelectron detector where the source happens to be internal to the detecting medium. We are therefore concerned with the development of the materials themselves, most notably silicon and germanium, and hence with the semiconductor device revolution that unfolded in the late 1940s and 1950s and indeed continues in the present day. Silicon is the second most abundant element on the earth's surface (oxygen is the first). Germanium is a by-product of burning coal and the smelting of copper, tin, and zinc ores. The availability of either silicon or germanium raw material was never an issue. However, producing detector-grade material has been a significant issue. The question of silicon or germanium as the material of choice has oscillated with silicon being the more stable (but more difficult to purify) and germanium having the better semiconducting properties (but more sensitive to surface instabilities and temperature). The room temperature silicon surface barrier particle detectors from 1960 onward, together with room temperature Si(Li) particle detectors from 1962 onward, started a revolution in charged particle spectrometry. Liquid nitrogen–cooled Ge(Li) detectors, from 1966 onward, then liquid nitrogen–cooled HPGe detectors from 1970 onward, did the same for γ-ray spectrometry. Commercial availability lagged 2 or 3 years behind the pioneering work. The matching of liquid nitrogen–cooled Si(Li) detectors to the silicon junction field effect transistor (JFET), in 1964–1966, revolutionised energy dispersive x-ray spectrometry (EDXRS), expanding the applications via energy dispersive x-ray fluorescence (EDXRF), around 1969, and energy dispersive x-ray microanalysis (EDXMA), around 1971. Applications moved from nuclear physics into virtually every corner of science and, in some cases, art (e.g., identification of pigments in paint) as well.

It is interesting to place the development of crystals of 'pure' material for detectors within the timeline of the development of the more ubiquitous devices such as the bipolar transistor, JFETs, MOSFETs, and integrated circuits (ICs). However, we will observe that the material requirements are far more stringent for radiation detectors than for most of these devices, and a significant evolutionary branching occurred around 1960. This was also because one technology was moving toward microminiaturisation and the other toward larger thicker devices with greater radiation sensitivity and stopping power. This branching led to lithium-compensated material (a technique now unique to semiconductor x-ray detectors) and later very high purity silicon and germanium (again unique to radiation detectors).

11.2.1 Search for Materials before World War II

After the successful development of the gas proportional detector in the 1930s, it was only natural to look for a successor and toward the semiconductors, which were gaining importance as World War II loomed. John Orton gives an excellent account of the history of semiconductors in his book [1]. The advantages, such as an inherently better energy resolution (lower ionisation energy), greater stopping power, compactness, and ruggedness of a solid-state version, were obvious from the start.

It was established early in the nineteenth century that there existed a category of materials that fell into neither insulators (such as sulphur) nor conductors (metals, such as silver). Michael Faraday noted the negative temperature coefficient of silver sulphide in his observations published in 1833. This is in contrast to the positive coefficient for all 'conductors'. Willoughby Smith reported the observation of a telegraph clerk (a Mr. May, 1872)—of the reduction in resistivity of a selenium sample when illuminated with light. Although selenium in its purest form is not a semiconductor, its compounds ZnSe and CdSe are. A hint of things to come appeared in 1874 when Karl Ferdinand Braun [2], a 24-year-old Ph.D. graduate of the University of Berlin, studied the electrical behaviour of various metallic contacts to sulphide crystals such as galena and pyrites and discovered that, unlike metal-to-metal contacts, the current–voltage characteristics were *nonlinear*. In short, they formed a *junction*. At about the same time, Werner Siemens realised that the photovoltaic effect in selenium and copper oxide rectifiers represented a direct light-to-electrical energy conversion. Arthur Schuster found similar results for contacts of clean copper to copper oxide. We now know that under certain circumstances, these oxides can be made to behave as semiconductors and that these were early observations of 'localised junctions'. Braun was convinced that the effects were attributable to the properties of his point contacts, which would later be referred to as 'cat's whiskers'. The first patent for a cat's whisker was by the Indian radio pioneer Jagadis Bose [3] in 1901. This brought it to the attention of the electronics community and Braun, in 1904, applied point contacts to the problem of detecting the recently discovered electromagnetic radio waves (Heinrich Herz, 1888). The cat's whisker acted as a rectifier and filtered out the high frequency radio carrier, passing the audio frequency signal. No one knew why at the time. Braun used contacts to crystals such as PbS, SiC, Te, and Si. Crystals of PbS seemed to give the best results, but silicon from Westinghouse Electric Company also gave good results. This predates the use of vacuum tubes (John Fleming started experiments on them in 1904). These were prone to failure, had a high power requirement, and produced heat. Crystal devices seemed a possible alternative. However, the cat's whisker was eventually superseded by the triode vacuum tube (Lee de Forest 1906), which was taken up by AT&T (the Bell Telephone Co), USA, in 1912.

The founder of the Bell Telephone Co., Alexander Graham Bell, was the son of a Scottish professor, and in 1870 the family emigrated to Canada. Bell had entered the amplifier arena via his interest in deaf aids (both his mother and wife were deaf) and the unit of logarithmic power ratio, the deciBel (dB), is named for him. His patent for a telephone was issued in 1876, and he created the Bell Telephone Co. the following year. In 1885, the company was renamed American Telephone and Telegraph (AT&T), and in 1925 AT&T set up the Bell Laboratories at Murray Hill, NJ, to focus on basic research. Marvin Kelly, director of research at Bell Laboratories, instituted a research program on semiconductors in the 1930s.

AT&T had a solid monopoly on telephones but was eventually broken up by an antitrust committee in 1984. AT&T (now Alcatel-Lucent Technologies) kept the long distance systems and Bell Laboratories at Murray Hill.

Among the other large companies that were to play a part later in our story were General Electric (GE) and Radio Corporation of America (RCA). The history of GE goes back to Thomas Edison, who set up the Edison Electric Light Co. (Menlo Park, CA) in 1878. In 1889, Edison consolidated all his companies as 'Edison General Electric Co.', and in 1892 the company merged with Thomson-Houston Electric Co. RCA was founded in 1919 by Owen D. Young as a retail arm of Edison GE for radio sales. RCA moved into semiconductors in 1948 under the leadership of David Sarnoff (president, 1930–1971). In 1986, Edison GE acquired RCA (which was then owned by NBC News) to form GE.

The cat's whisker rectifier was to have a new lease of life during World War II when the vacuum triode rectifier was found to be useless for the high radio frequencies needed for the development of radar. This was because of the long electron transit times involved, whereas in the cat's whisker not only is the capacitance lower but the carriers only have to travel distances of the order of ~μm or less and could operate up to 3 GHz. Details of this interesting story can be found in the work of Garratt [4]. Edwin Hall (of the Hall effect) had been studying the effect of transverse magnetic fields on conducting metals back in 1879, but Koenigsberger and coworkers were the first to study the effects in semiconductors between 1907 and 1920. In 1912, Baedeker showed that the Hall effect in CuI depended strongly on 'doping' with iodine, which influenced the free charge carrier concentration. Thus, the Hall effect allowed the behaviour of free carrier density to be distinguished from free carrier mobility.

The sensitivity of certain materials to x-rays goes right back to x-ray discovery. In 1895, Wilhelm Röntgen working at the University of Wurtzburg, Germany, noticed that certain crystals scintillated when near a conducting vacuum tube, thus establishing the existence of x-rays. Further investigations with Joffé [5] from 1913 to 1921 showed that the electrical conductivity of insulating crystals increased when they were in the vicinity of radioactive sources. In the late 1920s, investigations of the charge induced in biased crystals at low temperature were being carried out in at the University of Gottingen, North Germany.

11.2.2 1930s: Theoretical Understanding

In 1932, Karl Hecht [6] published a detailed work correlating this induced charge on insulating crystals, such as silver chloride, with the bias applied and the localised position of illumination of the crystals. He was able to show that the results were consistent with the charge induced being proportional to the average distance travelled by the released electrons before being trapped, thus anticipating Shockley and Ramo's work (see below). He described this in the form of an equation, now known as the Hecht equation (see Section 1.7.4), which was revived many years later to investigate the $\mu\tau$ product in potential detector materials.

In the early 1930s, a picture of semiconductors was beginning to form, and in 1931 Alan Wilson presented to the UK Royal Society the first fairly adequate theory of their properties. Although his explanation of diode rectification was later shown to be wrong, his ideas were still being assimilated long after, even during the creation of the first transistors at Bell Laboratories in 1947. A major obstacle to progress was that the practical samples of silicon and germanium being used were not pure or indeed crystalline. Wilson even made the statement that silicon should be regarded as a metal! A satisfactory explanation of diode operation came in 1938 when Nevill Mott [7] at Bristol University, UK, proposed a barrier potential model for rectification. Independently, Walter Schottky [8] at Siemens in Germany and Alexander Davydov [9] in Russia also published barrier models. By the start of World War II, there was a basic understanding of the properties of optical absorption, photoconductivity, and the photovoltaic effect in semiconductors. Also, around this time, two theoretical landmarks for the basic understanding of detectors were established by scientists with very different backgrounds but uncannily similar names (Ramo and Fano). Although their work was not directly concerned with semiconductors, these names were destined to be common parlance in discussions on semiconductor radiation detector physics. In 1939, Simon Ramo [10] devised a method of computing the signal induced by charge moving between two conductors with an electric field between them. To be fair, William Shockley [11], working at Bell Laboratories (of whom we will hear much more later), had also devised a less elegant but equivalent method a year before, although it is now commonly known as Ramo's theorem—but more correctly, the Shockley–Ramo theorem. As discussed in Section 1.7.2, the significance of this theorem to radiation detectors is that the charge generated by ionising radiation need only to have moved some finite distance in the field to register a signal on the detector contact. The consequences for detector technology were both good (trapping of charge can to an extent be tolerated for mere 'detection') and bad (trapping gives rise to underdeveloped signals or 'background' in spectroscopy). Trapping–detrapping also gives rise to noise. The importance of the Shockley–Ramo theorem is that it gives a basis to calculate the signal precisely if all the dynamics of the charge motion in the detector are known. For the special case of a semiconductor where space charge is also present, Jen [12] showed that the Shockley–Ramo theorem was still valid. To bring the use of the theorem up to date, in 1963 Cavalleri et al. [13] showed that it was the fraction of the *distance* travelled along the track (rather than the *potential*) that defined the induced charge signal. Controversy still surrounded the application of the theorem until a more rigorous treatment was given in 1971, again by Cavalleri et al. [14]. The latter reference is sometimes referred to as the 'extended Shockley–Ramo theorem'.

Simon Ramo was born in Salt Lake City, UT, USA, in 1913, the son of Russian and Ukranian immigrants who ran a small store. When he wrote his paper, he was leading the electronics research at General Electric, Schenectady, and may be better known generally for his microwave research and the development contribution to the electron microscope (EM). In 1946 he became director of research at Hughes Aircraft, Los Angeles. It is rather ironic that although, as we will see later, both GE Research Laboratory and Hughes Research Laboratories were to be actively involved in semiconductor detector development, he played no part, having left in 1953 in frustration of the management style of Howard Hughes. He set up a new company, the

Ramo-Wooldridge (RW) Corporation, which became the lead contractor for the U.S. Air Force program largely responsible for the development of the Atlas Rocket (1958). RW became incorporated with Thompson in 1958, forming the TRW Corporation.

Ugo Fano published the results of calculations on the ionisation yields in matter with particular reference to gas proportional counters in 1943 [15]. In a later paper [16], he calculated the *fluctuations* in the number of charge pairs and showed that it was related to the Poisson variance by a factor that he represented (fortuitously) by the symbol F, and since this seminal paper it has always been referred to as the 'Fano factor'. As explained in Section 2.4, its importance lies in setting a limit (the 'Fano limit') on the *resolution* achievable for a detector constructed of the active material. The significance of F for detector resolution (in connection with gas proportional detectors) was pointed out to Fano by R.G. Sachs, as he confesses in his pioneering paper [16]. The Fano factors for silicon and germanium were not calculated or measured until 20 years later, and the measurement and interpretation has always led to great controversy (see Section 2.4).

Ugo Fano was born in Turin, Italy, in 1912. His father, Gino Fano, was a professor of mathematics at the University of Turin and specialised in differential geometry. Gino is remembered, for example, by the 'Fano plane' in the theory of musical scales and Fano varieties of multidimensional spaces. Ugo gained his Ph.D. in mathematics at Turin 1934 and fled from Fascism to the United States in 1939 and eventually worked in the Physics Department of Columbia University, where he wrote his paper. He moved to the National Bureau of Standards (NBS) in Connecticut in 1946 and wrote other papers on the same topic. He remained at NBS for 20 years and died in 2001.

11.3 Development of Materials during World War II

World War II had several distinct effects on progress toward the control of semiconductor technology. The race to achieve viable radar systems and a better understanding of nuclear radiation spurred on the semiconductor technology associated with detection in both these endeavours. Most of the progress came from North America and benefited from an influx of scientists and engineers fleeing Europe in the years building up to the war itself. The United States and the UK collaborated on the development of radar, and Denis Robinson (in the UK) had read the German literature of Alfred Thoma and Hans Hollman and concluded that the microwave receiver had to include the cat's whisker as rectifier in the heterodyne mixer. In 1942 Herbert Skinner, also working in the UK, had found that a tungsten whisker contacting a silicon or germanium crystal was the most promising from the point of view of the ruggedness required in warfare, and attention swung to these materials. It is interesting to note that the American, Frederick Seitz, in his monumental book, *Modern Theory of Solids* (published in 1940) [17], had not given silicon or germanium a mention. However, he later gave a historical account of the cat's whisker and the role of silicon [18] in electronics. Soon, purified and doped silicon were developed at the University of Pennsylvania and Dupont Company in the United States, whereas Purdue University (Lark-Horowitz's group, see below) and Sylvania Electric Products Inc., Pennsylvania, concentrated on germanium.

The cat's whisker action was understood to be due to the diffusion of electrons forming a localised 'inversion' of the semiconductor surface, effectively forming a point 'Schottky barrier' contact. This does, incidentally, illustrate how great care should be taken in making any pressure contacts directly to a semiconductor surface. But the variability in the performance of the crystal rectifiers sparked off efforts devoted to the growth of pure single crystals. As an early example of doping, it was discovered empirically that a 0.002% addition of boron reduced the resistivity by a factor of 100 (by increasing the hole carrier density). Also by 1943, Germany had seen the advantage radar gave to the Allies. In February 1943, a British Sterling bomber was shot down in Holland and revealed to the Germans its in-flight radar (they may have already had short-range radar). Herman Goring ordered a thorough analysis and instigated an attempt to catch up with the technology. They were, of course, familiar with the theories of Schottky and Mott, and Herbert Mataré [19] was put to work on the problem in Poland at the Telefunken factory. One of the materials investigated was germanium supplied by a Luftwaffe research team in Munich that included Heinrich Welker. After the war, the Allies teamed up Mataré with Welker and others to work on higher purity germanium rectifiers at a French Westinghouse subsidiary. To follow this story up, Mataré filed a patent in April 1948. As a result, the French Westinghouse company had their version of the junction transistor, the 'Transistron', in production but the French government and Westinghouse failed to capitalise on their technical advantage. The work was abandoned within a short time because the French government withdrew its funding in favour of an intensive program to develop atomic energy. In 1952, Mataré returned to Germany and founded the company Intermetall, one of the first manufacturers of transistors outside Bell Laboratories in the United States and the first to demonstrate a transistor radio (a year before TI). Welker moved to Siemens and became a pioneer of III–V semiconductors and head of R&D in 1969.

11.4 1940–1960: Crystal Counters

In 1945, a graduate student, P.J. van Heerden, working at the University of Utrecht, Holland, published his thesis [20] on new 'crystal counters', the first crystal detectors with any practical application. The work was significant but was only exposed to the wider scientific community toward the end of the decade [21]. He discovered that if contacts are put on various crystals and a field applied, a signal was generated when it was exposed to ionising radiation. Although the ions produced in the solid were quickly trapped, they did move a finite distance and they did induce the signal predicted by the Hecht equation and the Shockley–Ramo theory. The detector consisted of a natural crystal of silver chloride with silver contacts deposited on them and cooled by liquid air. The amplifier circuit used was similar to those used for the conventional pulse ionisation chambers. He was able to demonstrate signals from electrons, α particles, and x-rays. The detectors quickly degraded because of the trapped space charge causing 'polarisation' but could be recovered by reversing the field or annealing the crystal. There were manufacturing problems and in the end they could not compete with the new scintillation counters, and no further development took place at Utrecht. But it did prompt work elsewhere.

Besides AgCl, other crystals were tried such as CdS [22], ZnS [23], and diamond [24]. They consisted of small specially selected crystals (<1 cm^3), coated on opposite faces with a conductor. A bias was supplied with a very high value load resistor across which a signal was taken to an amplifier as shown in Figure 11.1.

By the time of the First Symposium on Scintillation Radiation Detectors [25], at the University of Rochester, USA, in 1948, crystal counters accounted for about half of the papers [26]. Furthermore, the recognized expert on solids, as mentioned, and close friend of Bardeen (Bell Laboratories), Frederick Seitz [27], presented a paper on the possibility of a solid-state 'Geiger counter' using the avalanche process. Various other reports appeared in the journal *Nucleonics* and the *Proceedings of the Institute of Radio Engineers* (IRE) by their protagonist, Robert Hofstadter [28,29]. These considered the alkali metal halides but surprisingly not silicon or germanium. Writing in 1958 in the first edition of his book, William J. Price [30] (U.S. Air Force), listed the advantages of crystal counters or, as they were sometimes called, 'homogeneous' or 'bulk' counters. These were linearity, response time, resolution, insensitivity to magnetic fields (unlike scintillation counters), and the relative insensitivity to gamma and

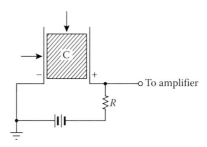

Figure 11.1 Early crystal counter.

neutron background. He mentions also their light sensitivity and their use, combined with a scintillation detector such as CsI, to give a signal/noise ratio several orders higher than the scintillator alone. The disadvantages are listed as the short operating lifetimes (polarisation), small output signal (this was before the invention of the FET), radiation damage, and dependence on ambient conditions. This prompted him to venture that crystal detectors would have little future because of the greater advantages of scintillation detectors. But this was the first solid-state detector to *directly* convert ionising radiation energy to an electrical signal (unlike the scintillation counter that converts the energy to light and then, via a photomultiplier, to an electrical signal) and does not differ in principle from the 'insulator' compound semiconductor detectors being used and further developed today (see Chapter 10). The only difference was the primitive nature of the *material*. In his second edition of his book (1964), he added a chapter on semiconductor detectors and had to admit how wrong he had been in 1958!

Another form of crystal detector was also developed, as reviewed by Taylor [31], also of the U.S. Air Force, in his book of 1963. These used material that had been manufactured especially for the purpose. They had diffused contacts and were known as 'bulk detectors' to distinguish them from the junction detectors, which we will describe later. They differed from the crystal detectors by using a large (>1 cm^3) grown single semiconductor crystal rather than polycrystalline insulator types. GaAs appeared to be one of the most hopeful materials at room temperature. It was commented 'intrinsic Si (even it were available) would not be useful at room temperature and germanium is not useful even at liquid nitrogen temperatures. The surfaces must be very clean and they must be used in the dark in a dry atmosphere or vacuum'. But it was found that n-type Si doped uniformly with gold could be used virtually as an insulator at low temperature. The gold gives deep levels that pin the Fermi level midgap and produce near 'intrinsic' silicon. At 120 K, the resistivity was ~10^{15} Ω cm. However, the gold obviously increased carrier trapping. Davies [32], working at Knolls Atomic Power Laboratory (KAPL) in 1958, evaporated gold onto gold-doped n-type silicon at room temperature, making rectifying contacts. This was rather analogous to the lithium-doped detectors that were to come 2 years later. He also made ohmic contacts by further diffusing gold into the surface at 400°C–500°C. However, the lifetime for holes was only 10^{-9} to 10^{-7} s, and the crystals had very strong polarisation effects. Philips in Eindhoven, the Netherlands, still had a commercial interest in these 'conduction' detectors in 1961 [33] and Jim Mayer (at Hughes) [34] still discussed them in his review of 1962.

Crystal detectors were the forerunners not only of the junction detectors but continued to be of interest to the present day with the WBGS 'semi-insulator' detectors, with particular interest in their operation at room temperature. Different examples such as mercuric iodide, lead iodide, silicon carbide, and diamond, have gone in and out of fashion over the years.

11.5 1940s: Role of National Laboratories, Bell Telephone Laboratories, and Other Corporations

The national laboratories were set up in the United States during World War II largely for the development on the atomic bomb (the Manhattan Project), and they collaborated with various universities on associated radiation detector research. For security reasons, Chalk River, Montreal, was chosen during the war as a British–Canadian Atomic Energy Laboratory. It was headed by John Cockroft (co-inventor of Cockroft-Walton accelerator) and employed a number of scientists from the British Commonwealth. After the war, Cockroft moved back to the UK to set up the Atomic Energy Research Establishment (AERE) Harwell, which he considered the 'daughter establishment of Chalk River'. Chalk River and its workers (including Bromley, McKenzie, Goulding, Waugh, Ewan, Tavendale, Fowler, Malm, McMath, Martini, and Sakai) from all parts of the world, were to play a major part in semiconductor development and some, in the later establishment of commercial interests. Chalk River also worked closely with the RCA research laboratories, Montreal (Williams, Webb), in the early days.

Before the war, much pioneering work had been carried out in Germany at the universities (Gottingen, Wurzburg) and the Kaiser Wilhelm Institutes, notably Munich and Heidelberg. After the war the latter became the Max Planck Institutes (MPI), but it was many years later before real pioneering work, for example, on silicon drift detectors (SDDs) (Section 11.26 and Chapter 9), was again carried out in Germany. After the development of the atomic bomb at Los Alamos and Oak Ridge, TN, nuclear physics became very prestigious. For example, Ernest Lawrence, inventor of the cyclotron particle accelerator, received a large increase in funding at Berkeley, CA, for further work, largely through the backing of General Groves, the leader of the Manhattan Project. Furthermore, the building of the first nuclear reactor in Chicago in 1942 and the demonstration of controlled nuclear fission by Enrico Fermi, caused a great deal of interest and prompted extra funding for nuclear physics and radiation measurement. The search was on for a more efficient medium than the gases of proportional counters. Despite this, as we have seen, the first progress toward semiconductor radiation detectors came from Europe in the form of crystal detectors.

The war effort also established the concept of research *teams*, working cohesively on scientific problems, rather than individuals working independently. In the United States, two national laboratories had been set up on a crash basis to develop the atomic bomb. These two, Argonne near Chicago, and Oak Ridge in Tennessee, were joined shortly after the war by the civilian nuclear research center, Brookhaven National Laboratory (BNL), Long Island, NY. These research centres, including Chalk River in Canada, collaborated with and recruited from peripheral universities such as Chicago, Purdue, Michigan State, Battelle Institute, Columbia, Princeton, MIT, Harvard, McMaster, and Toronto. Other laboratories involved in the atomic bomb development were at Berkeley (LBL), California, which operated in Los Alamos, NM, where the bomb testing took place. Apart from these, government contracts had helped to set up purpose-built industrial laboratories such as RCA, GE, and KAPL, Schenectady. The latter was operated by GE (since 1946) to research power reactors. Most importantly, in terms of semiconductor development, was the Bell Telephone Co. Laboratories, Murray Hill,

NJ, a stone's throw away from Brookhaven. The nature of the Bell Laboratories was also to have an influence. Fortuitously, rather than trying to keep the products of their work proprietary, Bell Laboratories licensed out their inventions to other industries and enabled such companies as Texas Instruments (TI) to grow rapidly. Some of the very best researchers were concentrated at Bell Laboratories, and there was a fairly loose attitude toward the management of the research carried out. For example, because of the unpredictable behaviour of silicon (it was more difficult to purify), one of the first directives was to cancel work on silicon in favour of germanium. Somehow one of the researchers, Russel Ohl, kept working on silicon. In 1939, he had made progress in realising that the unpredictability was attributable to imperfections and impurities that were a source of free electrons in the crystals. The silicon work of Russel Ohl was later shifted to the solid state group led by London-born physicist William Shockley and American chemist Stanley Morgan. Their group included John Bardeen (a quiet theoretician from Wisconsin) and Walter Brattain (a quiet experimentalist and ex-rancher from Oregon). They, with Shockley, were destined to discover the junction or bipolar transistor in 1947 and receive the Nobel Prize for physics in 1956. The story is a fascinating one and has been told many times, for example, see Riordan and Hoddeson [35] and Shurkin [36].

11.5.1 Field Effect Transistor

What Shockley and his colleagues were really seeking to make was the FET. This can be regarded as a simple solid-state version of the triode vacuum tube that had been so successful. The realization of the FET ran parallel to, and was essential to, the development of the modern solid-state x-ray detector. The idea was to use a potential applied to the surface to form a depletion layer and in this way control the resistivity of the semiconductor. In 1926, a patent outlining similar ideas had been filed by Julius Lilienfeld [37]. Lilienfeld was a Polish professor of physics at Leipzig University who emigrated to the United States in the same year. Figure 11.2 shows the defining figure from his patent. A schematic of this arrangement is shown in Figure 11.3.

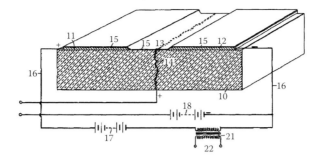

Figure 11.2 FET design drawing by Lilienfeld, 1926, taken from his patent. (From Lilienfeld J.E., Method and Apparatus for Controlling Electric Currents, US Patent 1745175, 1926. With permission.)

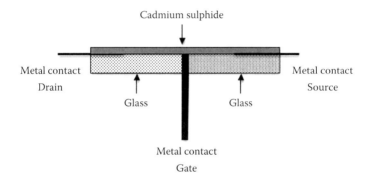

Figure 11.3 Schematic showing Lilienfeld's FET design.

Lilienfeld's proposal was to apply a voltage between the two horizontal metal contacts and to modulate the current flowing in the thin layer of amorphous copper sulphide by applying an oscillating voltage to a third metal electrode. From the text in the patent, it is clear that Lilienfeld was aware that by modulating the conductivity of the copper sulphide layer he could achieve signal amplification. The potential applied to the metal would either attract or repel electrons in the surface of the copper sulphide, creating a region with either a lower or higher conductivity. This region would either increase or decrease the current flowing through the copper sulphide. Today, we would say that, with the correct polarity, the applied voltage would partially deplete the copper sulphide. Lilienfeld's pioneering work was followed by other work that is admirably detailed in the book by Morris [38].

There is no evidence, however, that Lilienfeld's device was tested, but its invention and its claims establish the structure and the application of the device. Lilienfeld further refined his concept in a patent [39] 2 years later and extended it to cover lead oxide. The more detailed manufacturing and assembly techniques described suggest the possibility that such devices were made but possibly did not work. Another patent of what was essentially an FET, was filed by Oskar Heil [40], a German engineer working with New Zealander Ernest Rutherford at Cambridge in the UK in 1934. Further contributions also came from both Robert Pohl and Rudolph Hilsch working at Gottingham University, Germany in 1938. It was these patents that were to later thwart Shockley from filing his own patent for the FET, although there is no evidence that the earlier workers actually made a device that worked. It was the high density of *surface states* that screened any field effect and prevented any practical devices from working. These were the same surface states that were also to plague semiconductor radiation detector manufacturers over the years.

What Shockley and his colleagues actually stumbled upon was the germanium junction transistor, a device that depends on the *diffusion* of charge rather than the *depletion* of it. Germanium had been produced during World War II by Purdue University for microwave detectors based on the cat's whisker. In the Bell Laboratories process, water was used for cleaning and this dissolved any surface oxide that had formed on the germanium. Had they used silicon with its stable natural or 'native' oxide, they perhaps would have seen its beneficial effects regarding the inhibiting of surface states and might have invented the MOSFET earlier. As it was, there was little understanding of the germanium transistor (which used as its basis *minority carrier injection from an emitter*) at the time, except that it *worked*. Because of the problems of water condensation in his vacuum cryostat, Brattain decided to immerse his devices in an organic liquid. By good fortune, this had the effect of reducing the surface states considerably and made it possible for him to observe true field effects. The effect worked better on germanium and he decided to use GeO_2 for his 'FET'. However, as we have observed, the oxide washed off and he ended up with *two* contacts on the germanium crystal. He had made the first bipolar junction transistor. The date was Christmas eve 1947. The world had entered the age of semiconductor electronics.

The role of Shockley in the invention has been debated many times. It seems Shockley had become bored with the way the work was progressing and was not involved in the actual invention. Brattain, in particular, therefore resented Shockley rewriting history and taking center stage. It is true that Shockley was obsessed with the concept of the FET and his drive and determination to achieve it brought the somewhat unexpected junction transistor. He also wrote an important work, *Electrons and Holes in Semiconductors* (1950) [41], which became standard reading. After later sidelining Brattain and Bardeen from further work on the junction transistor, Shockley and Pearson were stimulated to revisit the FET and they successfully demonstrated the 'field effect' in 1948 [42]. Shockley was instrumental in the invention and in 1952 gave a classic theoretical description of the FET [43]. Soon after, Dacey and Ross [44] manufactured and tested such devices at Bell Laboratories. The first commercial FET was probably the 'Tecnetron' manufactured by the French Thomson-Houston Co. (CFTH) in 1958, designed by a Polish scientist, Stanislas Teszner. A picture of such devices is shown in Figure 11.4 taken from *Semiconductors* by Hyde (published in 1965) [45].

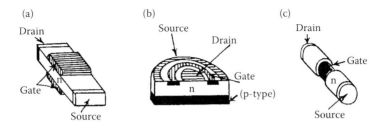

Figure 11.4 Details of different FET designs from Hyde's book. (a) FET with bar-type geometry, (b) FET with redial geometry (pre 1964) and (c) FET with cylindrical geometry (Tecnetron, 1958). (From Hyde F.J., *Semiconductors*, Mcdonald & Co., London, 1965. With permission.)

Also shown in this figure is a device (b) with radial symmetry. This is the forerunner of the design of many of the modern low-noise, low-capacity FETs (see Figure 1.48). The early semiconductor detectors were being used with thermionic valve amplifiers, but the emerging developments on junction transistors and FETs were soon to replace these components (see Section 11.15).

The Tecnetron did not progress well in the markets. Its transconductance was low and its gate leakage was high. However, it did establish a product line, and a company called Crystalonics (Cambridge, MA) produced the next product in the United States. This product (C610) was made from silicon, and we believe it was the first product to be based on the planar process using silicon dioxide. Its pinch off voltage was 20 V and its transconductance was between 100 and 400 μ Siemens at a drain voltage of 24 V. The gate current was less than 100 nA.

A patent was issued to John Wallmark of RCA in 1957 that used an insulating layer beneath the gate electrode. Chemically treating the surface to produce a film of hydrogenated germanium monoxide formed the insulating layer making the first MOSFET. The FET story is taken up again in Section 11.10 after 1959, when Martin 'John' Atalla showed that thermally grown oxides stabilised silicon surfaces and planar technology was established. In the early 1960s, RCA was to pioneer silicon bipolar metal–oxide–silicon (MOS) devices. Only with these advances could anything like the modern low-noise silicon JFETs and MOSFETs be achieved. Their implementation in x-ray detector amplifiers around 1964 is discussed in Section 11.15.

11.5.2 p–n Junction

As-grown p–n junctions in silicon and germanium were first noted at Bell Laboratories in the early 1940s. In an effort to produce a single crystal, Russel Ohl noticed that when silicon was melted in a vertical tube and slowly frozen from the top down the resulting rod showed a very large photovoltaic effect. The current through a sample depended on heat and light. When he showed this to Brattain, the latter concluded that there was some sort of barrier at a physical junction in the crystal that had different impurities on either side. They had stumbled on a p–n junction and the basis of all transistors and junction diode radiation detectors. When the rod was etched in nitric acid, the junction delineation was physically observable. The phenomenon was later explained by the difference of the segregation coefficients of boron and phosphorus impurities in the melt, the phosphorus being effectively swept to the bottom of the rod. For transistors, the doping level of the base should be low, the mobility of the injected minority carriers should be high, and their lifetime in the base should be high. The first demand required high-purity material, the second required a perfect single crystal, and the third demanded both high purity and high crystalline perfection. In 1942, Karl Lark-Horowitz of Purdue University (Lafayette, IN, USA) decided to work on germanium, not silicon. His reasoning was again simple: germanium had a much lower melting point and would be easier to purify. Karl Horowitz (Lark was his wife's name) came from Vienna and took the position of professor at Purdue in 1928. He set a long tradition in the study of semiconductors and radiation detectors at Purdue. His group reported doping to obtain p- and n-type material identified by the Hall effect and various types of photo effects. They also supplied the germanium to Bell Laboratories to make the first transistor, but they themselves seemingly missed the significance of injected minority carriers and transistor action.

With attention swinging to germanium in 1942, the mining company EaglePicher found an outlet for the germanium oxide powder that had been an unimportant by-product of their tin production. Some of this found itself in the hands of Randy Whaley, a graduate student at Purdue University. The university had received a grant from the Office of Scientific Research and Development to improve the crystal rectifier and microwave detectors. Working with a trial-and-error approach, the Purdue group eventually discovered that group III dopants produced p-type material and group V produced n-type material. They also discovered that n-type material made the better rectifiers. Another student, Seymour Benzer [46,47], developed crystal rectifiers that could withstand very high reverse bias (exceeding 100 V). In the end, Purdue, Bell

Laboratories, Massachusetts Institute of Technology (MIT), and the University of Chicago were all collaborating in producing good germanium.

11.6 1950s: Radiation Detection

In 1948, Bardeen and Brattain [48] and Shockley and Pearson [42] were describing the high resistivity region under reverse bias of the point contact to germanium as the 'exhaustion layer'. In the UK, it was referred to as the 'barrier layer', but now it is usually referred to as the 'depletion layer'. Shockley and Pearson [42] used Poisson's equation to estimate the depth of this layer and realised that it might be ideal as a relatively charge-free volume to detect the ionisation due to radiation and hence operate as a counter much like the bulk crystal counters and gas proportional counters. A year later, at the suggestion of Shockley, Kenneth McKay [49] used an n-type germanium point diode to verify that the junctions were indeed radiation sensitive under reverse bias. McKay observed pulses on a cathode ray oscilloscope (CRO) and noted that the pulse height from polonium α particles saturated as the bias was increased. He also used a commercial Western Electric Type 400B diode that had been cut open and the wax impregnation removed. Western Electric was the manufacturing branch of Bell Laboratories, or rather the parent company, AT&T. He noted that perhaps with a different geometry, a p–n junction diode might be a better choice, providing the capacitance could be kept small. This was a turning point in the concept of the semiconductor radiation detector. McKay was soon followed by Lark-Horowitz and his team [50] in 1950. They were able to demonstrate pulses from individual β particles and increased the junction resistance by cooling. They even suggested that neutron transmutation might be used to produce artificial barriers of greater volume. McKay continued and in 1951 reported work [51] using a point contact n–p–n junction and α particles to measure the lifetime of the holes produced. Later, he reported work using a specially *grown* germanium n–p junction [52] to measure the energy to create one electron hole pair (ω) as 3.0 ±

0.4 eV, remarkably close to the presently accepted value for germanium. As he pointed out, previous measurements using point contacts were invalid because of interfacial states and the possibility of avalanche giving an unknown multiplication factor. His report is something of a classic in that he recorded the individual pulses from the germanium. His circuit and the CRO display are reproduced in Figure 11.5.

The development of pulse-height analyzers or 'kick sorters' was in its infancy in 1951. Dennis Wilkinson and Otto Frisch

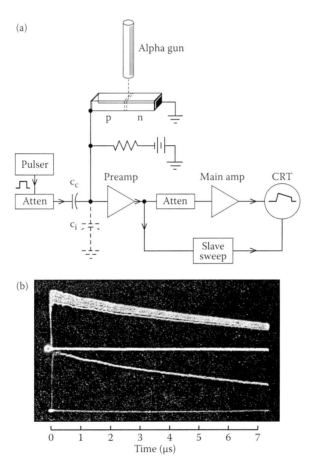

Figure 11.5 McKay's original circuit and oscilloscope response. (a) Circuit for detecting alpha particles using a germanium p-n junction as detector and (b) cathode ray oscilloscope traces resulting from detected alpha particles. (From McKay, K.G., *Phys. Rev.*, 84, 833, 1951. With permission.)

were working on them at the University of Cambridge, UK, and no commercial ones were available until 1952 (a 20-channel Atomic Instruments product built to an Oak Ridge National Laboratory (ORNL) design). Atomic Instruments later became Baird-Atomic. If McKay had used one, he would have been the first to obtain an energy-frequency histogram or 'spectrum' from a semiconductor detector by many years. His sample was a single crystal of 1–20 Ω cm resistivity germanium of only ~1 mm² cross section ~20 mm long. The junction was somewhere midway and the barrier width only ~5 μm at a few volts bias. It may be noted that he was also probably the first to inject a test charge for calibration purposes. He shone α particles from polonium (5.3 MeV) into the depletion region through the surface inversion layer parallel to the junction. He realised also that most of the charge deposited outside the barrier *diffuses* into it with little trapping or recombination. He took the maximum pulse height for his measurement assuming that there was some incomplete charge collection (ICC) for the smaller signals. He went on to surmise that the 'lost' energy ($\omega - E_g$) was in fact transferred to the crystal lattice (as phonons). Walter Brattain [53] continued to be involved in the effects of radiation on germanium. McKay later worked with McAfee [54] on the avalanche diode detectors again using α particles in silicon and germanium. Their measurements for ω for silicon (3.6 eV) and germanium (2.9 eV) were again remarkably close to the presently accepted values.

Thus, it was realised that the depletion region formed by the barrier had the ideal properties of a detecting medium, being high resistivity and relatively devoid of free carriers, like an insulator. All practical silicon and germanium detectors from this time on used reverse-biased specially formed p–n junctions. The purer the germanium, the deeper the depletion layer, the more sensitive the detector, and the lower the noise (capacitance). There was suddenly a great deal of interest from the nuclear physics community in the new 'junction detectors'. Work was quick off the mark, for example, in Russia [55], East Germany [56], United States [57], and UK [58]. The race was on to obtain better material for transistor manufacturing and purer single crystals for radiation detectors with deeper depletion at reasonably low bias. The early texts on semiconductor radiation detectors emphasise this interest in nuclear physics and in particular nuclear structure physics [30,31,59–61].

11.7 1960s: Story of Silicon and Germanium

Silicon and germanium detectors represent more than 99% of the world market for x-ray and gamma-ray spectrometers. Silica (Latin *silicis*, flint) was prehistoric man's first tool and its oxide was used in glass manufacture. Amorphous silicon was known from Humphrey Davy and Gay Lussac's time in the early 1800s. In 1854, Deville succeeded in preparing crystalline silicon. Mendeleev had predicted the existence of germanium in 1871, realising that it would have similarities to silicon and naming it 'ekasilicon'. It was first isolated from argyrodite in 1886 by Clemens Winkler, who named it after his homeland, Germany. All the work during World War II and transistor work at Bell Laboratories had been carried out with polycrystalline germanium. Shockley believed that as the junctions were of such small dimensions, polycrystalline germanium would suffice and saw no reason to support germanium crystal growing. However, in 1950, Gordon Teal persuaded him otherwise, and he, with John Little, reported a crystal growth method [62] in the same year. William Pfann [63], working at Bell Laboratories in 1952, put the case thus: 'Of what value is a detailed study of some substance which contains unspecified impurities or crystalline imperfections that later are shown to have a critical bearing on its behaviour'. Growing crystals was traditionally carried out pulling a carrier vertically from a melt of the substance. This was the method laid down by the Polish scientist Jan Czochralsky in 1918 and by Percy Bridgman, who was working at Harvard at about the same time, and is still known as the Czochralsky or Bridgman method. In 1950, Teal and Little and Rob Hall, who was working at GE [64], reported the pulling of single crystals of germanium from the melt and using directional cooling to achieve the segregation of impurities. By this means, they could achieve levels of dopant concentration of $\sim 10^{14}$/cc. Teal moved to TI in 1952, and improvements were steadily made in material and design with silicon junction transistors demonstrated by TI in 1954 with a 10-MHz cut-off frequency. Rob Hall was studying electron–hole recombination in germanium in 1955 [64].

Using the segregation idea observed in the freezing process that gave rise to the p–n junction described above, Pfann [63] introduced 'zone refining' to purify the material in 1952. This used one or more molten zones that are moved through the ingot at about 10 cm/h. This swept the impurities along and reduced the impurity level to below 1 part in 10^{10}. An improvement to the zone refining method was made by Henry Theuerer [65], in the same year, by using a vertical silicon ingot supported at both ends and scanning the floating molten zone down it. Silicon material manufactured in this manner is referred to as 'float zone' (FZ) silicon. The surface tension of the molten silicon is sufficient to stop it from running away and hence it never comes in contact with the outer vessel that had been causing contamination of the crystal. In practice, the process was usually carried out in a pure helium atmosphere. Unfortunately, the FZ refining method was not so effective in removing boron (the segregation coefficient of 0.8 is too close to unity) and produced p-type material (but which, as we will see later, could be lithium drifted). The next advance was to 'seed' the molten zone at its starting position using a small single crystal, a method used to obtain lifetimes of several milliseconds in germanium (see Cressell and Powell [66], Marconi, UK, 1957). The careful control of growth rate and temperature gradients also paid dividends.

By 1955, EaglePicher was supplying 95% of all raw germanium used in research and production in the United States. There was now a great deal of commercial interest outside Bell Laboratories in growing the silicon and germanium crystals. Obviously, the large vacuum tube companies such as RCA, Sylvania, GE, Phillips, Westinghouse, and Philco, were interested, as were chemical companies such as Monsanto, duPont (Lombardo, Atkinson, Kelermen) and Merck, in Danville, PA (Benedict). Also, Hughes and some small start-up companies such as TI, Fairchild and Transitron were investing in the new semiconductors for transistors. Some of the large companies such as GE (Rob Hall), Phillips Corp., NY (Johannes Meuleman), RCA, Montreal (McIntyre, Williams, Webb), Westinghouse

(Church), and Harshaw (Nixon, Schmidt, Stewart) were also interested in manufacturing the new junction radiation detectors. The latter had purchased Molechem (Baum), manufacturers of surface barrier detectors in 1964. Merck was the largest supplier of silicon for detectors by 1960, but duPont and Dow Corning also supplied silicon and Sylvania and Semimetals, germanium. By the mid 1960s, specialist suppliers of crystals such as Semi-Elements Inc. were also offering silicon and germanium and other materials specifically for solid-state detectors.

Outside North America, in Germany, the long established company Wacker-Chemie GmbH started high-purity silicon production in 1953. Hoboken (based near Geel, Belgium, now 'Umicore Hoboken') started by processing germanium from the impurity metal in the refining of raw copper ore by the mining company Union Miniere (Vieille Montagne). Philips, Eindhoven, in the Netherlands, was also an early producer and involved in detector manufacturing. In the UK, Ferranti had been involved in the radar war effort, and moved into transistor manufacture. They became the first European company to produce a silicon diode (1955). Plessey, also heavily involved in the war effort, took an interest in semiconductors and got a kick-start with a licensing agreement from Philco in 1951. By 1957 they were manufacturing transistors under the name 'Semiconductors Ltd'. At first, these were germanium, but in 1958 they decided to switch to silicon. In 1954, GEC (Wembley, UK; not to be confused with GE, Schenectady, USA) were making germanium ingots of 1–2 kg and high-quality ingots up to 5 kg by 1957 and probably led the world in growing germanium at this time. In 1961, they merged with their main competitor Mullard (part of Philips) and formed Associated Semiconductor Manufacturers (ASM) Ltd. They were making n-type germanium for transistors in Stockport, UK. A.R. Powell was working on germanium at AERE, Harwell [67], and, Billig (from Germany) [68] was growing ingots at Associated Electrical Industries (AEI), based at Aldermaston, UK. By 1962, AEI had made the first Ge(Li) detector [69] and, by 1969, a good 32 cm^3 planar Ge(Li) from the material [70]. At the end of 1966, Harwell established a 2-year contract with GEC Research, Wembley, for p-type germanium with a vacuum puller for 3-in crystals set up under S. Bradshaw. They also supplied germanium to Chalk River, Canada.

With the advent of lithium compensation [71] in 1960, which is discussed further in Section 11.14, the problem of boron impurity in both silicon and germanium was largely eliminated. However, other impurities and dislocations in the materials still had to be minimised. The first germanium grown especially for Ge(Li) detectors was probably produced by Adda and Benson [72] at Bell Laboratories, Allentown, in 1968. The detectors were made and tested by de Witt and McKenzie [73] at Murray Hill. At the second Gatlinburg Conference (1969), Webb et al. [74] of RCA reported that the ASM germanium gave the best resolution Ge(Li) detectors of samples taken from four sources. However, the ASM subsidiary of GEC was broken up in 1970. By then, all the centres involved with the lithium drifting of germanium were obtaining material from where they could, for example, from Sylvania, Hoboken, Vieille Montagne, ASM, and Semimetals, but achieving low yields. At some point, Sylvania lost the process altogether [75]. Hoboken Czochralski–grown germanium seems to have been most favoured. Chalk River reported yields for the period 1967–1969 of less than 9% for 'very good' crystals. Lawrence Livermore Laboratories reported similar yields. The problems included slow drifting and trapping. The cost was also a factor ($9/g for good material). Thus, many centres—Chalk River, Berkeley, Livermore, ORNL, Lucas Heights (Australia), and Belgrade— were setting up to grow their own (mainly Czochralski) material. The advent of high-purity germanium (see Section 11.22) in 1970 would change all that.

With the switch over from germanium to silicon transistors, the problems of obtaining silicon for lithium drifting were not so acute. Resistivities of 100–1000 Ω cm p-type material could be drifted. The high resistivity silicon for junction detectors was ~10 kΩ cm p-type and was supplied by companies such as Wacker (Germany) and Topsil (Denmark). At first, the high resistivity material occurred more or less by accident. In the later passes of purification, a p–n junction was sometimes produced somewhere along the crystal and the high resistivity material was obtained either side of it. For depletion depths of greater than 2 mm, purer material was required.

11.8 Other Materials

As mentioned in Chapter 10, CdTe detectors had been studied since around 1957, and early references were given in the Asheville Conference (1960) Proceedings (p. 194) [76]. Large crystals had become available from Philips, Eindhoven (work of D. de Nobel), since 1959. After the successes with both Si(Li) and Ge(Li) detector manufacturing, the question remained: could drifting (lithium or some other interstitial dopant) be applicable to other materials such as GaAs and CdTe? It had been shown [77] that lithium acted as an acceptor in GaAs, and efforts were made to make detectors at a number of centres. The most successful seem to have been the Russians [78,79] at the Ioffe Institute around 1966, where they lithium drifted *n-type* CdTe to make γ-ray detectors with 200-μm-deep depletion. They achieved 5% resolution measured with Cs-137. Their conclusion, however, was that the defect density in the material needed to be reduced drastically.

By the mid 1960s, Hughes Research Laboratories had had programs to look at CdTe, the most promising of the compound semiconductors. Their progress may well be best summarised by a one paragraph report by Jim Mayer [80] at the Gatlinburg Conference in 1969. 'We have just seen the first photopeak from CdTe using 400 keV γ-rays after four to five years!' HgI_2 was known as a photoconductor from 1959, but the first results as a γ-ray and x-ray detector did not come until the early 1970s [81,82]. For further details on the history of HgI_2 and other WBGS, see Chapter 10. Commercial exploitation of WBGS did not begin until the 1980s (see Section 11.18).

11.9 1960s: Progress in Detector Manufacture and Spectroscopy

The discovery of x-rays by Wilhelm Röntgen in 1885 was one of the most unexpected happenings in science: the regular way that energy levels in atomic structures varied with atomic numbers, giving a predictable means of identifying elements by their *characteristic* x-ray emissions, one of the most beautiful. This only became apparent with the x-ray fluorescence spectra collected from many elements by Henry Moseley working under Ernest Rutherford at Manchester University, UK, around 1913 and 1914 [83]. In relating the x-ray lines to atomic number, he is generally regarded as the founder of XRS, but he was unfortunately killed in World War I at the age of 28. By the 1920s, these characteristic x-rays for most elements had been recorded using wavelength dispersive crystal spectrometers. Also, one of the most intellectually satisfying phenomena was the way this atomic quantisation effect was repeated at higher energies with the strong nuclear force in the atomic nuclei replacing the Coulomb force. The γ-ray spectra identified the nuclides present and gave an insight into nuclear structure.

The problem with semiconductors before the advent of lithium compensation was the lack of sufficient active material depth both for the photoelectron and the x-ray ranges. The x-rays themselves are absorbed exponentially in the material with the 50% intensity falling in the range 2 μm to 2 mm for 1–30 keV x-rays in silicon and 100 μm to 2 cm for 30–1000 keV γ-rays in germanium. Despite the advent of surface barrier contacts for x-ray 'windows' (see Section 11.13.2), the low energies were restricted by noise threshold to >~6.5 keV. The high energies were restricted by efficiency, but despite this there were attempts made to detect photopeaks without lithium compensation (see Section 11.9.2).

11.9.1 Centres of Activity

The year 1960 was being approached with reasonable supplies of single crystal silicon and germanium, and the race was on to develop radiation detectors for x-ray and γ-ray spectroscopy. In 1958, John Walter started making germanium surface barrier detectors at ORNL, work that was presented more generally in 1960 at several seminal conferences. The organisers of the Seventh Scintillation Symposium in February 1960 in Washington were persuaded to include some papers on semiconductor detectors. This was the first time that part of a conference had been dedicated to them, and Jim Mayer [84], who was then at Hughes Research Laboratories, was requested to review the status of semiconductor detectors. It was here that Blankenship [85] presented details of the manufacturing of surface barrier detectors at ORNL and his now well-known nomogram for silicon junction detectors that gave the parameters of the material necessary for adequate depletion. This meeting was followed by two meetings in Tennessee, first at Asheville in September [76] and then Gatlinburg [86] in October 1960. Workers from Europe were represented for the first time at both these conferences. The main players in North America were the government laboratories at Brookhaven (Miller, Donovan), working closely with Bell Laboratories (Gibson, Buck, Brown, Wagner, Radeka, Llacer), Lawrence Livermore National Laboratory, California (Armantrout, Smierkowski); ORNL, Oak Ridge, TN (Walter, Borkowski, Dabbs, Blankenship, Fox); Argonne National Laboratory (Mann, Haslett, Bilger), Batelle Laboratories, Washington state; and Chalk River, Ontario Canada (Bromley, McKenzie, Tavendale, Ewan, Malm, Dinger, Sakai, Fowler, Williams, Webb, Martini, McMath). In the private sector, besides Bell Laboratories, there were some other 'big boys', GE (Schenectady), Hughes, (Los Angeles), Phillips, Harshaw, Amperex (subsidiary of Philips, Eindhoven, the Netherlands), Union Carbide, and RCA (Montreal, Canada) (Figure 11.6).

Figure 11.6 A selection of 'Logos' from large companies involved in this emerging industry.

Also there were the universities—Purdue University, LaFayette, IN (Mayer); University of California, Berkeley, CA (Goulding, Jaklevic, Elliot, Madden, Hansen, Lothrop, Stone, Walton, Malone, Landis, Pehl); University of California, Stanford, CA (Zulliger, Aitken, Drummond), California Institute of Technology (Nordberg, Emery); University of Pennsylvania, Michigan State University, and Yale University (Bromley). Some other U.S. universities were also manufacturing for their own use, for example, Western Reserve, Northwestern, and McMaster in Canada. Other work was also taking place in universities outside of North America: Paris (Amsel, Baruch, Swierkowski, Coche), Leningrad (Ryvkin, Matveev, Alferov), Prague (Tauk), Giessen (Roemer), Karlsruhe (Meyer), Bir mingham UK (England, Hammer), Manchester UK (North-rop), Liverpool UK (Sareen). There were also several small start-up companies in the United States consisting of a few people. Commercial interests often were in geographical clusters around the source of their technology and key employees. Nucleonics companies Tennelec, Oak Ridge (Fairstein, a spinoff from ORNL), and Sturrup Nuclear/Canberra, Middletown, CT. Detector manufacturers included ORTEC (another spinoff from ORNL, Oak Ridge, TN), Technical Measurements Corp. (TMC; Newhaven, CT), Nuclear Diodes Inc. (Illinois), and Electro-Nuclear Laboratories (Mountain View, CA). Briefly, even well-established companies such as Duncan Electric in Lafayette (whose founder, Thomas Duncan, was an early benefactor to Purdue University) were involved. There was also Solid State Radiations Inc. (SSR) in Los Angeles, which was set up by ex-Hughes people and nuclear physicists to address the needs of nuclear physics, space applications, and later medical applications. Intertechnique was also well established in France. These and other significant companies in the history are discussed in Section 11.18.

Nuclear physicists were the users and were involved traditionally with the development of particle detectors such as gas proportional counters, spark chambers, wire chambers, and scintillation counters. They also knew the desired characteristics of the new generation of detectors and were heavily involved. However, it did prompt H.E. Wegner of Los Alamos Laboratories to remark at the Asheville conference

Figure 11.7 Company 'logos' for start-up businesses appearing in the market.

that 'A nuclear physicist constructing a semiconducting detector is like a carpenter fixing an automobile with a crowbar'. This trend in development and operation of such specialised industries was to be repeated many times, including, it may be said, the present authors! However, Bertolini and Coche concluded in their excellent book, *Semiconductor Detectors* (1968) [87], with different sentiments: 'We have a personal conviction that Experimental Nuclear Physicists can make superlative semiconductor detectors, irrespective of the effort involved, provided the Solid State Physicists can supply adequate crystals'.

In the wake of the launch of the Russian satellite *Sputnik* (October 1957), President John Kennedy made his famous speech in 1961 declaring the determination of the United States to 'put a man on the moon by the end of the decade'. This rekindled the university, government, and industry cooperation experienced previously in the war years. Money poured into space flight research and low-power miniaturised devices. Together with military interests, space flight boosted activity in semiconductor research in the United States in the 1960s. Semiconductor detector manufactures in United States benefited indirectly (better material, transistors, and FETs), but some of it found its way directly to the manufactures of small silicon junction radiation monitors. SSR [88], supported by contracts with the Aeronautical Systems Division of the Wright-Patterson Air Force base in Ohio, manufactured Si(Li) particle detectors that flew on a U.S. Navy satellite in June 1961. The Discoverer series of satellites had already flown with a linear array of junction detectors forming a β-ray spectrometer [89] to study the Van Allen belt. Bell Laboratories was involved in supplying silicon junction detectors made from the highest available resistivity silicon (10–20 kΩ cm) for the Explorer and Telstar [90] satellites in 1964 and was also involved in supplying Si(Li) particle detectors [91] for space applications in 1966. The cost of the early commercial semiconductor detectors was high, and many centres of research around the world were encouraged to at least attempt to make their own.

Outside North America in the late 1950s and early 1960s, there was little money available and much less activity. However, detectors were being investigated by users in France at CEN Saclay (Koch): CSF (Compagnie Generale de Telegraphie sans Fils) at Puteaux was one supplier. CSF, headed by Maurice Ponte (a Ph.D. physicist), was both large (15,000 employees) and research led (1400 engineers). The company became part of Thomson-CSF in 1968. The French Thomson-Houston itself, Compagnie Francaise Thomson-Houston (CFTH), was originally a sister company of GE in the United States, as was the British Thomson-Houston (BTH), which became AEI in 1959. 'p–i–n' diode detectors were manufactured by CFTH in France (Maillard, Bobenrinck). From CFTH, the modern Thomson Group evolved, demerging in 1999 to form Thomson Multimedia and Thomson-CSF. La Radiotechnique (RTC) at Caen was a French subsidiary of Philips and supplied Ge(Li) [92] and n-type silicon detectors in the United States through Amperex. Johannes Meuleman, one of their chief workers, later moved to the Phillips Corp. (New York, USA). Compagnie Lignes Telegraphiques et Telephoniques (LTT), at Conflans, Sainte Honorine, was mentioned by both Georges Amsel and Lydie Koch in 1960 as supplying detectors. For the history of Intertechnique, see Section 11.18.

In the UK, there was activity at the Atomic Weapons Research Establishment (AWRE) Aldermaston (Davies, Northrop, Gunnerson, later Muggleton and Ellis), AEI Aldermaston (Freck, Wakefield), AERE Harwell (Dearnaley, Whitehead, Blamires, Howes), as well as the Services Electronics Research Laboratory (SERL) (Hilsum, Moncaster, Gibbons, Simpson, Northrop). In the Industrial sector, besides AEI, which, as mentioned, produced the first Ge(Li) in 1962, there were a number of nucleonics companies. Elliott Bros. (London) was an independent long established company that entered into the semiconductor detector business making surface barrier detectors. Isotope Developments was founded in 1950 by a team of scientists from Harwell, to design and manufacture radiation measurement equipment on a site in Reading and later became part of the Elliott Nucleonics Division. They were taken over by GEC in 1967 (see Section 11.18). Mullard (ASM) had been growing germanium for transistors since 1959 [93] and by 1970 had a program to make Ge(Li) detectors with AERE Harwell and AWRE Aldermaston [94].

In West Germany, by the mid-1960s work was being carried out at Karlsruhe (Meyer), Hamburg (Andersson-Lindstrom, Zausig), Technical University Munich (TUM), (Bernt, Eichinger), Max Planck Institute (Staeudner), and Heidelberg University (Kemmer). They were using Hoboken germanium and Wacker-Chemie silicon. Staeudner was later manufacturing Ge(Li) detectors for Nucleonic and Electronic GmbH (Nucletron), in Munich. Josef Kemmer was later to be involved with the introduction of lithography and planar processing with his own company, Ketek (see Section 11.25).

In Italy, Euratom at Ispra (Bertollini, Cappellani, Restelli) used Montecatini silicon (after a takeover, this company became Montedison), and Emilio Gatti was starting work at the Milan Polytechnic, particularly on the front–end electronics. In Japan, groups were late in starting and were mainly based at Rikkyo University, Tokyo, where Ge(Li) detectors were manufactured. The work by Ishizuka, Husami, and others mainly appeared in Japanese reports between 1968 and 1974. In Australia, Ge(Li) detectors were being made by Alister Tavendale and George Ewan at Lucas Heights near Sydney, using Sylvania germanium. They also spent time later working at Chalk River, Canada.

Activity behind the Iron Curtain is more difficult to assess because of the Cold War censorship that existed then. Soviet scientists were only rarely allowed to travel and were discouraged from publishing in non-Soviet journals. Even those papers that were translated into English were not widely read in the West. There were strong teams in Russia (Alferov, Ryvkin, Konovalenko, Maleev, Yuchkevich, Airapetyants, Vitovski, Vul), and in East Germany (Vavilov).

11.9.2 Semiconductor X-Ray Spectrometers

Because of the intense activity in 1960–1961, it is difficult to ascertain which group was the first to observe a spectrum showing a photopeak using a semiconductor detector. Previous attempts had only seen the low energy Compton edges but the Bell Laboratories–Brookhaven collaboration [95] first reported the detection of γ-ray photopeaks in June 1960, using a silicon shallow diffused junction detector with a depletion depth of about 0.8 mm. This was also reported at the Asheville Conference [76]. Their Co-57 spectrum is reproduced in Figure 11.8.

Soon after, Ms. Lydie Koch [96] and colleagues at CEN Saclay, France, reported a γ-ray peak using a silicon p–i–n diode manufactured by CSF. This work was also reported at the Asheville Conference. They irradiated the diode *parallel* to the junction to increase the detection efficiency. In this way, they could get more than 1 mm depth of depleted silicon in a 1 kΩ cm n-type Si wafer (8 mm × 1.1 × 0.28 mm). Her spectrum for the 279-keV γ-ray from an Hg-203 source is reproduced in Figure 11.9.

The statistics appeared rather poor but from the maximum pulse amplitude, they were able to deduce ω ~ 3.53 eV. This agreed well with the measurement of Blankenship and Borkowski [97] (3.50 ± 0.05 eV) and Mayer [98] (3.5 eV) for α

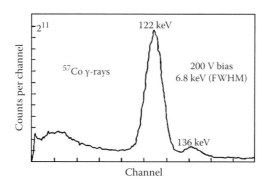

Figure 11.8 Co-57 spectrum using a shallow, silicon diffused junction detector, 1960. (From Dabbs, J.W.T., Walter, F.J. (Eds.), *Semiconductor Nuclear Particle Detectors*. (Asheville Conf. 1960), NAS-871, Washington, D.C., 1961. With permission.)

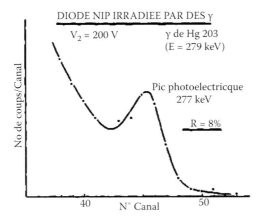

Figure 11.9 Hg-203 spectrum using a silicon p–i–n diode, exposed parallel to the junction, 1960. (From Koch, L. et al., *Comp. Rend.*, 252, 74, 1960. With permission.)

particles, and the prediction by Shockley [99] at the Prague meeting in 1960 that ω depends essentially on the semiconductor energy gap and the total phonon energy emitted. The idea of increasing the efficiency using transverse radiation was also used later by Harry Mann [100] and others at Argonne and was even used in germanium for 160 MeV protons [101] from the Harvard cyclotron. Side-entry detectors with increased energy efficiency range (e.g., [102]) have been described many times since.

From the discussions at the Asheville Conference, Mayer may also have already observed photopeaks using his new Si(Li) detector in 1960, and Baily et al. [103] and Mayer et al. [104] reported these in April and May 1961. Their result for Cs-137 (662 keV γ) is shown in Figure 11.10.

Soon after, Mann et al. [105,106] at Argonne reported the 14.6- and 662-keV photopeaks from Cs-137 with both room temperature and liquid nitrogen cooled Si(Li) detectors. Their best resolution was ~9 keV FWHM (full-width half-maximum) at 662 keV (see Figure 11.11).

It was realised by Nybakken and Vali [107], who were working for Boeing in Seattle, WA, that silicon detectors might be useful in Mossbauer studies. In a quite pioneering work in 1963, they constructed a vacuum chamber with a beryllium window and borrowed a guard-ring Si(Li) detector from SSR

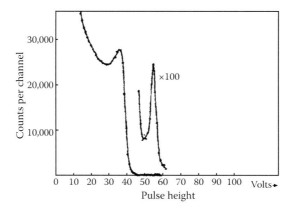

Figure 11.10 Cs-137 using a Si(Li) detector, 1961. (From Baily, N.A. et al., *Rev. Sci. Instrum.*, 32, 865, 1961. With permission.)

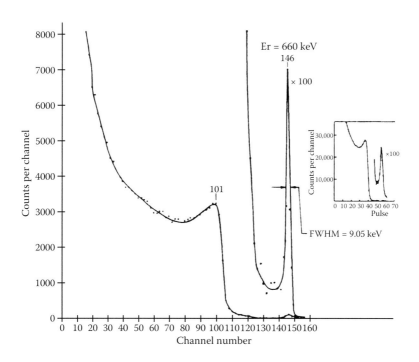

Figure 11.11 Cs-137 with a cooled Si(Li) detector, 1962. (From Mann, H.M. et al., *IRE* NS-9, 4, 43, 1962. With permission.)

(1 mm drift depth) and a preamplifier from ORTEC. They went so far as to cool the FET (in fact, the whole preamplifier), probably being the first to do so. Using a pulser, they estimated the noise to be 2.7 keV (see Figure 11.12).

Mayer remarked at the Asheville Conference that germanium had not been investigated yet but should have better efficiency. As mentioned previously, Freck and Wakefield [69], working at AEI in the UK, were to confirm this prediction in 1962 by making a 1.5-mm lithium drifted depth Ge(Li) detector cooled with liquid air in a vacuum cryostat. Their spectrum for Cs-137 is reproduced in Figure 11.13.

Toward the end of 1962, Webb and Williams [108], at RCA, Montreal (in a program with Chalk River), had drifted germanium to a depth of 5 mm and demonstrated much improved spectra. Figure 11.14 shows their Co-57 spectrum.

They cooled with liquid nitrogen and warned that the detectors 'would need redrifting each time they are operated'. Tavendale and Ewan [109] at Chalk River, Canada, followed soon after with an 18-mm-diameter, 8-mm-deep Ge(Li), and they were using them for nuclear spectroscopy. Fox [110] and others at ORNL also made small planar Ge(Li) detectors cooled by immersing them in liquid nitrogen in a small glass dewar. By 1966, small encapsulated Ge(Li) detectors were

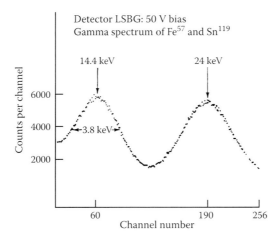

Figure 11.12 Spectrum of Fe-57 and Sn-119 using a guard ring Si(Li), 1964.

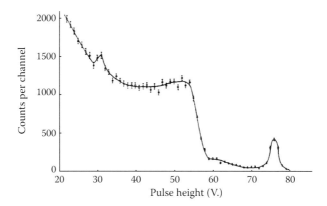

Figure 11.13 Cs-137 using a cooled Ge(Li) detector, 1962. (From Freck, D.V., Wakefield, J., *Nature*, 193, 669, 1962. With permission.)

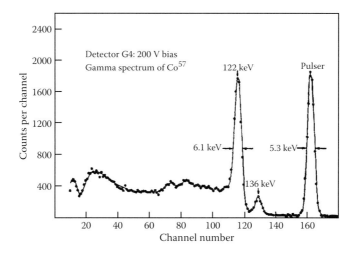

Figure 11.14 Co-57 Spectrum using a 5-mm Ge(Li) cooled to LN temperatures, 1963. (From Webb, P.P., Williams, R.L., *NIM*, 22, 361, 1963. With permission.)

being commercially manufactured. Figure 11.15 was accumulated with a Philips, Eindhoven [111], detector and illustrates the vast improvement over scintillation counters taken with the same specimen (Ho-166m).

Manny Elad, working toward a Ph.D. degree in 1965 at the University of California, Berkeley, did pioneering work with Michiyuki Nakamura [112,113] on FET preamplifiers (see

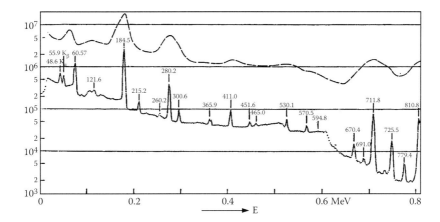

Figure 11.15 Ho-166m spectrum using a scintillation counter (top curve) and a cooled Ge(Li) detector, 1966. (From Hofker, W.K., *Philips Tech Rev.*, 27, 323, 1966. With permission.)

Section 11.15). He worked on an amplifier with the first-stage FET cooled in the cryostat alongside a small Ge(Li) crystal (made by Guy Armantrout at Livermore) and a Si(Li) crystal (made by Lothrop and Smith at Berkeley). One of the problems recognized by this group was that of vibration or 'microphony' caused, in this case, by the boiling liquid nitrogen. Figure 11.16

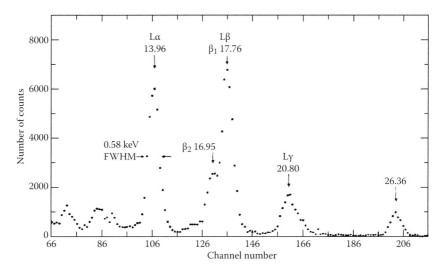

Figure 11.16 Am-241 spectrum using a Si(Li) and a cooled FET, 1966. (From Elad, E., Nakamura, M., *IEEE* NS-14, 523, 1967. With permission.)

shows the Am-241 x-ray spectrum from a 5-mm-diameter Si(Li) after the cryostat was allowed to boil dry.

The reported resolution was 700 eV FWHM at 6.4 keV, but he surmised that this could be brought below 500 eV once the microphony problem could be solved. By 1964, semiconductor detectors for nuclear physics were well established with Ge(Li) detectors leading the way. But even in 1966, Si(Li) detectors were only being used as particle detectors, and gas proportional detectors were still preferred for x-rays (<100 keV). Si(Li) detectors for x-ray spectrometry were dismissed by Dearnaley and Northrop [61] as 'virtually useless, as its photoelectric absorption coefficient is only one fortieth of that of germanium'.

In retrospect, the gestation time of EDXRS using Si(Li) detectors was very long. McKay and others had laid the seeds at the end of the 1950s, but maturity in the form of applications did not come about for another 16 years. After the 1961–1966 spurt of development, the evolution of a useful Si(Li) x-ray detector at the end of that decade was incremental and painfully slow. This progress will be discussed later in terms of the new applications of EDXRF and EDXMA and the companies involved. However, technical progress had been made in five major areas: the understanding of surface states and their role in reverse bias diode leakage currents, surface passivation to reduce and stabilise the surface state density, contacts (in particular thin surface barrier contacts), lithium compensation, and low noise amplifiers. We will now look at these areas in detail.

11.10 Surface States and Nature's Gift of SiO_2

The Russian physicist Igor Tamm [114] had predicted the existence of surface states in 1932. Shockley [115] also published theoretical papers on surface states in 1939. But no one really anticipated how important these would be in practice. It was only the presence of surface states that first prevented the Bell Laboratories group from realising a working FET and similarly the related surface leakage currents that were dominating the noise in the new junction radiation detectors. Bardeen had calculated that it was sufficient for only one in every 1000 surface atoms ($\sim 10^{16}/m^2$) to have a surface state to completely mask any predicted field effect for an FET to work. The density of interface states certainly needed to be an order of magnitude lower than this. He considered how a voltage of 100 V applied across 0.1 μm of n-Ge on a quartz plate could influence surface charge and increase its conductance. Treating it as a field-plate capacitor the fractional change was calculated to be ~1% (it is 30% higher for germanium than silicon because of the greater dielectric constant). However, they could not measure anything at Bell Laboratories. The explanation, given by Bardeen [116], was that the electrons induced in the germanium are trapped by *surface states* and the presence of charged surface states caused conducting surface 'channels' in the germanium (see Section 5.1.9). For example, a positively charged state would attract a channel of electrons to the surface trapped in a two-dimensional well that could contribute to the conductivity of the surface. In n-type Ge, this was an n-type accumulation layer and in p-type Ge an n-type inversion layer.

The solution for Bell Laboratories was to get away from surface effects by making the carriers diffuse across an intermediate region in the bulk germanium made by two p–n junctions in the one single crystal, one forward biased and one reverse biased. This was the junction transistor, a bulk semiconductor device with all its parts easily manufactured by controlling doping during crystal growth. Full credit should go to the crystal growers, Morgan Sparks and Gordon Teal, using double doping Czochralski methods. The wide base meant long transit times, and the cut-off frequencies were low. The first junction transistors from Raytheon were marketed in 1953 with 1 MHz cut-off frequency (but note that the French Westinghouse 'transistron', already mentioned, came out earlier). Gordon Teal moved to TI in 1952, and improvements were steadily made in material and design with the first silicon, rather than germanium, junction transistor demonstrated by TI in 1954 with a 10-MHz cut-off frequency. These silicon transistors were in much demand by the military because they did not suffer thermal runaway, and it gave TI a lead of about 2 years. They grew rapidly between 1954 and 1958. Bell Laboratories came back with a 'mesa' design of transistor that was demonstrated in germanium in 1954 and in silicon in 1955 (with a 120-MHz cut-off frequency). The diffusions were through masks on the top, defining a 'mesa' or 'table region' (named from the geological features in Western USA), and this led the way to the true *planar* diffused transistor (see Figure 11.17) and later to integrated circuits in 1959. The 'planar' description refers to the fact that all the diffusions and contacts were made to the top plane surface leading to side-by-side development and *wafer* technology for manufacture.

One of those working on transistors at Bell Laboratories, Walter Brown [117], was studying the change in resistance of the surface inversion layers of n–p–n transistors and was coming to the conclusion that the increase of surface resistance was a function of the magnitude of the inversion layer and the field *normal* to the surface. Studies were also made on the effects of water and alcohol treatments on 'pinching-off' of surface currents when the resistance was high. More theoretical work was carried out by Schrieffer [118], also at Bell Laboratories, showing that the surface mobility of carriers could be considerably less than the mobility in the bulk. Brown's paper on surface channels acknowledges Sparks and Smith for the making of his experimental germanium transistors. Brown later worked closely with Laurie Miller [119] and Paul Donovan of Brookhaven on silicon p–n junction radiation detectors under the auspices of the U.S. Atomic Energy Commission, as did Walter Gibson [120] later. To reduce the

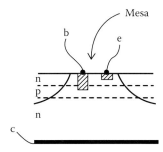

Figure 11.17 n–p–n mesa transistor defined in planar geometry. e, emitter; b, base; c, collector.

surface leakages and noise, it was essential to reduce the density of surface states and having done so render the surfaces stable. In short, a good surface passivation process was required.

As it turned out, the crucial element in planar technology was SiO_2, the native oxide of silicon. Unlike the oxides of germanium, it is very stable, being essentially an amorphous layer of quartz glass. It is resistant to most acids [but not hydrofluoric (HF)] and etches and possesses very good dielectric properties (moderate dielectric constant, high breakdown potential, and low dielectric lossiness). As an added bonus, the grown oxide forms an interface with silicon with very low surface state density in contrast to the unstable germanium oxide interface. Aside from protecting surfaces, it acts as an excellent passivation and very significantly acted as a barrier to the diffusion of dopants. Carl Frosch had discovered the formation of a thick oxide on silicon earlier in 1955 by accident at Bell Laboratories (water contamination in a diffusion capsule), but the full implication had not been appreciated at the time. This work was reported later in 1957 [121]. However, 'John' Atalla [122], also working at Bell Laboratories, is credited with finding a way to grow the oxide on silicon in a controlled manner. The silicon surface must be very clean and the oxide grown to a thickness of ~100 nm at ~1000°C in the presence of water vapour. As in Bell Laboratories' mesa diffused transistor, the process used diffusion through a mask but the oxide film on the silicon now formed the mask. The precursor to this was the silkscreen technique used, for example, to produce the proximity fuse for munitions during the war years. The diffusion automatically moves laterally in the silicon and under the oxide, as well as into it. This means that the junction where the diffusion meets the silicon is *under the oxide*. It acts as a near-perfect protection and passivation of the junction. The discovery of the thermal oxide technology eventually realised the JFET, MOSFET, and IC technology. The planar process itself is credited to a Swiss scientist Jean Hoerni (patent dated March 1962 and titled 'Method of Manufacturing Semiconductor Devices') working at Fairchild, California. Many people argue that this patent represents the process that revolutionised the electronics industry. A multitude of new devices were now possible. More than one device, or different devices, could be made on the same substrate. Junction protection, and hence reliability, was significantly improved. Hoerni detailed the advantages and pointed out that, although primarily aimed at silicon, it was not restricted to silicon.

The path to realising a working FET needed first of all the invention of the concept. This is credited to Lilienfeld. It also needed the understanding of surface states. This is credited to Bardeen. Finally, it needed the planar process with controlled diffusions and its brilliant use of the SiO_2 dielectric layer, and a reliable method of growing it is credited to Hoerni and Atalla. The theoretical treatments were credited to Schottky, Mott, Shockley, and others. The FET revolutionised the amplifier circuitry used for radiation detectors. It provided the first rugged compact low-noise sensor for the miniscule charge signals developed in the sensor. The noise was further reduced by cooling the FET and, in many cases, the detector semiconductor crystal also. For ultralow noise, it is still yet to be replaced by any other device. We have seen that the mainline silicon chip industries had found their saviour in thermally grown SiO_2. This was not applicable for germanium, but could it be applicable to silicon detector crystals? It was indeed tried at Bell Laboratories [123,124] and Berkeley [125–127]. However, the high temperatures required for the oxide growth (~1000°C and 120 atmospheres in steam) was a problem. The lithium diffusion, lithium compensation, and delicate surface barrier contacts were clearly destroyed by such treatments, and the diffusion of such undesirables as Cu and Au at these temperatures would degrade carrier lifetimes. The oxide therefore had to be grown on the silicon before any processing, and any reprocessing further down the line was out of the question. Furthermore, it was found [127] to be necessary to coat the oxide with epoxy or silicone to obtain the required added stability over several months. Although oxide growth as used in the mainline industries set a standard against which detector passivation could be measured, an alternative had to be found. Again, evolution was causing a branching from the mainline automated semiconductor industries. The advantages of lithography would materialise only many years later, with p–i–n diodes, SDDs, charge-coupled devices (CCDs), and the like, but now at least the behaviour of the surfaces was better understood, thanks to certain individuals such as Bardeen, Brown, and Buck [128] working at Bell Laboratories and theoretical analysis by Statz [129] at Raytheon.

11.11 Processing and Passivation

To quote McKenzie [130] in his review of 1979: 'There are almost as many recipes for making semiconductor counters as there are silicon and germanium crystals or, at the least, detector fabrication facilities'. We could add now that most detector fabrication centres probably had (and still have) more than one process (e.g., for backup or different applications). The basic processes were laid down in the 1960s and evolution has in many cases not taken them far. Commercially, low yields (20%–40%) were tolerated and process changes were regarded as risky. Each center had its own preferred techniques and had found forms of surface passivation suited to the particular characteristics of the surfaces left after their etching and quenching stages. In general, the silicon etches and cleaning processes (based on the RCA cleaning formula first used in 1965 and still withstanding the test of time) were carried over from the semiconductor transistor work. Most used variations on the process given by Dearnaley et al. [131,132] in 1961 being essentially etches by CP4 (3 parts conc. nitric acid to 1 part 48% HF acid) or CP4A (5 parts conc. nitric acid to 3 parts 48% HF acid and 3 parts glacial acetic acid). 'CP4' derives from 'Camp Process No. 4' etch, one of the formulae developed by Paul Camp at RCA in 1954. They recommended that the etchant be precooled in ice to achieve a mirror finish. The acetic acid was used as a buffer and the etch process was quenched in deionised (DI) water either at room temperature or ice. In some cases, particularly germanium, where any water was deemed to be highly undesirable, the quench consisted of copious amounts of methanol. The 'top hat' and inverted 'T' geometries (see Section 11.12) for planar detectors were favoured because they were thought to give better exposure and more control during the etching [133].

The diffused, ion-implanted, or surface barrier contacts had to be masked during the etching. There were three methods in common practice: Picein (Apiezon) black wax (a carryover from the encapsulation of transistors), masking tape [e.g., PVC or polytetrafluoroethylene (PTFE; Teflon)], or polypropylene clamps. An etching cell using PTFE clamps to seal the surfaces from the etchant was proposed by Albert Muggleton [134] (AWRE, UK) in 1965. This avoided the time-consuming painting and curing of wax and waste, in the form of expensive high-grade solvents and methanol. Although researchers and manufacturers often changed or combined techniques to suite, the West Coast companies USA (see Section 11.18) seemed to favour wax, particularly on the smaller planar crystals. The etching could take place with agitation in a PTFE 'basket' as carried out at ORTEC. It was quite a technical advance when it was discovered that a wax solution (usually in TCE) formed strong surface tension bonding with both silicon and germanium and with PTFE. The crystals could then be firmly cemented to 'spatulas' for more controlled etching, quenching, and cleaning. Furthermore, with the aid of a metal point, the serial number could be scribed through the wax and etched onto the crystal wall or front contact rim. The wax was very reliable as long as it did not get hot during etching. The tape method was generally used over larger areas (coaxial Ge(Li) and HPGe crystals). Any impurities in the adhesives (or wax itself in the case of Picein) did not appear to be a problem.

The tendency was for the etched surfaces of silicon and germanium to go strongly n-type mainly because of the ubiquitous water vapour. Hence, the remark of Statz (Raytheon) at the Asheville Conference (1960), 'I just generally tend to avoid water vapour. Period'. Quoting Statz again from the same Conference, 'It is a good idea to make the surface layer of different conductivity type from the bulk'. This was an inversion layer, that is, a junction, isolating the surface film from the bulk. Surface states were also important in transistor manufacture, and Bell Laboratories had carried out extensive studies. Buck [76,135,136] reported many variations of surface treatments and their influence on surface states and leakage current and, at the Gatlinburg meeting in 1969, on surface noise [137]. But as noted earlier, treatments were also dependent on the material as well as the basic etch and quench.

The general consensus forming in the 1960s was that a final rinse in TCE and methanol left surfaces clean and slightly n-type. Subsequent oxidation, in the absence of water, then drove them toward p-type and hence toward neutrality or as it was termed 'flat band' condition. This could be achieved with HF acid vapour or dips, using the diode current/voltage characteristics described by Buck and others to ascertain the surface states and how they were changing. The process could be restarted to some extent by using an HF dip to remove the oxide and DI water rinses, if it was deemed necessary. At some predetermined point a passivation could be applied to leave it usually slightly n-type.

There were many arguments at the time over whether a passivation coating was really necessary and there is no doubt that some devices stayed good after many transfers without any passivation. From a commercial manufacturing point of view a physical passivation in the form of a coating seemed desirable for the sake of repeatability and protection. Of course processes as described demanded individual crystal attention and the necessity of small batches. This was acceptable because the products were highly specialised and could demand a high unit price. Production was usually only counted in hundreds of detectors per year. Evolution was taking the semiconductor detector manufacturing industries away from the mainline mass production of the silicon chip to small labour-intensive cottage industries. The large companies were therefore eventually to lose interest, making way for small start-up companies.

Silicon grows a native oxide but this is only ~4 nm thick. It could be made thicker by oxidising in water vapour at a reasonably low temperature (~95°C). This was the method used by one of these small start-up companies, Simtec [138] (see Figure 11.18), in the mid 1960s.

The native oxide could also be replaced or augmented by a low temperature application of a glass-like material. Reasonably successful attempts were made, as can be noted from the 1960 Asheville Conference. Williams (RCA), in his paper, refers to a 'blue sheen from silicon glass' and Mayer in discussions says: 'The arsenic glass we used at Hughes Research Laboratories has drawbacks but is the best we have found'. Grainger (also

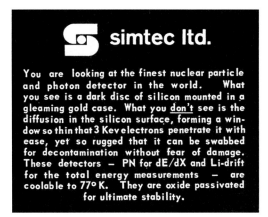

Figure 11.18 Simtec's claims regarding diffused contacts and oxide passivation.

at Hughes Research Laboratories) [139] at Gatlinburg in 1960 bemoans the flaking or cracking of the 'protective arsenic glass'. This form of passivation was used, but As could be a spectral contaminant. Glass coatings on silicon substrates, for example, by decomposition of siloxanes, were documented in the semiconductor industries since 1962 [140]. These were also being used at Hughes in 1965 [141]. In 1966, Goulding [142] remarked, 'One commercial manufacturer produces oxide passivated Si(Li)s but the oxide is not thermally grown'. This was probably TMC. In the same year, Dearnaley and Northrop [61], referring to passivation, wrote 'unfortunately there is not as yet either a complete understanding of the process nor a really satisfactory way of dealing with it'. Buck [137] summarised the various ways of depositing SiO_2, from RF sputtering to anodisation in solution.

The attention of most companies switched to conformal coatings that could be painted on and cured at moderate temperatures (<100°C) and could be removed for reetching if necessary. There are many examples. Mayer [76] said at Asheville (1960) that Apiezon (Picein) wax worked reasonably well, whereas Fox at ORNL used epoxies [143]. Lothrop at Berkeley [144] used polyurethane varnish ('Marvethane' by Sherwin-Williams). Xylene [145] was known to drive surfaces toward p-type, and xylene-based paints such as 'box-car red' (named from its use in model railways) were an early choice, for example, by Nuclear Diodes. Various 'photoresists' used in lithography had been developed, for example, by the Dow Corning Glass Company, for the mainline silicon industry. Other examples were Kodak KMER [146] and OCG 747. These various coatings could be used to drive the surfaces p-type or n-type and were also found to be useful passivations. By doping them before application with ionic materials such as iodine [143] solution, arsenic, or lead compounds, there was some further control of the surfaces (see Section 5.1.9). However, in certain applications such as EDXRF and total reflection XRF (TXRF), such heavy elements could be fluoresced by x-rays or photoelectrons. During 1967–1968, Sher [147], working at Simon Fraser University, British Columbia, experimented with evaporated calcium fluoride coatings and de Witt and McKenzie [148], working at Bell Laboratories, with calcium chloride quenches on germanium. Many workers [149–152] found that a quick rinse in hydrogen peroxide (a preferential etch along certain crystal planes) reduced the surface leakage current in germanium.

We have seen that SiO_2 is difficult to apply at low temperature, but the monoxide, SiO, is easy to thermally evaporate, and Dinger [153,154] at Chalk River first suggested, almost a decade later, that this was a good passivation for germanium surfaces. It is quite likely that the film converted to SiO_2 at least partially in the process [155]. This recipe caught on and was widely used. It was also applicable to silicon. For germanium, it was the passivation of choice but a dispute arose over the ownership of the process. It resulted in challenged patents, threat of lawsuits, and much ill will throughout the industry. There is still a question as to whether the technique is owned by any particular company. Dinger's passivation was widely used at least up to the development at Berkeley [156] of a sputtered amorphous germanium coating. Some control of the surfaces was accorded by a degree of hydrogenation (driving them toward n-type). Amorphous germanium or GeO was also used at AERE Harwell by John Howes (private communication).

The uniqueness of HF-treated silicon surfaces, making them hydrophobic, was recognized by Buck [157] back in 1958. Over the years, it was appreciated that, in the process, the dangling bonds at the surface were terminated by hydrogen, and this constituted a form of passivation. Hydrogen passivation using atomic hydrogen at 200°C was first tried by Benton et al. [158] in 1980 at Bell Laboratories and a few years later by Chabal [159] and Ubara [160]. Chabal and others have reviewed these techniques in their work, *Handbook of Semiconductor Wafer Cleaning Technology* [161]. The passivation can best be achieved with a series of HF acid dips of diminishing acidity ending in DI water as described in Section 5.1.10.

11.12 1960s: Evolution of Detector Geometries

The region where the p–n junction comes to the surface is a source of leakage current and noise, and it was Shockley [162] in 1954, who first realised that contouring of the silicon surface could be used to reduce the surface potential gradient. In the late 1950s, the French company LTT, as noted earlier, was making diffused junctions for silicon p–n detectors used at the University of Paris Van de Graaf accelerator. Here, Amsel et al. [163] were defining a circular window area in the 5-μm-deep Ga diffused surface by etching away a shallow shelf forming a 'mesa', similar to the concept of the mesa transistor (Figure 11.17). Figure 11.19 shows their design.

Using a 100-channel RIDL analyzer, they obtained heavy ion spectra and a resolution on 6-MeV α particles of ~40 keV at

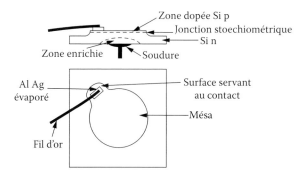

Figure 11.19 'Mesa' design applied to a Ga-diffused silicon junction detector. (From Amsel, G. et al., *NIM*, 8, 92, 1960. With permission.)

room temperature. Presciently, in May 1962 Harry Mann [164], noted in reference to the manufacture of Si(Li) particle detectors at Argonne that 'it would appear that the geometry of the exposed edge of the junction is a crucial factor in reducing noise as well as leakage current'. He did not elaborate on this, and this important observation does not appear to have been further investigated by them. Gerald Huth appreciated the value of a chamfer or 'top hat' geometry for controlling leakage current in 1963 [165,166] while working at the GE Space Sciences Centre on avalanche photodiodes (APDs). These required very high fields and bias voltages (1–2.8 kV) to achieve the avalanche process, and they used contouring to reduce the surface current and noise. His designs are shown in Figure 11.20.

By 1966, Huth [167] was recessing a 'window' into the p$^+$ side of his detectors (see Figure 11.20). This further reduced leakage currents (electron injection) and noise as discussed at Gatlinburg (1969) [137]. The recess protects the delicate junction from damage and contact pressure and was later used to good effect to protect the surface barrier junction on planar silicon, Si(Li) and germanium crystals. APDs were being widely used in EDXRF applications in the mid 1960s [168]. Working with Cassidy of AEC, Huth was able to detect 5.9 keV x-rays from a Fe-55 source with a resolution of 2.4 keV [169]. It was realised that the APDs should be particularly good for soft x-rays and, in fact, Huth at the 1969 Gatlinburg Conference [137] reported the detection of Al Kα x-rays (1.48 keV) with 30%–40% efficiency up to a temperature of 100°C. In the discussion at the end of Buck's paper on surface state noise, Huth remarked, 'The surface sensitivity of semiconductor devices has been studied by many people over the past 10 years. It was solved in the transistor area by the planar process (SiO$_2$). For higher resistivity material, such as used in high voltage rectifiers, it was solved by contouring. For the bulk charge densities of interest in Si(Li)s (~10^{11}/cm^3) the difficulty of control of the surface charge becomes severe. I do not believe that such control can be reproducibly achieved and one must, therefore, depend on other techniques, for instance, the pinching off of surface channels'. In this, he was correct. Cylindrical detectors *can* be made with near-perfect surface charge conditions as described by Brown, but to do it reproducibly in a production environment

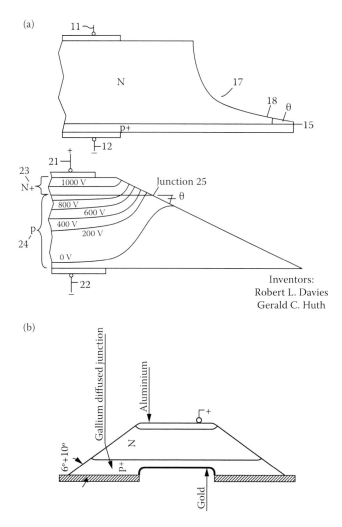

Figure 11.20 Use of 'top hat' geometry to control leakage currents. (a) 'top hat' geometry design for controlling surface leakage current showing equipotentials and (b) recessed entrance window. (From Huth, G.C. et al., *Rev. Sci. Instrum.*, 34, 1283, 1963; Huth, G.C. et al., US Patent 3491272, 1963; Huth, G.C. et al., NYO Tech. Report 3246-8, 1966. With permission.)

without resource to very time consuming and expensive equipment and laboratories we have, even today, to use contouring.

Si(Li) detectors were being fabricated at Bell Laboratories in 1964 by drifting lithium from a 1-cm^2 spot on flat wafers of twice this size, and there was considerable outward spreading during the drifting. It appears that it was the suggestion of Laurie Miller, as quoted by Coleman and Rogers [170], working at Brookhaven, that the drift profile could be contained if a 'top hat' geometry was used for the drift. Around this time (1963 to 1965), before his graduate studies at Stanford University, Jorge Llacer [171–173] was working in the Instrumentation Division at Brookhaven on semiconductor surfaces, and he must be credited as the first to fully appreciate the significance of Brown's work at Bell Laboratories, that is, that the magnitude of the field *normal* to the surface was instrumental in increasing the surface channel resistance. The drifting method of Miller fortuitously provided the means for turning the surface normal to the field lines by contouring (see Figure 1.59, Section 1.16.2). Llacer was using a platinum probe with a light source scanning the surfaces and studied the surface current variation with bias. This incidentally clearly showed that the depletion in a deep Si(Li) crystals was from the p-type region (x-ray window side), that is, the bulk material after lithium drift is slightly n- (or 'ν'-) type. Figure 11.21 shows the results for some of his scans [173].

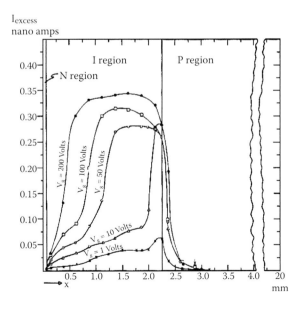

Figure 11.21 Scanning walls of a Si(Li) with a potential probe and a light source. (From Llacer, J., *IEEE* NS-13, 1, 93, 1966. With permission.)

He also surmised that the surface current was mainly
generated in the high field region at the junction boundary
between the p-type silicon substrate and lithium-compensated
silicon. He realised that by curving the *surface outward*, with
the usually n-type surface left after etching, the field could be
turned normal to the surface and pinch-off the troublesome
surface currents very significantly.

By 1966, Miller at Brookhaven was manufacturing 2-mm-
thick surface barrier Si(Li) detectors with a top hat geometry and Hayashi [174], and others at Bell Laboratories were
encapsulating such detectors for use in particle telescopes on
space flights. These detectors also incorporated grooves in the
Li diffusion to act as 'guard rings' to reduce leakage current.
Fred Goulding's group at Berkeley pioneered this concept in
the early 1960s [175–177]. They were studying surface leakage current and its associated noise, and this was one way
of controlling it on a cylindrical detector as shown in Figure
11.22. Although it was at first just a means of studying surface

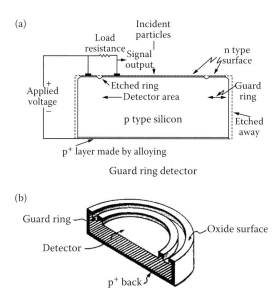

Figure 11.22 Use of a guard ring on a Si diffused contact detector.
(a) Design and biasing of a guard ring on a p-type silicon detector and
(b) section showing physical construction with outer guard ring. (From
Hansen, W.L., Goulding, F.S., UCRL 7436, 1960. With permission.)

currents, it was later used to minimise them, particularly when devices were required to operate at elevated temperatures or to act in anticoincidence to reduce spectral background.

They were also homing in on to the same top hat geometry with what they called a 'shelf structure' [178–180]. This work was also reported at the Gatlinburg Conference in 1969 and was followed by a patent [181] on the geometry. The mechanical shaping (grinding or drilling) to form the mesa on the thick Si(Li) crystals by the Berkeley group was usually carried out after the lithium drifting [182]. The alternative to the top hat was a deep groove or 'inverted T', also sometimes referred to as the 'Woo geometry' [183], after Ed Woo, who pioneered it at TMC/Kevex in California. Thus grew up two distinct planar geometries, both used commercially to this day. Figure 11.23 [173] shows the planar, top hat, and groove geometries.

Surface barrier contacts (see Section 11.13.2) were first applied to lithium-drifted detectors for the cleanup drift. For example, Mayer [184] applied them around 1962. An added advantage of the geometries described above was that the edges of the contacts could be terminated by applying the

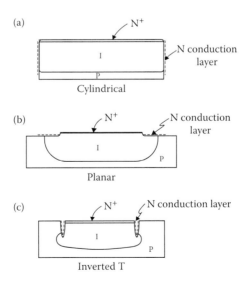

Figure 11.23 Top hat and grooved geometries. (a) Planar geometry, (b) top hat geometry and (c) grooved geometry. (From Llacer, J., *IEEE* NS-13, 1, 93, 1966. With permission.)

metal contact over the undrifted–drifted p–n boundary, thereby avoiding the edge breakdown problems indicated by Fox and Borkowski [185]. Ian Campbell [186], working at the University of Guelph, Canada, in 1971, devised a clever method of determining the profile of the active volume of detectors *in situ* using coincidence techniques and the annihilation radiation from a Na-22 source. Figure 11.24 shows the result on scans of an ORTEC planar Ge(Li).

Figure 11.24a shows the sensitive volume as measured by the technique and Figure 11.24b displays the dimensions of a similar crystal. This clearly delineates the drifted volume

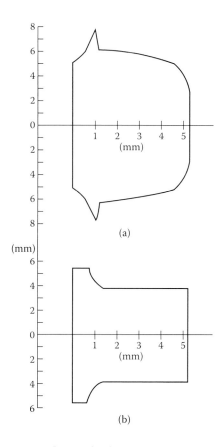

Figure 11.24 Sensitive volume of a Ge(Li) detector measured using Na-22 scans and a coincidence measurement method. (a) Results from scanning the crystal using a coincidence technique and (b) dimensions of a similar crystal. (From Campbell, J.L. et al., *NIM*, 92, 237, 1971. With permission.)

under the 'brim' of the top hat geometry. The geometries usually had cylindrical symmetry, but there were exceptions. For diffraction applications, it was important that all the diffracted x-rays from the sample were collected. The active area had to be minimised to optimise the energy resolution, and this led to rectangular strip or oval-shaped Si(Li) crystals. ORTEC, for example, was manufacturing crystals with active area 1.5 by 19 mm [187], and EDAX and Scintag were to use oval ones for special applications.

Planar Ge(Li) detectors, as we have seen above, used these geometries, but there was great pressure to increase the *volume* to achieve higher efficiency for very high energy γ-rays. It was clear that the planar geometry giving only ~5 mm depth was not appropriate for large active volumes. Miller et al. [188] had suggested a 'wraparound' (or coaxial) geometry for a Si(Li) detector in 1963, but this lent itself better to the large germanium ingots being produced a few years later. A coaxial geometry Ge(Li) was finally developed by Alister Tavendale and Howard Malm [189,190] at Chalk River, and a patent [191] was filed in January 1965. This coaxial geometry was also pioneered by Hans Fiedler [192], working at McMaster University, Canada. The detector consisted of a cylinder of p-type germanium with the lithium-diffused junction on its outer surface. They could be 'closed' or 'open'-ended and could have an undrifted p-type core or a borehole as illustrated in Figure 11.25 [191]. In practice, the early coaxial Ge(Li) detectors were rhomboid in shape, reflecting the shape of the crucible used to grow the germanium ingot.

The American Nuclear Society awarded Tavendale and Ewan of Chalk River the first Radiation Industry Award in 1967 for their work on the large-volume coaxial Ge(Li) detectors, which revolutionised nuclear spectroscopy. However, this geometry had the disadvantage that the electric field was no longer uniform but, rather, logarithmic and was weakest in the outer regions (which represented the majority of the volume) and at the corners. The bias voltage had to be as high as possible in order to avoid charge trapping in these regions. In 1967, de Bruin and Hoekstra [193] in Delft, Holland, used coaxial drifting but then removed the junction on the walls as a method of achieving deep (7 mm) cylindrical Ge(Li)

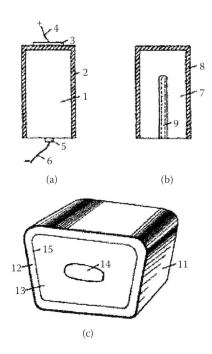

Figure 11.25 Early coaxial Ge(Li) detectors. (a) Closed end, (b) closed end with p-type core and (c) rhomboid structure, double open ended. (From Tavendale, A.J., USPatent 3374124, 1968. With permission.)

detectors. Coaxial geometry was also tried for Si(Li) detectors by Tavendale [194] while he was at the Australian AEC in 1966. This was not successful, mainly because of the rapid contraction of the p-type core in the final stages of drift. Coaxial geometry Si(Li) crystals were again attempted at the University of Koln, Germany [195] (see Figure 11.26), with some success and at also at Brookhaven [196] in 1969.

At Koln, the drifting was carried out from a 1-mm-diameter ultrasonically drilled hole and the detector operated at 1.5 kV bias. At Brookhaven, the drifting was carried out using a 19 mm diameter cylindrical silicon ingot from Monsanto that was then sliced into 2- to 3-mm-thick wafers. The x-rays entered one of the large area 'open' ends of the coaxial Si(Li) with no metal on the window; thus, the x-rays entered the transverse logarithmic field directly. It was argued that because of the increased penetration of x-rays into silicon, surface effects could be minimised. These detectors worked surprisingly well

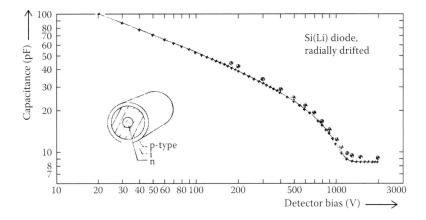

Figure 11.26 Radially drifted Si(Li) detectors. (From Kuhn, E.W. et al., *Proc. 1st European Symposium on Semiconductor Detectors for Nuclear Radiation*, Eichinger, P., Keil, G. (Eds.), 127, Munich, 1970. With permission.)

particularly for high energy x-rays, giving a resolution of ~300 eV FWHM at 6.4 keV for a 250-mm^2 active area. The capacitance was only 0.7 pF, which illustrates the advantage of the transverse field, and had the idea been pursued, we might have had the equivalent of the SDD 20 years earlier! Similar coaxial or 'bullet'-shaped detectors have also been designed for silicon and other materials more recently [197]. Alternative side groove geometries were also experimented with, and in 1975 Rob Hall at GE [198] patented a 'multiple side groove' geometry for germanium shown in Figure 11.27.

This idea was later briefly resurrected by Japanese workers [199,200] for HPSi detectors. However, despite these

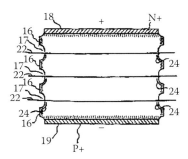

Figure 11.27 Germanium detector with multiple side grooves. (From Hall, R.N., US Patent 3825759 to GE, 1972. With permission.)

experiments, the top hat and groove geometries for silicon and germanium planar x-ray detectors and the coaxial geometry for large germanium γ-ray detectors were destined to become the commercial norm.

11.13 Contacts

There are three main types of contact in use today—diffused, surface barrier, and ion-implanted—and each can be used as both rectifying and ohmic contacts. As we have discussed in Chapter 4, these can also be modified or 'engineered' by multi-layering. We will now discuss these roughly in the chronological order in which they were developed.

11.13.1 1950s: Diffused Junction Contacts

As we have seen, mesa transistors were being made by diffusions through masks in the mid-1950s. The diffused junctions were termed 'alloy' junctions and were first made by the then called 'fusion' of the impurities with germanium. Such junctions were made for rectifiers at GE Schenectady and described by Hall and Dunlap [201] in 1950. Because the impurity concentration and its gradient could be high, the junction could be located very close to one surface. This had implications if the surface was to be used as a detector of light (a photodiode) or particles. Moreover, the large impurity gradient gave rise to high reverse bias (>100 V) before breakdown occurred. The first use of diffusions as window contacts for large area photodiodes was probably carried out by the Russians, Alferov, Ryvkin, Konovalenko, Tuchkevich, and Uvarov [202], in 1955. These detectors were p-type germanium 1–1.5 mm thick with active areas up to 10 mm^2. The same detectors were also shown to be sensitive to 55 keV x-rays [203]. In 1959, Goetzberger and Shockley [204], at the newly formed Shockley Transistor Corporation, showed that depositing B_2O_3 and P_2O_5 in an oxidising atmosphere at high temperature could make very uniform diffusions of boron and phosphorus in silicon. The diffusion rate from the oxide to the silicon was high and by successive removal of the oxide with HF and fresh depositions, very uniform junctions of about 0.4 μm could be achieved. Also, by this time, the French company LTT was making diffused junction particle detectors for Amsel et al. [163] at the University of Paris. Workers at Bell Laboratories mentioned in February of 1960 [205] that a 'paint on' diffusion of phosphorus from a solution of P_2O_5, which was then heated to 950°C, was effective. The full recipe was given later at the Asheville Conference, and Mayer [84] at Hughes Aircraft described 25 mm² shallow diffused junction detectors with resolution <60 keV on 6-MeV α particles. They could easily be made thinner than ~1 μm, and Blankenship [97], working at ORNL, claimed a depth of 0.1 μm, proclaiming that the 'window problem was solved'. However, there were problems of 'multiple peaking' or 'ghost peaks' in spectra (see Sections 3.4 and 11.19.4), a hot topic at the Asheville Conference, and it was clear that for low-energy particles and x-rays, the diffused junction was not an ideal window contact. Diffused junctions are still used for the windows of photodiodes but for x-ray and particle detectors they have been superseded by surface barriers or ion-implanted contacts. They are still used as back contacts of course, for example, on Si(Li) detectors, because they are used in the drifting process and for their ruggedness.

11.13.2 1950s: Surface Barrier Contacts

In parallel with the diffused junctions, surface barrier contacts were being developed. They were, however, more difficult to make successfully and were not (and still not) completely understood. In a report [206], and a series of lectures consisting of 170 pages at the Belgrade High Energy Physics Conference in August 1965, Fred Goulding [207] devoted only a few lines saying 'the lack of basic understanding of the mechanism leads to serious problems. We will not devote any further discussion to this type of device'. Surface barriers can still be difficult to make, can be unstable, and are now unique to the radiation detector industry. They were developed from the metal point contact or cat's whisker, which in some respects would seem to be a good choice for detectors. The junction was a thin surface inversion layer, and its small area around the point contact made for low capacitance. But being of small area, it made a very insensitive radiation detector. What was required was to extend the shallow junction deeper and over a broader area. In 1950, Seymour Benzer [208], working at Purdue University, described germanium photodiodes made by evaporating thin films of gold (and other metals) onto a high purity n-type germanium crystal. These were the first 'surface barrier' contacts, and he immediately filed a patent (see Figure 11.28).

In January 1951, Moore and Herman of RCA [209] patented a 'semiconductor signal translating system', which could obtain signals from a germanium gold surface barrier detector from a modulated electron beam. This was followed in 1952 by Pantchechnikov [210], also at RCA, who succeeded in making photocells by the evaporation of gold onto clean etched germanium. This simple method of producing a 'surface barrier junction' over an extended area had the added advantages of giving a thin uniform 'window' for radiation and also of not involving high temperatures that could affect the charge lifetime of the material. In further work at Purdue, Simon [211] obtained signals from evaporated gold on n-type germanium surface barriers in 1955, and Mayer and Gossick [212] verified their signal to energy linearity 1 year later. A simple design was soon adopted in particle detector manufacturing

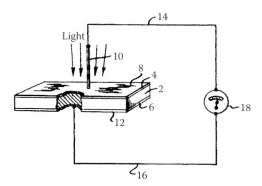

Figure 11.28 First surface barrier contact. Labels refer to following: **2**, bulk n-type germanium; **4**, etched germanium surface; **6**, back contact; **8**, thin evaporated metal film; **10**, metal contact wire to film **8**. (From Benzer, S., US Patent 2622117, 1950. With permission.)

by Mayer, who made surface barriers up to 16 mm^2 area at Purdue and by Walter et al. [213] at ORNL.

In his 1979 review, McKenzie [214] showed the construction of the early Ge surface barriers (see Figure 11.29).

The choice of gold was rather arbitrary, influenced by the industrial manufacturing traditions and ease of evaporation. Many workers [213,215,216] showed the importance of the diffusion of oxygen across the gold contact, forming surface states and 'hardening' the rectifying characteristics of the detector. Indeed, by cleaving silicon under vacuum and subsequent exposure to oxygen, it was shown that an interface oxide was *essential* for a surface barrier contact as shown in Figure 11.30.

Russians workers [217] at the Ioffe Institute in Leningrad had showed that the performance of germanium surface barrier detectors improved significantly with cooling. The first application as a practical particle detector was by McKenzie and Bromley [218] using a cooled 2 mm by 2 mm germanium gold surface barrier and particles from the 3-MeV Van de Graaf accelerator at Chalk River in 1959. Cooled germanium surface barrier detectors were routinely used for a while as particle detectors in nuclear physics experiments, but the cooling was inconvenient. So in 1959, Chalk River initiated a contract with RCA Research Laboratories in Montreal to attempt to make silicon p–n junction detectors to operate at room temperature. According to McKenzie's review, results of the first attempts by Webb at RCA were not very encouraging, showing no resolution at room temperature. But in the mean time, Chalk River had bought some 100 Ω cm n-type Si crystals and experimented with a Au surface barrier contact. Soon

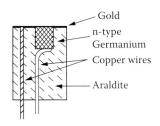

Figure 11.29 Germanium surface barrier detector. (From McKenzie, J.M., *NIM A*, 162, 49, 1979. With permission.)

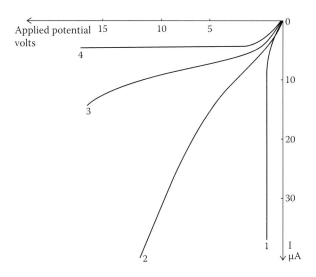

Change of the reverse characteristics vs.
time, after the air contact
(1) in vacuum,
(2) 10 min after the contact with air,
(3) after 20 min,
(4) after 30 min.

Figure 11.30 Formation of a gold surface barrier contact and its dependence on oxygen. (From Siffert, P., Coche, A., *IEEE* NS-11, 3, 244, 1964. With permission.)

their technician, R.J. Toone, who made most of their detectors, had successfully made the world's first silicon surface barrier detector operating at room temperature [218]. The silicon surface barrier detectors had far superior resolution for α particles compared to the diffused junction detectors. Soon the Chalk River group [219] was reporting 16 keV resolution on 6-MeV α particles, which is respectable even by today's standards. McKenzie also recalls that the preamplifier was a crude three-transistor unit with no feedback that Bromley had copied from *Popular Science* (magazine)! After an informal meeting at Bell Laboratories arranged by Walter Brown, things moved quickly. At the American Physical Society Meeting in Cleveland Ohio in December 1959, both McKenzie [220] and Nordberg [221], who was working at the California Institute of Technology, presented their results on silicon surface barrier detectors. The latter made detectors with areas up to 50 mm^2 and obtained a resolution of about 60 keV on 5.3-MeV α particles. Mayer [222], now at Hughes Research Laboratories, also published results with silicon and germanium surface barrier detectors in 1959. Cooling silicon surface barrier detectors reduced the leakage current and noise considerably but did not improve the intrinsic resolution significantly [223]. Because the silicon detectors operated adequately at room temperature, germanium surface barriers were not pursued vigorously after 1960.

A detailed manufacturing process for these detectors was given by Dearnaley et al. at AERE Harwell [132] in 1961. By then, n-type silicon was becoming available, and this was necessary for the Au surface barrier contact. It has a lower resistivity than p-type silicon as the electron mobility is three times that of holes (hence, for the same resistivity and bias, the depletion depth is greater) but the surface naturally grows an oxide and develops a p-type inversion layer. As we have noted, the oxide is very thin and the junction barrier is virtually at the surface. The tail of the metal wave function extends through the oxide and acts to give a virtual 'Schottky barrier', so that there is some dependency on the metal, but the surface states that exist at the interface make the barrier heights far from those predicted by the Schottky/Mott theories. The barrier height is modified by the screening of the crystal, wholly or partially, from the externally applied field, and the detailed physics was not fully understood. Nickel was also used for a surface barrier contact by 1963 [31]. Andrews [224] had shown that a good low noise noninjecting ohmic rear contact was made by the evaporation of aluminium. This was used by Dearnaley [132]. However, he also showed that if the detector was fully depleted, this contact could become rectifying and injecting unless the silicon was lapped, so providing a degenerate surface. This suggested that aluminium might make a good rectifying surface barrier on *p-type* silicon, and such detectors were first made by Inskeep [225,226] at the University of Tennessee in 1963 and were later widely adopted commercially by ORTEC. They were, in fact, more rugged and less light sensitive than the gold surface barrier detectors. The dimensions of surface barrier detectors were similar to diffused junction detectors, but the applied bias was reversed to deplete the bulk.

The edge termination of the surface barrier on the silicon or germanium surface is very sensitive because it is here that the high fields exist and tend to spread laterally [185]. As a result, one of the problems was the electrical connection, usually to the bias. The edges were protected by doped epoxy resin. As noted above, in the case of Si(Li) crystals, the groove or top hat geometry provided a good method of terminating the surface barrier by evaporating the contact contiguously over the uncompensated periphery. A recess also protected the surface, and the crystal could be safely handled by the low resistivity edges.

11.13.3 Engineered Surface Barrier Contacts

From the start, it was clear that the surface barrier was complex. It was basically a Schottky metal–semiconductor barrier but drastically modified by the interface states that existed in the presence of the native oxide and was variable according to the surface history. The question was: can the charge collection characteristics of the natural barrier junction be improved upon?

The first thoughts along these lines probably dated from around 1960 with the work on CdS diodes by Carver Mead [227] (later of VLSI prominence) at Caltech. These ideas were extended generally to semiconductors by Arthur Bramley [228]. The idea was to improve the diode characteristics by interposing a very thin insulator or degenerate semiconductor between the metal and semiconductor surface. Ten years later, Mead and others [229] were involved with colleagues at Caltech, including Jim Mayer, studying the 'window thickness' of silicon surface barrier detectors. They rightly predicted that charge backdiffusion was largely responsible and assumed in their calculation that the metal acted as an absolute sink for the charge that reached it. However, they added: 'In silicon surface barrier detectors, there is an oxide layer between the metal and the silicon surface which, if sufficiently thick, would present a barrier to the penetration of carriers into the metal… the surface preparation techniques could have a large influence on the amount of charge loss and hence the window thickness'. Furthermore, it was thought that subsequent deficiencies of oxide might be the root cause of instabilities (leakage current, noise, and spectral peak tailing) and that an oxide might also provide some sort of passivation.

Anodic oxidation of the surface led to resistive contacts, but George Amsel [230] reported in 1961 that the evaporation of tin oxides before the evaporation of gold onto the etched silicon produced reliable surface barrier detectors. Amorphous germanium was advocated by a group at Birmingham, UK [231], as a non-injecting *back* contact on silicon under a metal evaporation in 1971 and by the Berkeley group [232] on HPGe 6 years later. The latter also showed experimentally that they could detect α particles (with reduced resolution) through a contact of 0.5 nm amorphous germanium under a film of aluminium on HPGe. In 1974, Llacer [233] was convinced that he could chemically grow a thick oxide 'possibly an anhydrous amorphous monoxide' on HPGe before applying Pd to make a surface barrier contact. He found that the detector using this contact gave high reliability with no adverse effects. He did not report on ICC and low-energy spectral performance. Very little work was done in this area until 1985, when a Japanese group [234] reported excellent stability and line shape on 60 keV x-rays at room temperature using an n-type silicon surface barrier with a very thin film of tungsten oxide under the gold contact. However, nobody had studied the effects of such intermediate thin films on ICC for ultralow energy radiation. This was subsequently studied by the present authors at Link Systems, UK, in the 1980s (as discussed in Sections 4.6 and 11.20.2).

11.13.4 1960s: Ion Implantation

The early semiconductor detector manufacturers largely pioneered ion implantation, and the experience gained from this work contributed considerably to the success of LSI in the general semiconductor industries toward the end of the 1960s. Russell Ohl and others made the first studies at Bell Laboratories between 1952 and 1955 [235–237]. Ohl bombarded a polycrystalline silicon surface with He ions of 200 eV to 30 keV to modify the characteristics of a point contact junction. They offered no theoretical explanation for the observed effects that included an increase in the reverse bias breakdown voltage. The first application for an α-particle detector was made in 1962 by Alvager and Hansen [238], who carried out a phosphorus implant in 9 kΩ cm p-type silicon at Argonne. They obtained 25-mm^2 area junction diodes giving a resolution of 75 keV for 6-MeV α particles. This was followed by Ferber [239] implanting boron into n-type silicon. At the Ninth Scintillation Counter and Semiconductor Detector Symposium in 1964, Martin [240] from the Ion Physics Corporation, presented the first comprehensive work on ion-implanted junction detectors. In 1968, Laegsgaard [241] used the Aarhus University isotope separator, in Denmark, to implant n-type silicon with boron through a mask. The silicon was supplied by Haldor-Topsoe A/S, Denmark (the engineer Haldor Topsoe founded the subsidiary, Topsil A/S in 1958) and, with the help of Walter Gibson, seconded from Bell Laboratories, they succeeded in making good position-sensitive strip detectors. Ion implantation of silicon is now used extensively for silicon particle detectors and boron implantation of n-type coaxial HPGe detectors is routine. For very thin widows for x-ray detection with Si(Li) and HPGe crystals, the surface barrier contact is still preferred. However, special techniques for ion-implanted contacts have recently been developed in particular for CCDs and SDDs with impressive results (see Section 4.5).

11.14 1960s: Lithium Compensation

The impurities in silicon and germanium were the limiting factor to the depletion depths available and therefore the energy sensitivity limitation for radiation detection.

Attempts were obviously made to purify the materials, germanium being easier than silicon as noted earlier. The surface barrier detectors were limited to 3–12 keV x-rays at best. It seemed that the pleas of T.R. Kohler of Phillips at the Asheville Conference in 1960 for 'a detector for 1–50 keV x-rays, the range of particular interest with good resolution' (for elemental analysis), were still unlikely to be easily satisfied. Similarly, Borkowsky (ORNL) stated, 'Goulding mentions the need for fast timing devices for HE Physics. But I think that the bigger need is for γ-ray detectors. We are talking 100,000 to 100,000,000 devices working at 10 V bias for Civil Defence'. Even today, silicon ingots of high enough purity (>20 kΩ cm) are not widely available, at least not enough on which to base commercial detector manufacture.

This 'material problem' was almost completely removed with work from an unexpected direction in 1960. Lithium diffuses readily into silicon interstitially (unlike boron and phosphorus, which displace the silicon atoms). The diffusion rate had been studied at Bell Laboratories using p–n junctions since 1953 [242]. This work was unrelated to work on radiation detectors: after all, for detector-grade material, *they* were trying to *remove* all impurities. Work continued at Bell Laboratories on the interactions of various 'dopants' in germanium, including lithium. By the mid 1950s, Reiss, Fuller and Morin [243] had concluded that Li ions paired with singly charged acceptors such as Ga. Bell Laboratories picked up on this for solar cells. They used techniques that allowed diffusion without the conversion of the p-type germanium to n-type, thus achieving partial 'compensation'.

Erik Pell, working at GE in New York, was also studying the mobility of ions such as Li^+ in silicon. He first described 'lithium ion drift', that is, the movement of the positive ion in an applied electric field, in this case from lithium-diffused silicon n–p junctions at elevated temperatures, in his classic paper of 1960 [244]. He used it to study the Li precipitation on impurities and defects in silicon, such as interstitials or vacancies [245]. He had considered its use for high impedance transistors as a possible application but did not realise, at first, the potential of the technique for producing deep depletion layers for radiation detectors. The drifting and pairing of the mobile Li^+ ions to impurities (such as O, B) and vacancies, coupled with a copper-staining technique [246] (see Section 5.1.3) that delineated the drifted volume, was a sensitive probe for studying their distributions in silicon crystals.

It was soon realised that this pairing completely transformed the electrical properties of silicon, automatically producing large volumes of effectively near-intrinsic silicon, a process that became known as 'lithium compensation'. The transformation was as physically significant as the interstitial carbon in iron to make steel and the crosslinking by sulphur in latex to make vulcanised rubber. The compensated material, 'Si(Li)' or 'silly', could be used in the manufacture of thick depleted material, and there was an immediate response by most laboratories working on semiconductor radiation detectors. It is debatable which laboratory first successfully produced a working Si(Li) device because of the extreme competition. Berkeley (Elliott) and Hughes (Mayer) are usually credited as being among the first. Harry Mann (Argonne) [247] credits Jim Mayer as being the first to report such detectors at the Asheville Conference at the end of September 1960, but although Pell himself did give a paper on ion drift there was no paper presented on Si(Li) detectors as such. Pell talked of the advantages of the process but added 'this is hypothetical as no practical detectors have thus far been made by the ion-drift technique'. He added that the obvious advantage is 'deep depletion and there may be 'self healing' with respect to resistivity changes caused by radiation damage'. There was no mention of drifting germanium. Jim Mayer, who had been a student under Karl Lark-Horowitz at Purdue University but was now at Hughes Laboratories, declared in the post-Pell paper comments that 'we have made some junctions of this type. One advantage is deep depletion at low bias'. There is also a note added in proof in the printed proceedings: 'Mayer's detectors had 1.5 mm depletion and show good sensitivity to β and γ rays. They operated a 1-mm-deep device at 20–65 V without any change in β resolution. One problem was the window dead layer that used the heavy doping technique and gives ~50 μm silicon dead layer. He believed that this problem could be overcome with 'proper techniques'".

Work at Argonne was started after Mann had visited Mayer and Elliott. There were competing commercial interests in the form of Hughes and SSR Inc., which had ex-Hughes employees including Stephen Friedland. Mayer [248] presented his work at the International Atomica Energy Agency Belgrade Conference on Nuclear Electronics in May 1961 and filed a patent for Hughes on lithium compensation for detectors the very next day [249]! However, these patents and most subsequent ones on detector manufacture seem to have had little impact. The work was more generally reported in the journal *Review of Scientific Instruments* [250] (paper received in April 1961), and they were drifting depths up to 5 mm [251]. First attempts at Harwell in UK [252] were not so successful.

Jack Elliott [253,254] published the Berkeley results in 1961, and Harry Mann [255] presented the Argonne results at the Third Annual Nuclear Engineering Education Conference at Argonne in January 1962 and later in the year at the IRE meeting, as did Blankenship [256], who was at ORNL. The Hughes paper [251] addressed the practicality of using Si(Li) detectors for gamma-ray spectrometry, for which germanium is obviously better suited. One must keep in mind that the Si(Li) particle detectors at this time were operated at room temperature and cooling with liquid air or nitrogen, although improving resolution, was not convenient. The equivalent gamma-ray detectors from Ge(Li) material, which had to be cooled, followed from the work at AEI Aldermaston UK by David Freck and James Wakefield [69] in 1962 as already discussed. Unfortunately, despite their publication in the prestigious journal *Nature* (but hardly mainline for this type of work), and the filing of various patents [257], this latter work was rather overlooked and was certainly not fully capitalised on by AEI or the UK in general. Ge(Li) detectors were subsequently further developed elsewhere.

At the Ninth Scintillation and Semiconductor (now added to conference title) Detector Symposium in Washington in 1964, Tavendale's [258] paper was the only one on Ge(Li) detectors. Fred Goulding put this down to surface problems, but McKenzie [214] thought that the Ge *material* was more important and Chalk River was lucky to have a good supplier (ASM/GEC from Europe). Two years later, Davis and Webb [259] at RCA also concluded that some material was 'good' but could not ascertain why. Surface treatment studies at Bell Laboratories and elsewhere also seemed dependent on the material.

Because the main application of semiconductor detectors was for nuclear physics experiments, there was considerable interest in their behaviour in severe radiation environments. Also, work at Bell Laboratories was moving toward putting encapsulated Si(Li) detectors on satellites, and there was concern about the radiation belts they would encounter. The early work in 1964 was pessimistic because of the relatively weak fields obtainable in Si(Li) detectors at room temperature. Coleman and Rogers [170] found considerable ICC after only 10^5 R dose of Co-60, but the detectors were restored after further drifting. The irradiation of Si(Li) particle detectors with protons and α particles was studied by Mann and Yntema [260] at Argonne. They noted that because the net space charge of ionised impurities is very low, any space charge due to trapping makes them very susceptible to the field being distorted, resulting in regions of ICC. There was a loss of resolution and increase of leakage current after $10^8/cm^2$ particles, and after $10^{10}/cm^2$ particles multiple peaking occurred. Again, they could be restored after a period of further drifting. This implied that the low field regions were attributable to lithium-lattice defect centres. For other types of detector, there was more optimism as Dearnaley and Northrop [61] noted in 1966: 'Fortuitously, whatever the resistivity and type of the initial silicon, irradiation increases the resistivity. This is because the simultaneous production of donor and acceptor levels pin the Fermi level midgap in the same way that Au doping does'. But like the Au-doped crystal detectors (see Section 11.4), the charge trapping increased. Radiation hardness in semiconductor detectors and semiconductors in general has, of course, continued to preoccupy many workers in areas such as high-energy physics (HEP), space missions, and military applications (see Section 4.9) ever since.

We will now examine the techniques developed for lithium compensation in more detail. Because of the difficulty in removing boron contamination from silicon, most of the silicon available in the 1960s was p-type. This was fortuitous because lithium drifting is only effective for p-type material where the acceptors are compensated by the Li^+ ions. Pell [244] carried out his diffusion work drifting into silicon of resistivity 10–100 Ω cm at 125°C to 175°C in air but covered it to reduce photocurrents, heat loss, and dust settlement. He pointed out that if the resistivity is too low we may lose the junction because of lithium precipitation, and if it is too high it may become 'intrinsic' at the drift temperature, causing the junction to disappear and preventing drift. Mayer [34] later confirmed the problem of drifting with too low a resistivity (<50 Ω cm) by showing that the capacitance of the resulting Si(Li) crystals increased with storage at room temperature over time. According to Pell, with a 1 kV/cm field across a cylindrical sample at 125°C, the drifting of 10 mm depth took ~25 days. Some workers continued to drift in air [178,256,261–263], whereas Jim Mayer [264] and subsequently Laurie Miller [188] at Brookhaven and Harry Mann [247] at Argonne, used liquid drifting methods: silicone fluid at 40°C–204°C in the case of Mayer and a fluorocarbon liquid at ~175°C in the case of the latter two. The liquid was thought to give more uniform temperatures and enable drifting at higher power. Because the drift time varies as the cube root of the power, the liquids allowed faster drifting (they claimed 15 W gives a 4-mm drift in 28 h). Liquids were also thought to help to keep the surfaces clean and stable, but this was debatable. Dearnaley and Lewis [262] pointed out that forced cooling of the surfaces probably gave rise to a radial variation in temperature. They preferred to drift in air in an enclosed region, and both drifting in air and liquids are still used successfully. There were even attempts [265] to drift in various gases such as nitrogen and helium doped with, for example, iodine in an effort to control the surface states.

It was realised that impurities and dislocations in the silicon could impede the drifting process, and Kuchly et al. [266] pointed out that the silicon should also be oxygen-free because LiO complexes could form. The early workers [164,178] drifted whole slices of the silicon ingot, albeit of smaller diameter than those available today (usually less than 8 mm diameter), and there was little attempt to confine the drift laterally. This only came later with the introduction of the grooved or top hat geometry to contain the drifted volume as described earlier. After diffusion, a shallow 'mesa' was formed by masking an area on the diffused side and etching for about 6 min. After the slice was drifted, the front was slightly recessed by etching and a gold contact evaporated over it to form a surface barrier. After chemical treatment and passivation, each slice was ready to use as a single radiation detector crystal. The leakage current during drift was flat or increased only very slowly, indicating that its origin was not thermal generation in the drifted volume (which would cause it to increase linearly) but surface leakage current. This was also indicated by the magnitude of the current, which was far greater than that expected from just thermal generation of carriers in the bulk [188]. The groove and top hat geometries had the effect of slowing the initial drift in its final stages but helped considerably to reduce the surface leakage currents, particularly during the separate 'cleanup' drift (see below), where the drifted material already extended around the groove or brim of the top hat. The progress of the lithium drift was monitored by a copper stain of the profile, which was a destructive sampling test, or by plotting the capacitance versus bias (C–V plot), at room temperature. As discussed in Section 5.1, during drift the volume approaches the 'intrinsic or insulator' condition and a C–V plot begins to level off. The voltage at which this happens (full depletion) is lowered as the drift temperature is lowered (see Figure 5.9, which was taken from Mayer [264]).

Miller started with FZ-Si of 100 Ω cm to 3 kΩ cm resistivity, drifting at 150°C. This relatively high temperature had the advantage of rapid drift, but the higher bulk leakage currents caused 'overcompensation' with lithium making it n-type. This meant that the field was not uniform at the operational temperatures of the detectors. The field distortions due to the space charge current during initial drift were minimised by drifting at high fields as pointed out by Mayer. Miller used circuitry to deliver constant power during drift and maintained the voltage while the liquid cooled (over about 1 h) at the end of the drift. In this way, he was aware that a 'cleanup' drift was taking place to some extent at lower temperatures. Mayer had pointed out that as currents are thermally generated in the bulk, the density of electrons and holes increases linearly across the drifted volume. As discussed in Section 5.1.7, it is this uneven distribution of free charge in the volume during drift that gives rise to an uneven compensation or 'graded junction' and distorts the field. The solution was to redrift (or 'cleanup') at a lower temperature for a prolonged time after the initial drift to 'level' this distribution. It also has the effect of 'levelling' the C–V curve, and the process has been called levelling. It was not until the noise and spectral background had been reduced in detectors much later that the marked improvements made by a cleanup drift could be appreciated. Early workers were content to just let the crystal cool slowly under bias in a single lithium drift.

In 1964, Goulding's group [267,268] at Berkeley were using automatically controlled drifting that adjusted the temperature such that the leakage current was constant. The temperature reduced from ~130°C to ~120°C during drift. The heating was removed when the resistance between two points on the surface being approached by the drift increased, but the bias remained on to accomplish some degree of cleanup. This was a first step towards their drifting under complete computer control [269] in 1989.

A separate cleanup drift was first reported in the drifting of germanium by Tavendale [190], who, after an initial drift at 70°C–80°C in chloroform, redrifted at 20°C–25°C in a silicone oil bath for 48 h. Cappellani et al. [270] at Ispra, Italy, substituted pentane for chloroform and drifted at a lower temperature (46°C) with a shorter cleanup period. In 1969, Lauber [271], working in Sweden, theoretically formalised the drifting process and suggested the requirement for a separate cleanup drift at a lower temperature for silicon, but he remarked that for germanium a cleanup drift was already usually carried out at around room temperature. Gradually, workers [269] started to settle on 1–2 kΩ cm silicon starting material with an initial drift with a ramp up (to protect the junction from breakdown) to 110°C and ramp down to 80°C (to give a cleanup). AERE Harwell [262] settled on a simple way of detecting the completion of the drift. By drifting up to a lapped metallised degenerate surface on the p-type substrate, a 'punch-through' of the lithium drift was indicated by a sudden increase in bulk leakage current due to the disappearance of the p–n junction. The lapped surface gave electron injection into the bulk, and this increase in current could be used to trigger the heater to turn off. This method was preferred because lithium 'punched-through' after different drift times and copper staining could then be used to confirm this punch-through [178] on each individual crystal (see Figure 5.3). This method then necessitated the formation of a new p-type junction contact, usually a p^+ Au surface barrier, to support the field during a quite separate cleanup drift at 80°C–100°C. This could be continued as long as thought necessary, but usually ~20 days. Commercial companies were using a separate cleanup drift, but of course details were always proprietary. The idea of lapping back to the compensated silicon after the initial drift and forming a surface barrier contact by etching and applying a gold contact both for cleanup drift and the final thin widow contact, seems to have come to several workers [188,262,263,272] at the same time around 1963. The Yale group [272], under Prof. Al Bromley, drifted rectangles of p-type silicon slices to within ~1 mm of the front contact and ground a recess through the uncompensated material into the drifted material (see Figure 5.18). This geometry avoids the weak peripheral field regions associated with contouring geometries (which were adopted universally later). A similar geometry was used by workers at the TUM [273] in the late 1960s, but because of the success of contouring in controlling surface leakage currents, the idea was not pursued by commercial companies.

Regarding the lithium drifting of germanium, we have already seen that Freck and Wakefield [69] were the first to make a successful Ge(Li) γ-ray detector in 1962. Because of the lack of a stable native oxide and a higher dielectric constant, the germanium surface states were more of a problem than for silicon and drifting had to be done in a dry atmosphere or liquid (Freck and Wakefield used an oil bath), albeit at a much lower temperature (60°C–80°C) than silicon. Above such temperatures, the uncompensated material is flooded with thermally generated carriers and the junction disappears. The Berkeley group [267] also drifted germanium, but made no attempt at a cleanup drift. A drift depth of 6 mm could be achieved in ~150 h. The window was at first a simple contact (such as gallium–indium) to the undrifted p-type germanium in the case of planar detectors and to the lithium diffused outer surface in the case of coaxial detectors. A relatively thin window (~6 μm) was achieved by Tavendale by drifting into an alloyed aluminium contact. But the process used for making surface barrier contacts to Si(Li) detectors was also applicable, and such detectors were being made commercially by 1965 by SSR [274]. Ge(Li) γ-ray detectors were also commercially available from a few other manufacturers (ORTEC, RCA, PGT, Nuclear Diodes, Philips, Nucletron).

The main disadvantage of the Ge(Li) detectors was the fact that not only did they have to operate at liquid nitrogen temperatures, but they also had to be stored at −50°C or lower otherwise the lithium distribution was disturbed. Accidental warm-up could be disastrous. Figure 11.31 shows a typical cryostat using Union Carbide gravity feed dewars in 1964 from Berkeley [275,276].

Efficient liquid nitrogen reservoirs, or 'dewars', were commercially available, for example, from the Linde Division of Union Carbide, New York, and Eugene Miner [277–279], working at Berkeley, had published useful cryostat designs for Ge(Li) and Si(Li) detectors between 1965 and 1967 along with the commercial sources for the materials and components that were being manufactured and used at Berkeley. The early dewars were of the gravity feed or 'chicken-feed' type, which delivered liquid nitrogen right to the cryo-head through an aluminium tube. It was known that detectors, Si(Li) detectors in particular, gave better results at higher temperatures than liquid nitrogen and internal heaters for the cryo-head

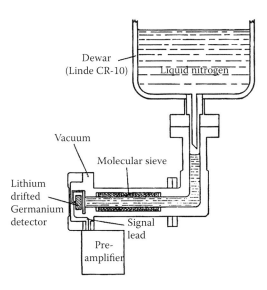

Figure 11.31 Liquid nitrogen–cooled gravity feed cryostat. (From Lothrop, R.P., UCRL-19413, 1969; Kuchly, J.M. et al., *NIM*, 47, 148, 1967. With permission.)

were installed [277]. Some early workers [280] adjusted this heater current to optimise the noise and resolution. It was then realised that it was the temperature of the FET that was important and a separate heater (usually a Zener diode) was fitted to the FET header heat sink. This led to some confusion, because it is difficult to vary these temperatures independently and the bubbling liquid nitrogen caused microphony, which dominated the noise. Although there may have been some temperature effects because of traps in the Si(Li) crystals, it was soon realised that the FET temperature dominated the noise if the microphony was reduced by isolating the vibrations caused by liquid nitrogen. The latter was achieved by cooling the cryo-head assembly by a solid copper cold-finger decoupled from the from the base of the dewar by a length of flexible copper braiding (see Figure 1.62b).

11.15 1960–1966: Amplifiers

The typical resolution of a cooled Si(Li) detector in 1964 was at best ~2 keV FWHM for low-energy x-rays. Amplifiers were vacuum tube–based but FETs had become commercially available, and the promise of FET-based preamps was recognized early. In 1962, Veljko Radeka [281] designed a negative resistive feedback preamplifier at Brookhaven using a Fairchild FSP-401 FET as the input stage. He also proved that it could be operated around liquid nitrogen temperature. Theron Blalock [282] at ORNL used a 2N2500 FET cooled to 125 K to achieve a noise in silicon of less than 1.6 keV. The early FETs that were used, such as the 2N2500, had considerable surface noise and when incorporated into preamplifiers were no better than the vacuum tubes available [283]. This situation was improved by the use of selected cooled low noise FETs. The best option seemed to be the TI 2N3823 (a metal–glass cased version of the Silex epoxy encapsulated 2N3819), the Motorola 2N4221, and later, the Union Carbide 2N4416, as these became available. The latter chip was, in fact, used for decades either from commercial sources (such as Union Carbide, TI, Motorola, TI Bedford UK) or, slightly modified (TI Bedford), from research centres such as Harwell UK. Blalock [283,284] investigated the new 2N3823 thoroughly in 1966 by plotting the noise as a function of temperature for 33 devices and found considerable variability as shown in Figure 11.32.

Although two very good FETs are shown (Nos. 26 and 27), it led him to comment, 'the noise characteristics vary widely and seem to be quite unpredictable from the electrical parameters'. He correctly predicted that the cause was temperature-dependent generation–recombination centres. The optimum temperature for good FETs was usually around 125 K. As mentioned previously, Nybakken and Vali [107] were probably the first to show the benefit of cooling the FET in a spectrometer in 1963, but Manny Elad and Michiyuki Nakumura [285], working at Berkeley 1965–1968 and investigating the yield of

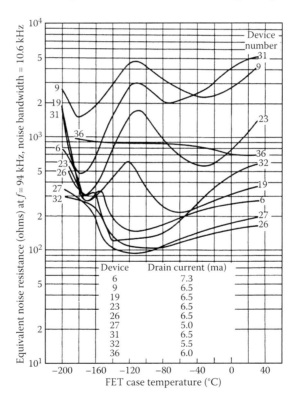

Figure 11.32 Noise as a function of temperature using different field effect transistors. (From Blalock, T.V., ORNL-TM-1055, 1965; Blalock, T.V., *IEEE* NS-13, 3, 457, 1966. With permission.)

the low-noise FETs, showed that they could be cooled inside the cryostat in close proximity to the Si(Li) or Ge(Li) crystal. He was able to achieve ~1-keV resolution on x-rays, soon reducing this still further in 1966 to ~700 eV [286] and even better [287] by 1967, when microphony was reduced. He used a 300-mm^2 Ge(Li) detector supplied by Guy Armantraut, who was working at Livermore. Even by early 1967, commercially available Si(Li) detectors did not use cooled FET preamplifiers. This prompted Harris and Shuler [288] of College of William and Mary, Williamsburg, VA, to take a standard 80 mm^2 TMC Si(Li) detector and replace the preamplifier with a Tennelec TC-130, where they inserted the same type of FET (2N3823) *into* the cryostat. This bold move resulted in a resolution of 460 eV on Fe Kα (6.4 keV) in contrast to the original 1.6 keV value. This was not only as a result of cooling the FET, but the reduction in input capacitance and more effective shielding of the FET. TMC was quick to respond to this with an improved cooled FET Si(Li) spectrometer.

11.16 1966–1971: Pulsed Optical Restore

We pick up the story around 1966. The high degree of sophistication achieved up to this period was largely as a result of extensive development efforts on high-resolution Ge(Li) detectors for nuclear physics at places such as Berkeley, Brookhaven, ORNL, and Chalk River. There was a period of slow painful progress with Si(Li) detectors involving the evolution of the crystal geometry, processing and passivation, and FET noise reduction, as already described, but by early 1969 a few spectrometers had resolutions as low as 250 eV, which would make the separation of elements above Ca possible. Figure 11.33, [112] shows a typical cryohead assembly from Berkeley from the mid 1960s, illustrating encapsulated FET with resistive/capacitor feedback.

Kandiah [289] had considered the problems associated with resistive feedback in high rate counting in 1966 and demonstrated a DC coupled arrangement with a dummy detector, where the average input current could be made zero by DC

Figure 11.33 Early cryostat front–end assembly showing detector, FET, and feedback components.

feedback to one of a pair of LED–photodiode (S1–D1, S2–D2) combinations as shown in Figure 11.34.

The peak distortion could be investigated using a random pulse generator (applied to one of the LED drivers) and was negligible up to a rate of ~10 kHz. The quiescent input current could be controlled with either one of the LEDs (depending on the current direction), and when the output exceeded a preset limit in any direction (due to quiescent current or the random pulse generator signal) the LED could be pulsed to restore it. This constituted 'limit restore'. The shot noise due to current in the photodiodes still added noise, and at Gatlinburg in 1969

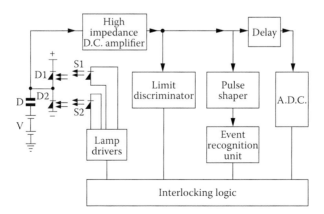

Figure 11.34 Replacing feedback resistor with optical coupling. (From Kandiah, K., *Proc. Conf. Rad. Meas. in Nucl. Power*, 420, IOP London, 1966. With permission.)

Kandiah [290] suggested pulsing the light onto a single photodiode only (D1 in Figure 11.34) for limit restore. This had long-term after-effects because of the slow charge in the undepleted regions of the reverse bias photodiode. Then, in a flash of inspiration, Goulding [291] realised that the drain gate of the FET itself, if exposed, could be used as an effective 'photodiode' as shown in Figure 11.35 [292].

He immediately patented the idea [293]. This obviously included as part of the modification, the removal of the FET chip from the header can and mounting it in a light-tight package (in attempt to prevent effects on the Si(Li) crystal) together with the LED. Fortuitously, this also eliminated the lossy dielectric glass of the FET header and reduced the noise still further. He had little problem 'decanning' the FET, but he did find that the commercially available LEDs that used clear epoxy lenses led to mechanical failures on cooling [294]. They achieved 180 eV resolution compared to 342 eV for the conventional resistor feedback preamplifier. Incidentally, the Si(Li) crystal itself can be also be used as a photodiode (D2 in Figure 11.34), a technique later used by EDAX when their crystal leakage currents were so low they were less than that of the FETs used. This gave limit restore in both current directions. Most other companies did not achieve such low crystal leakage currents, so they did not need the extra LED. The theory of dielectric noise had already been published by Veljko Radeka [295] while working in Brookhaven in 1968. By mounting the light-tight package close to the crystal, the stray capacitance

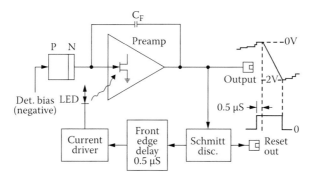

Figure 11.35 Pulsing the optical restore mechanism. (From Landis, D.A. et al., *IEEE* NS-18, 1, 115, 1971. With permission.)

and microphony was also reduced and, for the first time, resolutions improved to ~152 eV as demonstrated by Elad [296]. However, Elad still used continuous light feedback. Pulsed optical feedback using the FET as a photosensor was first described by Landis [297], who was working at Berkeley in 1971. He also showed that the limit on high rate counting was caused by light leaks from the package onto the Si(Li) crystal. The excited surface states could last 5 to 10 min (but shorter for germanium crystals). He surmised that this problem might be overcome, but the surface states excited on the FET itself might set the ultimate limits. The latter, as we shall see, was addressed by Kandiah by a modification to the FET structure (the optical modification). The progress toward pulsed optical restore was summed up by Goulding and Jaklevic [298] in 1971 with the following schematics (Figure 11.36).

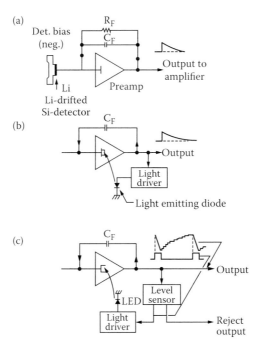

Figure 11.36 The evolution of pulsed optical restore as of 1971. (a) Resistive feedback arrangement, (b) 'signal by signal' restore and (c) 'limit' restore. (From Goulding, F.S., Jaklevic, J.M., UCRL-20625, 1971. With permission.)

In their figure, schematic (b) represents 'signal by signal' restore and schematic (c) limit restore. Similar techniques were soon applied to Ge(Li) detectors where resolutions of a few keV for γ-ray spectra were now achievable. The Berkeley group [299] had experimented with boron nitride (BN) ceramic, replacing the glass in a light-tight package. Kern and McKenzie at Bell Laboratories [300] also looked into the possibility of other ceramics and quantified the improvements using Western Electric glass, BeO, and Al_2O_3 ceramic encapsulations. It was found that the original glass could contribute more than 200 eV of noise at room temperature.

A number of alternative restore techniques were proposed [289,301–304] but the EDX detector industries largely gravitated to pulsed optical limit restore for the next 20 years and an FET package of sintered BN dielectric. The advantages of BN were its relatively low lossiness (see Table 1.5), ease of machining, low thermal expansion, affinity to epoxies, and ruggedness. Harwell, UK, however persevered with a PTFE package that, although theoretically superior (BN added a few eV of noise), was difficult to implement by Link Systems under commercial license. With the LED (usually GaP–IR) incorporated into the BN package, it was also relatively light-tight although it was sometimes necessary to impregnate the BN (e.g., with black picein wax). See Section 11.19.3 for examples of early FET packages.

11.17 Applications on the Horizon

Silicon surface barrier particle detectors and Ge(Li) gamma-ray detectors revolutionised nuclear physics. They replaced the commonly used magnetic spectrometers and scintillators, by virtue of their compactness, superior resolution, and fast timing capability. With the advancements, particularly with respect to noise and resolution, the applications moved from nuclear physics to *analytical techniques* such as neutron activation analysis using Ge(Li) detectors, Mössbauer studies, x-ray fluorescence (EDXRF) using Si(Li) detectors, medical tracer analysis and microanalysis (EDXMA) using Si(Li) detectors. The nature of the customers therefore changed. They were now metallurgists, chemists, biologists, medics, forensic scientists—in fact, the whole gamut of scientists and technologists worldwide. With this development, the nature of the industry changed from suppliers of solid-state detectors (SSR, ORTEC, TMC, Nuclear Diodes, RCA, PGT, AEI, Elliott Bros, LTT, Nucletron, Tadiran, and the like) to suppliers of complete analytical systems (TMC, Kevex, Nuclear Diodes/ EDAX, NEC, NSI, PGT, Link, etc.). Unlike nuclear physicists who were willing to cobble together their own system for their specific needs, the new customers just wanted a black box that gave them an answer. These changes were tracked by the establishment of small commercial interests starting in the United States but eventually moving wider afield.

The progress made in the development of semiconductor detectors for nuclear physics applications up to the mid-1960s was fairly openly reported. While the government laboratories such as Brookhaven, Berkeley, ORNL, and Battelle in the United States, Chalk River in Canada, AERE Harwell in the UK, and Saclay in France, were working on pure research and large companies such as Bell Laboratories, Hughes Laboratories, GE, RCA, and AEI were working speculatively, information was exchanged quite freely. This can be seen from the contents of the Asheville Conference, 1960 (organised by the National Academy of Science-National Research Council and hosted by ORNL–Union Carbide) through to the series of IRE Transactions (later IEEE) on Nuclear Science. The series of meetings now at the University of Michigan, Ann Arbor, began in Chicago 1964 and continued every 3 to 5 years at the University of Texas, Boston College, and subsequently at the University of Michigan, chaired by Prof. Glenn Knoll. Regular European conferences were rather slower to get off the ground. There were sporadic ones in Prague (1960), Belgrade (1961 and 1965), AERE Harwell (1961), and Versailles (1968). The first regular series started at the TUM in Munich in 1970 and continued roughly every 4 years, the sixth being held in Milan (1992). A separate international series devoted more or less to compound semiconductors and room temperature detectors was started in Strasbourg (1971) and has continued every few years and in 2008 combined with the IEEE at Dresden. The written proceedings of all these gatherings give an insight into the course of semiconductor detector development during the early years.

Once commercialism in the form of the much smaller spinoff companies specialising in semiconductor detector manufacture entered the scene (from about 1967 onward), much information became proprietary and the competition was fierce. What advances were published by then were mainly for the prestige value and advertising potential, and of course many pertinent details were duly omitted. There was a reticence even to patent in case too much was divulged to the competition. Detectors themselves often had seals that nullified any warrantees when broken. Before giving a snapshot of the companies involved let us look at the histories of the applications themselves.

11.17.1 EDXRF Analysis

XRS had already been developed using wavelength dispersive methods (using analyzing crystals or gratings), but this was a time-consuming *sequential* technique. Energy dispersive methods offered by gas and semiconductor detectors allowed analysis at all energies rapidly and *simultaneously*. Si(Li) detectors only began to be exploited commercially for this technology in earnest around 1966–1967, when a number of small companies took up the challenge. The two techniques are illustrated in Figure 11.37.

The upper schematic illustrates the application of the Bragg reflection law at an angle (a) using an analyzing crystal with atomic planes separated by a distance d. The diffracted x-ray beam has a wavelength λ given by $2d \sin(a) = n\lambda$, where n is an integer 1, 2,... (the 'order' of the diffraction). The XRF spectrum is built up by counting the x-ray intensity for different values of the angle (a) on a goniometer. The crystal is *wavelength* dispersive (WDXRF). The x-ray detector in this case can be low resolution (gas proportional or scintillation detector). The resolution of the spectrometer is governed by the selectivity of λ and is usually of the order of a few 10s of eVs. The lower schematic illustrates the use of a semiconductor detector to analyze all x-ray energies simultaneously in a fixed geometry. In this case the semiconductor crystal is

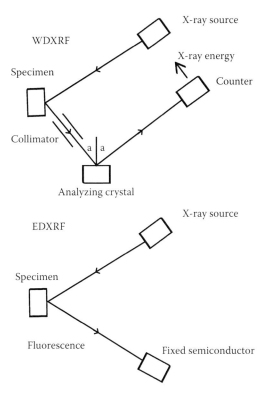

Figure 11.37 Arrangements of components in WD and EDXRF analysis.

energy dispersive (EDXRF). The resolution is governed by the semiconductor detector (130–200 eV) and although it is not as good as in WDXRF, this arrangement is much faster and cheaper. The x-ray source (x-ray tube or radioactive source) for WDXRF must be of a much higher intensity than that for EDXRF. X-ray tubes have the great advantage in that they can be turned off, but for WDXRF they need to be high powered and cooled. For an extensive treatment of XRF techniques in general, there are many specialist books available, for example, that by Beckhoff [305].

In 1966, Elad's FET preamplifier and Miner's cryostats were put to good effect by Harry Bowman [280] at the Lawrence Radiation Laboratory. He used two detectors, a Si(Li) made at Berkeley and a Ge(Li) made at Livermore, in an experimental radioactive source excited EDXRF arrangement. He noted that the efficiency of measurement of the excited x-rays from a sample was orders of magnitude greater than for a conventional wavelength dispersive spectrometer (WDS) system. Moreover, there was no requirement for a carefully machined goniometer resulting in a system that was much more compact, simpler to use and cheaper to manufacture. A battery-operated portable system was even predicted. But the *coup de grace* was that all wavelengths (energies) were measured simultaneously. Both quantitative and qualitative EDXRF analyses were demonstrated. He ended this seminal paper by suggesting the use of these semiconductor spectrometers for use on EMs. In further work at Berkeley in 1968, Bob Giauque [306] used an annular radioactive source for EDXRF and concluded that a 1% analytical accuracy was possible. The steady improvements taking place in the detector industries were having an impact. Resolutions better than 500 eV were offered in 1968 better than 250 eV in 1969 and approaching 150 eV in the early 1970s. Fred Goulding and Joe Jaklevic [307] compared the performance of Si(Li) detectors and Ge(Li) detectors for XRF. They concluded that trace elements were obscured by 'various Artefacts' in the Ge(Li) spectra caused by the detector itself: 'it appears that germanium detectors have little value for trace element analysis'. However, in the 1970s, GE, based in Philadelphia, was promoting a new HPGe-based EDXRF system and claiming analysis 'from fluorine to plutonium'! (see Figure 11.38).

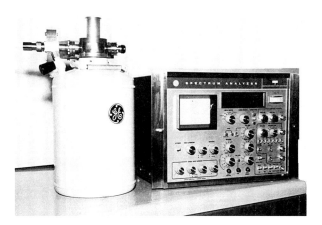

Figure 11.38 Early XRF system from General Electric using HPGe detector, 1970.

Improvements in pileup rejection and baseline restoration in the main amplifier further improved the quality of x-ray and γ-ray spectra. In 1971, the Berkeley group [308] took EDXRF to an extreme to which it has never really caught up with commercially since. They constructed a windowless system using a specially designed x-ray tube (see Section 11.20). Many different radioactive sources in different geometries were coming available in a safe sealed form [309] for excitation. This included the cyclotron-produced isotope Fe-55 with a beryllium window that was used as a reference source for low energy x-rays replacing the commonly used specification on the Fe Kα line (6.4 keV) that required some sort of separate excitation. The 5.9-keV MnKα line (which could now be resolved from the MnKβ line) would eventually (in 1981) be adopted as the IEEE standard [310] for resolution measurements of x-ray detectors. In fact, all resolution specifications for x-ray detectors from then on were understood to be FWHM values for the 5.9 keV MnKα line.

John Rhodes [311] (Columbia Scientific Industries, Austin, TX) used a 30-mm² by 2-mm-deep Si(Li) detector in a source excited analyzer for silver ores in 1967. Later, he summarised the literature for radioactive source excited EDXRF from 1966 to 1970 [312]. Of the 116 analyzers, only 17 used Si(Li) detectors, one used a Ge(Li) detector [313], two (both from Huth et al. [314]) used APDs. The remaining 96 used gas proportional detectors. Neither these nor the APDs could hope to compete with the x-ray resolutions being achieved with Si(Li)s or, for that matter, with Ge(Li) detectors. Rhodes countered the inconvenience of the APDs using liquid nitrogen by pointing out that being thus temperature 'controlled', they are now very stable. In Bertolini and Coche's book, *Semiconductor Detectors* (1968) [87], there are only two paragraphs devoted to the application of EDXRF—indicating that the technique was not well established then, at least from a European point of view.

11.17.2 Online and Portable EDXRF

Columbia Scientific Industries in the United States, Nuclear Enterprises (NE) in the UK, and Rigaku in Japan pioneered portable and online EDXRF analyzers between 1965 and

1976. They mainly used NaI scintillators or gas proportional detectors. There were some early trials of Si(Li) detectors using radioactive sources for online EDXRF. The pioneer in the United States was Kennecott Copper Mine, Salt Lake City, UT (1966–1971) [315], and in the UK, Warren Spring Government Laboratory [316]. Up to 1970 gas proportional counters or scintillators were preferred because of their ruggedness, room temperature operation, and relative insensitivity to vibration.

11.17.3 X-Ray Tubes

The modern x-ray tube had its origins in the hot filament high vacuum tube introduced by the American physicist William Coolidge in 1913. The principle is the generation of Bremsstrahlung radiation when an energetic electron beam is stopped in a metal target. In the 1960s, large, HV, power-consuming x-ray tubes were available commercially from, for example, Watkins-Johnson, Machlett, GE, and Diano Corp. However, there was not the great aversion to radioisotope sources in the public domain then as there is now, and these were rugged, reliable, and compact. In fact, TMC moved away from the tubes in favour of sources and, now, as Kevex [317], continued with source excitation. Meanwhile, Jaklevic at Berkeley [318] was pioneering small x-ray tubes specifically designed for Si(Li) EDXRF spectrometers, and these were soon commercially available. There was an eventual recognition that these were less hazardous than the high-activity (100 mCi or more) sources with accompanying legislative accounting and disposal problems, and at least tubes could be turned off! The manufacture of reliable vacuum-sealed x-ray tubes had a whole black magic of its own, just like gas detectors and Si(Li)s a whole and companies such as Kevex, NSI, and Nuclear Diodes never got into it themselves. Kevex acquired the x-ray tube division of Watkins Johnson in about 1981, and Link acquired X-Tech (a spinoff from Watkins Johnson) some time after this. Brian Skillicorn did pioneering work at Kevex, and for the technical problems relating to the production of compact, stable HV tubes for EDXRF, the reader may refer to his publications. Kevex was happy to supply Si(Li) detectors and x-ray tubes to other EDXRF System companies such as Columbia Scientific Industries, Tracor-Northern, Finnigan, and some time later, Link Systems. Tracor-Northern was based in Middleton, WI, and was formed from the old established company Tracerlab and the MCA company Northern Scientific. In 1974, Kevex collaborated with Tracor-Northern in producing a PDP-11 computer-based MCA EDXRF analyzer. Kevex's own analyzer was traded as Kevex-ray, later the 'ANALYST', and the NSI one as 'MECA', later 'Spectrace'. The Nuclear Diodes (now known as EDAX) analyzer was traded as 'EXAM'. In 1972, Nuclear Equipment Corporation (NEC) introduced 'Speediffrax', a combined EDXRF and diffraction system (see Section 11.8 showing a picture of Speediffrax). When the Naval Radiological Defence Laboratories (NRDL) in San Francisco closed in 1969, Paul Hurley, who was manufacturing Ge(Li) detectors [319], founded Quanta Metrix Corporation. The company manufactured EDXRF equipment using NEC detectors. ORTEC sold their Materials Analysis (EDXRF) Division to Jack Driscoll of HNU Systems (founded 1973, the name is based on $E = h\nu$), and they continued to use the terms SEFA (source excited XRF analyzer) and TEFA (tube excited XRF analyzer). Research at GKSS National Research Centre in Germany also led to some commercial enterprises (Seiffert, Atomika, Spectro).

11.17.4 Maturity of EDXRF

There had been considerable advancement in semiconductor detector manufacturing techniques in the first half of the 1960s. Now with the dramatic improvements in noise and resolution, theories could be tested and used to predict what (if any) further progress could be expected. For example, van Roosbroeck [320] had calculated a Fano factor of 0.12 for semiconductors that seemed in reasonable agreement with the measurements for silicon [321] and germanium [322]. By the 1970s, TMC (Kevex), Nuclear Diodes (EDAX), PGT, NEC, and NSI had complete EDXRF spectrometer systems based on Si(Li) detectors, and these were soon being used for useful EDXRF work [323]. The remaining improvements would have to come from further electronic noise reduction and crystal ICC reduction mainly carried out by the commercial companies. The first real demonstration of EDXRF for *trace* element analysis (1 ppm or less) was carried out in 1972, when the Berkeley group [324] used a Si(Li) detector closely coupled with a Mo transmission target x-ray tube. TXRF was first used by Yoneda et al. [325] in 1971. Berkeley was extolling the virtues of its guard ring detectors for their good charge collection and low background. With improved processing and collimation, guard rings were later shown to be unnecessary and—with the space constrictions in some EM columns—often inappropriate for EDXMA.

11.17.5 EDXMA

In EDXMA, the atomic excitations and subsequent relaxations are stimulated by an electron beam rather than by an X-ray source. Although this has disadvantages, such as the generation of Bremsstrahlung background radiation, it has several advantages, such as a highly focused beam. In EDXRF, the ionisation cross sections increase with the atomic number Z of the sample (photoelectric effect). In electron excitation (EDXMA), the cross sections decrease (because the atomic electron binding energies increase) thus favouring light element analysis. In the 1960s, such beams were already extant in the ubiquitous EMs, and elemental information was a powerful addition to the EM image. Two popular instruments, the scanning (SEM) and the transmission EM (TEM and STEM) were rapidly being used in research, and the large companies such as GE, Siemens, Philips, AEI, and Hitachi were the driving forces. In the SEM, an electron beam is focused onto the surface of a thick sample by magnets some distance above it. The x-ray detector is mounted into one of the already existing horizontal column ports in line with the sample stage. In the TEM, the sample is thin enough to transmit the beam to form an image below it. The energy of the beam is much higher and the focusing magnets much closer to the specimen. For the x-rays to escape from the specimen, the x-ray detector is usually arranged at a high takeoff (HTO) angle. Common arrangements for EDXMA on a SEM and TEM are shown in Figure 11.39.

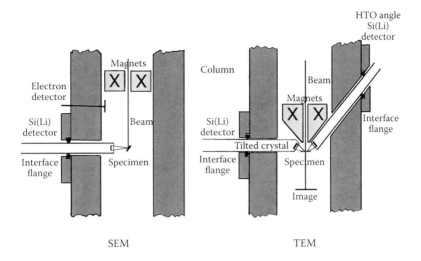

Figure 11.39 Possible detector mounting arrangements in a typical SEM and TEM.

The technical challenges for EDXMA presented by the TEM are greater than those for the SEM. The thin specimens give lower count rates, the space is very restricted, higher voltages lead to high energy electron backscattering, and there is danger of radiation leakage through the column interface port.

The history of the EM goes back to 1928, when Max Knoll and Ernst Ruska (Technische Hochschule, Berlin) carried out experiments based on earlier theoretical work by Hans Busch. Ruska made an operational electron microprobe (EMP) while working at Siemens, Germany, in 1931 (he only received the Nobel prize for this work in 1986). Siemens quickly filed a patent. Later, the electron beam was scanned with coils over the specimen area of interest in an SEM that was being produced commercially by Siemens in 1937. The first transmission beam instrument (TEM) was developed at the University of Toronto in 1938, followed by the first commercial TEM (from Siemens) the following year. It was realised that the electron beam spot on the specimen, besides producing an image from the electron secondary emission and scattering, was also producing characteristic x-rays (along with, unfortunately,

Bremsstrahlung background x-rays). M. von Ardenne proposed using a secondary electron collection electrode instead of a photographic plate, and Zworykin built one in 1942. The resolution was ~50 nm. The next obvious experiment was carried out in France. For his PhD thesis under the supervision of Prof. Guinier, Raymond Castaing [326] fitted a WDS to a TEM at the University of Paris in 1951, thus opening it up to analytical capability or electron probe microanalysis (EPMA). Researchers could now discover what their specimens were made of. Scanning techniques were developed at Cambridge University, UK, in 1957. In 1958 Tube Investments (TI) UK built its own SEM with a TV monitor, and in 1959 Cambridge Instruments (Microscan) collaborated with TI to produce SEMs commercially. Incidentally, the original Cambridge Scientific Instruments was founded in 1885 by Horace Darwin, the son of Charles Darwin. With Cambridge University, they improved the resolution and depth of field with their 'Stereoscan' system. Peter Duncumb, under the supervision of V.E. Coslett at Cambridge University, was the first to extend this to a scanning microprobe to image the sample in terms of x-ray emission (x-ray 'mapping') in 1957. He reviewed this and other early work in 1999 [327]. The first EPMA with gas proportional detectors was used by Dolby [328] in 1959. This was limited by the energy resolution (1.3 to 1.8 keV on Cu Kα) of the detectors. The first EDXMA on an SEM in 1963, also carried out by Duncomb [329], also used gas proportional detectors.

The advances made in Si(Li) detectors stimulated the first suggestion by Bowman and Hyde [280] in 1966 that they might be useful for EDXRF. With simultaneous elemental analysis and no moving parts or need for alignment, EDXRF had many attractions over the sequential WDXRF, and the technique was already well established using gas proportional detectors (Siemens had been supplying them on their TEMs since 1963). By early 1969, a few Si(Li) spectrometers had resolutions as good as 250 eV, which would make the separation of elements above Ca possible. The early work on EDXMA has previously been reviewed by some of the pioneers from the time such as Kurt Heinrich [330].

The first use of a Si(Li) detector for EDXMA is a little contentious, with a few companies claiming this accolade. The first publications described qualitative rather than quantitative work, merely showing the potential of the Si(Li) detector on electron column machines. Ray Fitzgerald (University of California, San Diego) [331], in an influential noncommercial collaboration, mounted an ORTEC Si(Li) detector of resolution ~600 eV on an Applied Research Laboratories (ARL) EMP. The support of John Walter and Rex Trammel of ORTEC is acknowledged. Elements down to Ca Kα could be identified, but it could not resolve Si Kα from P Kα. Figure 11.40 shows the detector used.

In Figure 11.40, the flange on the right of the unit is bolted onto the boss of a vertical LN dewar. By the time of the Fifth International Conference on X-ray Optics and Microanalysis (Tubingen, Sept 1968), the Si(Li) was reported as a tool for EPMA by D.B. Wittry [332]. The beam spot size and beam current in an SEM were much less than those in the EMP. This meant that WDS was no longer practical, but that some biological samples could be analyzed. The first reference of a Nuclear Diodes detector being used on an SEM is in April 1969 by John Russ [333], who was then a microscopist at JEOL (Japanese Electron-Optics Laboratories), Medford, MA, USA. By the October 1969 IEEE Meeting on Nuclear Science in San Francisco, Manny Elad [334] was with Nuclear Diodes and reported with Alan Sandborg and John Russ their use of a Si(Li) on a JSM U-3 and a Cambridge Instruments Stereoscan SEM. They did some simple x-ray mapping and even claimed to detect O Kα x-rays from a silicon dioxide sample with a 25-μm Be window. The O Kα peak is clearly resolved (Figure 11.41).

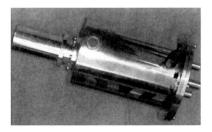

Figure 11.40 First recorded Si(Li) detector to be mounted on an electron microprobe. (From Fitzgerald, R. et al., *Science*, 159, 528, 1968. With permission.)

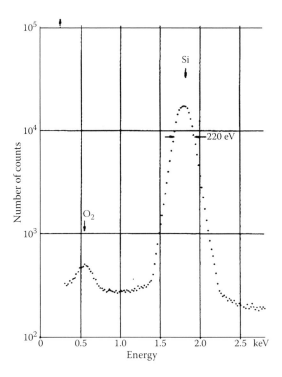

Figure 11.41 Early x-ray spectrum using a Si(Li) on an SEM and showing an oxygen line.

This claim was reiterated by Russ at the ASTM Meeting in Toronto, June 1970 [335]. In retrospect, it seems more likely that they were fluorescing an ice layer on the Si(Li) crystal.

Rather like the Asheville Conference a decade before, the Toronto ASTM [335] meeting was pivotal in defining future directions. ORTEC, Nuclear Diodes, Kevex, and PGT described their detector performances on EM columns. Nuclear Diodes [336] even had one on a Hitachi HU-11A TEM at Kennecott Copper Corp. (Lexington, MA). This work was the first to highlight the particular challenges faced when interfacing a Si(Li) detector to a TEM (Figure 11.42).

The columns had not been designed for this application, and they had to use the cold-trap port and change the specimens through the lead-glass window. Furthermore, they had radiation leakage (~3 mR/h) when operating at 50 kV. Add to

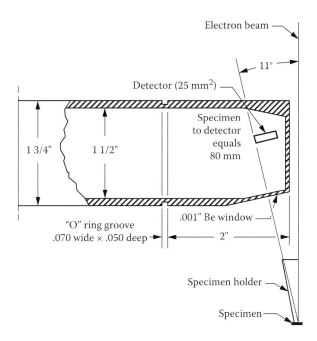

Figure 11.42 Interfacing a Si(Li) to a TEM. (From Bender, S.L., Duff, R.H., *ASTM Tech Publ.*, 485, 180, 1971. With permission.)

this the compactness of the specimen chamber and the necessity of a tilted Si(Li) crystal and one can see why many EDX companies shied away from interfacing to TEMs. Early in 1974, ORTEC introduced a complete data analysis system for EDXMA. However, this still looked like a cobbled nuclear physics system, and the detectors had large-diameter probes (see Figure 11.43).

Don Beaman and J.A. Isasi [337,338], working with L.F. Soloski of Dow Chemicals, Midland, MI, USA, must be credited as the first to assess the potential for *quantitative* microanalysis. Besides their own work, they also considered the quantitative work with Si(Li) detectors of other researchers [339,340]. They concluded: 'Even though one can expect superior accuracy with the WDS at this time, the outlook is favourable as is evidenced by the historical progress in accuracy improvement encountered in electron probe analysis'. They summarised this in a plot (Figure 11.44).

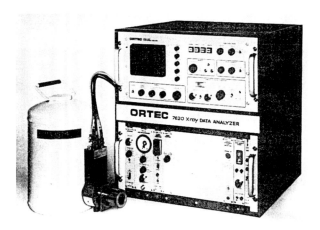

Figure 11.43 An early ORTEC system for x-ray microanalysis on an SEM.

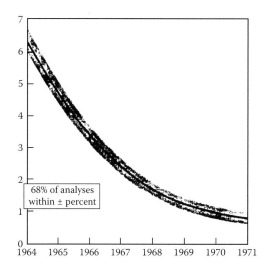

Figure 11.44 Reduction in scatter on measurements on electron probes as spectrometer performance improves over time.

This shows the scatter of measurements obtained for elements $Z > 10$ in alloys. This 'exponential' progress in the minimum detectable levels by EPMA is seen to continue into the 1990s as shown in Figure 11.45 [341].

This continued improvement in the minimum detectable limits is attributable not just to the detector improvement but

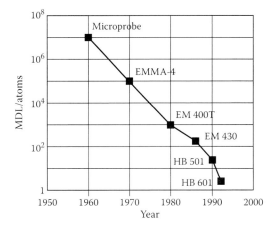

Figure 11.45 Minimum detectable limits improving with time. (From Cliff, G., Lorimer, G.W., *EMSA/MAS*, Boston, San Francisco Press, 1464, 1992. With permission.)

also to improvements in electron probes and STEMs and sample preparation.

11.17.6 PIXE

Particle-induced x-ray emission (PIXE) first became popular with the advent of particle accelerators (mostly Cockcroft-Walton and Van de Graaf) primarily designed for nuclear physics studies. By 1971, there were about 500 particle accelerators in the United States alone. Figure 11.46 shows the basic arrangement for PIXE using an accelerator proton beam with a Si(Li) detector.

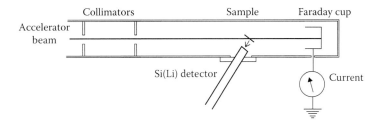

Figure 11.46 Detector used for PIXE analysis.

The Coulomb force on a proton as it penetrates a specimen is the same as that for an electron because their charges are of the same magnitude. However, because the proton is much heavier than the electron, its deceleration is much less. As a consequence, the intensity of Bremsstrahlung radiation generated is much less. In fact, this background radiation intensity is roughly inversely proportional to the rest mass squared of the penetrating particle. The 'particles' (ions, mostly protons) used in PIXE are massive compared to the electrons in an EM and produce much reduced spectral background and also less lateral scattering. The ionisation cross sections for charged particles are a maximum when their velocities match those of the orbiting atomic electrons most closely. Thus, whereas in EMs the energies are usually 10–50 keV, for PIXE the protons need to be 1–5 MeV. This is not, of course, a problem for, say, a typical Van de Graaff accelerator. Moreover, the ionisation cross sections increase with the mass (and energy) of the exciting particle and can be very large and, unlike photon excitation (photoelectric effect), increase for light elements. For these reasons, PIXE is useful for light element analysis, and low beam currents can be used over a very small spot size.

James Chadwick [342] (the discoverer of the neutron) was the first to study characteristic x-rays induced by α particle bombardment in 1912. The concept and application of PIXE was first developed at Lund University in Sweden. One of the first studies using a windowless Kevex Si(Li) detector was conducted by Musket and Bauer [343,344] using 100–350 keV protons at the Sandia National Laboratories, USA, in 1973. Figure 11.47 shows their PIXE spectrum for a graphite sample at 350 keV excitation.

Ron Musket later developed an α particle excited XRF product (see Section 11.20.1) after joining Kevex. The pioneers of true analytical PIXE work include T.B. Johansson [345] at Lund Institute of Technology, Sweden, Jerry Duggan [346] at ORNL, and Ian Campbell at Guelph University, Canada, who with Johansson, reviewed the topic in 1988 [347]. Campbell was also concerned with the detailed modelling of x-ray spectra from silicon and germanium detectors for PIXE applications (see Chapter 2).

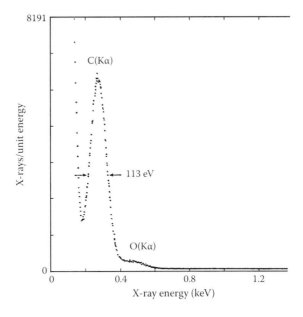

Figure 11.47 Carbon x-rays from graphite excited by 350 keV protons. (From Musket, R.G., *NIM* 117, 385, 1974. With permission.)

11.18 Companies

We are indebted to the following individuals for supplying information on the companies mentioned in this section:

Mervyn Hobden (AEI, e2v)
Roland Henck (Eurisys)
Harold Schmitt (ORTEC)
Ron Keyser (ORTEC)
Palmer Hotz (NEC)
Chris Cox (PGT)
Peter Milward (Nuclear Diodes, Link)
Hans Fiedler and Wieland Scholz (TMC/Kevex)
Orren Tench (Canberra)

    The history and relationships between commercial companies (see Figure 11.48) is less easy to follow in detail than, say, the careers of the people who worked in them. Dates of their establishment and demise sometimes have to be guessed

Figure 11.48 (**See colour insert.**) Examples of the Logos of Detector Companies.

and trading names sometimes changed seemingly on a whim. Documentation is less regulated than in the academic world, and much of what follows comes from claims of instrument manufacturers in their product literature, their websites, and various other documents. The appropriate caution must therefore be applied. The detector-making community has always been relatively small, and there has been a great deal of mobility between companies. Some of the original workers (see above) have kindly supplied information, but we offer advance apologies for errors and omissions. Corrections are, of course, welcomed.

Commercial interests were often in geographical clusters around the source of their technology and key employees. SSR started near Hughes Semiconductors, Los Angeles, CA. ORTEC and Tennelec started near ORNL, Oak Ridge, TN. Kevex, NEC, USC/NSI (later becoming Tracor-Northern), and Electro-Nuclear Labs, all emerged near Berkeley, and Stanford in Silicon Valley, San Francisco Bay area. Nuclear Diodes began near Fermilab, Chicago, and Argonne. Canberra/ Sturrup Nuclear, Bradley Semiconductors, and TMC began near Yale University, New Haven, CT. RCA Victor, Simtec, Aptec, Nuclear Radiation Developments, near Chalk River, McMaster, Montreal, Canada. PGT and Molechem (Hamner Electronics) began near RCA Princeton, close to Brookhaven, NY. In the United States, companies commonly had a strong Chinese representative on the technical side. Examples are SSR (Louis Wang), Kevex (Peter Lum, Ed Woo), NSI (Louis Wang), Electro-Nuclear Labs (Ken-Tang Chow), EDAX (Ray Yeh), PGT (Leong Ying).

Outside North America, the companies were again largely dependent on government laboratories and key universities. In France, there was LTT and RTC, established near Paris (Saclay). Enertec was founded near the Nuclear Research Centre, Cronenbourg, and the Louis Pasteur University, both in Strasbourg. In Germany, there was Nucletron (associated with University of Giessen but moved near Max Planck Institute, Munich), and in the UK, 20th Century Electronics (later 'Centronic'), Isotope Products, Isotope Developments, AEI, and Link Systems were all founded in the UK Home Counties not too far from AERE Harwell and AWRE Aldermaston. Philips

(Eindhoven, the Netherlands) was an early established electronics company (see below) that was to benefit from a good local supply of germanium (Hoboken).

Although a great deal of work on semiconductor detectors was taking place in Eastern Bloc countries by the late 1950s and early 1960s, there was, of course, no proliferation of commercial spinoff companies as there was elsewhere. The main centres were the Ioffe Institute, Leningrad (St. Petersburg), the Lebedev Institute, Moscow, the Institute of Nuclear Research, Swierk, near Warsaw, Poland, and the Riga Research and Development Institute for Radio-isotope Apparatus (RNIIRP) established in 1966 in Latvia. Only the latter resulted in commercial companies that still exist such as Ritec and Baltic Scientific (see below).

Despite Japan taking a leading role in commercial industrial semiconductor development, particularly after World War II and partly funded by the U.S. Government, they never seriously entered the semiconductor radiation detector world market. Detectors had been studied for many years at centres such as the Universities of Rikkyo and Tokyo (K. Husimi), but Si(Li) x-ray detectors had been manufactured only by the large EM companies such as JEOL and Hitachi (Horiba) and by Instrument companies such as Seiko. Even then, manufacture was mostly for use in their in-house systems and the home market.

In Germany, the growth of x-ray detector companies only really took off with the advent of the SDD developed by Kemmer and others at TUM, Garching, near Munich, Germany (see Section 11.26). Kemmer's own company, Ketek, formed in 1989, was followed by a spinoff company PNSensors. Ketek worked closely with Rontec, a Si(Li) detector manufacturer, in Berlin in 1996. Rontec was acquired by Bruker AXS in 2005. Vacutec, another German company, founded in Dresden in 1995, made Si(Li) detectors. They inherited a long experience in making semiconductor detectors under the company name Vacutronik dating back to the early 1960s.

Some companies formed symbiotic relationships with the large EM companies. Examples are Kevex with Hitachi, EDAX with Amray and Philips, Link with Cambridge Scientific Instruments, and both Oxford Instruments and Gresham Scientific Instruments with JEOL.

We now outline some of the main participants in detector evolution and the expansion of EDXRF, EDXMA and other x-ray applications roughly in the chronological order in which they were founded. Space, of course, limits us to just a sample of them. We include, for the sake of completeness, companies that specialised in compound semiconductor detectors although they were mostly active later (1980s onward). However, we will discuss the German SSD companies again in Section 11.26 that is devoted to them.

SOLID STATE RADIATIONS, INC.
2261 SOUTH CARMELINA AVE., LOS ANGELES 64, CALIF.

SSR. We have already mentioned the central role of SSR during the early years. The company had close connections with the University of Connecticut and Hughes Laboratories and was established in Los Angeles in 1960. The company was represented at the Asheville Conference in 1960 by Frank Ziemba, who had worked at the University of Connecticut and at Hughes Laboratories. By 1962, they had been joined by Louis Wang and by 1964 Michael Zatzick (also ex-University of Connecticut, Nuclear Physics). Michael Zatzick was involved with neutron and gamma-ray detectors. Stephen Friedland was a professor of physics at the University of Connecticut and became the manager of Radiation Physics at Hughes Aircraft Nuclear Electronics Laboratory before moving to SSR. His background was in biological applications of semiconductor detectors. Henry Katzenstein was president of SSR and was involved in designing electronic equipment for biophysics, radiobiological, and space exploration. The company specialised in lithium-drifted surface barrier silicon and germanium particle detectors for nuclear physics, space, medical, and biological applications, and was supported by research contracts from NASA. Wang left SSR in 1964 for Westinghouse Electric Company, California, but joined NSI (see below) around 1967. By 1964, SSR, along with ORTEC and RCA Victor, was also supplying liquid nitrogen–cooled planar Ge(Li) γ-ray detectors. SSR, to our knowledge, never entered the EDXRF or EDXMA arena, preferring to concentrate on the miniaturisation aspects for nuclear medicine. We shall see later the challenges of inserting detectors into the restricted space in SEMs and TEMs let alone into hypodermic needles and catheters! The company was still in existence in 1972 but became Quantrad Sensors and was bought by EG&G in the 1980s.

ORTEC. In the late 1950s, ORNL was concentrating on silicon surface barrier detectors, whereas others such as Chalk River and Bell Laboratories were mainly making diffused junction detectors. The work at ORNL was organised under Joe Fowler (Physics Division, including John Walter, John Dabbs, and technician David Peach) and Cas Borkowski (Instrumental Division, including Jim Blankenship and Richard Fox). John Walter was working with n-type silicon from Merck. The amplifiers were designed by Ed Fairstein and the nuclear physicists, Harold Schmitt and John Neiler were among the users of their detectors. It was clear that they could not make enough detectors, even for ORNL's own use, so the idea of an extracurriculum company was proposed. Neither Fowler nor Borkowski was supportive. Borkowski was afraid of the group breaking up and put a blanket prohibition on his group members participating or investing. However, the company (ORTEC) was founded in July 1960 by Harold Schmitt, who acted as president, and located at 901 Turnpike Building, Oak Ridge, TN. This represented one of the earliest examples of a technology transfer to the private sector. The team remained working at ORNL at first. Note that the detector specialist Jim Blankenship was blocked by Borkowski and was not involved. Walter, Dabbs, and E.O. Wollen joined ORTEC later. Ed Fairstein valued his independence and formed a separate company, Tennelec, Oak Ridge, in the same year. He provided designs for Infabco (founded by Ed Fairstein's technician at ORNL, Herman Hurst) to manufacture the electronics and Fairstein checked them. Two to three years later, ORTEC acquired Infabco (but not Tennelec, which later became a competitor).

ORTEC's first exposure to the market was at the Gatlinburg Conference [86] (October 1960), where they shared a booth with

Tennelec. Pricing ($40–200, depending on the quality) was based on that of the alternative detectors such as scintillators and the commercial diffused junction detectors. The orders came flooding in, the first from Rice University in October 1960, and it was clear that they would have to sever their links with ORNL. In May 1961, Tom Yount became the full-time president (Harold Schmitt became chairman). Bill Weiss (from GE, Cincinnati) joined as a detector guru in June 1961, and Bob Dilworth (electronics ex-ORNL) joined late 1961. John Neiler joined full time as technical director. They were also quick to acquire lithium-drifting technology. In 1967, ORTEC was acquired by EG&G, which brought in the germanium crystal pullers that it was using to manufacture transistors. The techniques for growing detector-grade material were developed at ORTEC where crystal growing was supervised by John Walter and Rex Trammell with Larry Darken (both also from ORNL).

In 1980, Walter left to set up his own company, Waltec, making surface barrier detectors, which was acquired by Tennelec in 1985. Rob Sareen (who had been working on surface barrier detectors at the University of Liverpool, UK) joined ORTEC in 1969, and in 1970 Dale Gedcke joined (from the University of Alberta, Canada). In 1972, Sanford 'Sandy' Wagner, who had been at Brookhaven working with Laurie Miller, joined them. 'Manny' Elad joined from Nuclear Diodes in around 1967. By 1977, ORTEC had also gained the services of Mario Martini (ex-Chalk River and Simtec), Tom Raudorf (from Simtec), 'TJ' Paulus, and Mike Bedwell (both from ORNL). ORTEC then had a formidable team and was renowned as *the* detector manufacturer. However, several key people were to leave (including John Walter, Larry Darken, Rob Sareen), and ORTEC later sold off its EDXRF division to HNU Systems. In May of 1999, EG&G purchased the Analytical Instruments Division of Perkin-Elmer. EG&G then changed its name to PerkinElmer, and ORTEC became part of the Analytical Instruments Division of PerkinElmer. As we saw, it was the first to demonstrate EDXMA but dropped out of both this and the EDXRF market. In 2005, ORTEC was purchased by AMETEK, Inc. and became part of the Advanced Measurement Technology subsidiary. It is still a major supplier of large HPGe detectors for γ-ray spectrometry, growing its own germanium crystals.

EG&G. EQ&G has a history dating back before World War 2 and was much involved in the war effort. Early history details are readily available elsewhere. The company was founded in 1934 (incorporated in 1947) by physicists from MIT, Edgerton, Germeshausen, and Grier. It soon became, among many other things, a well-established nucleonics company. In the early 1960s, it had bases in Boston, Las Vegas, and Santa Barbara. By acquisition EG&G was manufacturing diffused junction silicon detectors using a guard ring structure in 1965 in Boston, MA. It acquired ORTEC (see above) in 1967. In the process of reshaping the original EG&G, the Federal Systems Division was sold to The Carlyle Group in 1999 and used the EG&G name. In August 2002, it was acquired by United Research Services (URS) Corporation. The EG&G name is now only used in a few divisions outside the United States. EG&G, Santa Barbara was originally involved with radiation monitoring and protection using scintillation detectors under AEC contracts. Michael Schieber was growing HgI_2 by repeated sublimation at the Hebrew University, Israel, in 1973 and started a project with EG&G Santa Barbara around the same time. He worked with Wayne Schnepple and a little later Lodewijk van den Berg, both of EG&G. The latter flew on a space mission to grow HgI_2 crystals in zero gravity and now works at Constellation (Largo, FL), making HgI_2 material and detectors.

Philips. We have already discussed the involvement of Philips, Eindhoven, in the Netherlands with crystal detectors and Ge(Li) detectors in the 1960s. The company was founded in 1891, by Gerard Philips (who was incidentally a cousin of Karl Marx). It manufactured light bulbs and, later, vacuum tubes, x-ray tubes, and EMs. Its shield logo—representing radio waves and the ether—dates from the 1920s and although modified over the years has remained basically constant. The company grew and diversified in Europe dramatically. Its semiconductor activity started in 1953 and it was manufacturing Ge(Li) and Si(Li) detectors from 1965 to 1979. Its interest in x-ray detectors transferred to its U.S. acquisition, EDAX (see below). The detector personnel included L. Heijne (crystal detectors), W.K. Hofker (junction detectors), and J. Verplanke (HPGe detectors). Jan Verplancke later moved to Hoboken working with Walter Schoenmaekers on semiconductor detectors at Olen, Belgium (see Canberra below). In 1998, Philips Electron Optics merged with FEI Co., Eindhoven, manufacturing EMs.

Nuclear Diodes. Nuclear Diodes (see Section 11.9.1) was an early supplier of particle detectors and, by 1964, also Ge(Li) γ-ray spectrometers. Founded in 1962 by Charles J. Walsh with Alan Sandborg as vice president and detector expert, Nuclear Diodes was set up in Prairie View, IL, just north of Chicago, and became **EDAX** in 1972. This was a good choice of name and the term 'EDAX' became to be used as a generic term for the EDXMA technique. John Russ (from GE, San Jose and JEOL, Medford) and 'Manny' Elad (from Berkeley) joined in around 1969. EDAX was purchased by Philips (early 1980s) and consolidated with their headquarters in Mahwah, NJ. It became part of AMETEK in 2001.

Tennelec. Tennelec was founded as a nucleonics company by Ed Fairstein in 1960. Being a spinoff from ORNL, the company was based (like ORTEC) in Oak Ridge, TN. Tennelec merged with the company Nucleus in 1974. Its NIM modules achieved acceptance worldwide, and the company moved into the detector market with the purchase of Waltec in 1985 and the HPGe crystal growing facility of GE that they moved to Oak Ridge. For a time, Peter Hewka, who had worked with Huth on Si APDs and since the early 1970s on HPGe detectors at GE, continued to run the facility but Larry Darken (ex ORNL and ORTEC) took over around 1988. By 1989, Tennelec, along with ORTEC and Hoboken (Belgium), were the only suppliers of detector-grade HPGe material in quantity in the Western world. They entered the HPGe detector market with Pat Sansingkeow (ex-ORTEC), giving them a firm manufacturing basis, and Tennelec-Nucleus was purchased, mainly for its germanium production, by Link Scientific (UEI; see below) in 1989.

Princeton Gamma Tech (PGT). An historical description of developments of EDS by two PGT employees was given by Friel and Mott [348]. In 1965, Joe A. Baicker, a graduate of Columbia University (PhD in Nuclear Physics) and previously working for RCA (Princeton, NJ, USA), founded PGT as Princeton Research & Development Co. PGT manufactured Ge(Li) detectors and pioneered all-germanium (no scintillator) anti-Compton spectrometers and detectors for bore-hole logging. By 1975, Micha Harchol and Daniel Lister had joined. PGT claims to be the first commercial supplier of HPGe detectors (1973) and manufacturer of the first HPGe detectors flown in space. In 1983, Joe Baicker left and founded Radiation Data Co. (Skillman, NJ), which specializes in radon detection. PGT sold out to the Finnish company, Outokumpu, in 1985, and Joe Baicker retired completely from PGT in 1988. PGT's EDXMA business was bought by Bruker AXS GmbH in 2005.

Canberra. Some history of Canberra has been published [349]. Canberra Industries was founded in 1965 by Chuck Greer and Emery Olcott both graduates in management from

MIT (1963). It purchased the nucleonics company Sturrup, Larabee, and Warmers, which became Sturrup Nuclear (founded 1963) (Middletown, CT). Henry Webb, an Englishman and a technician at Sturrup, remained to design a NIM product line and he quickly became a very productive and innovative chief engineer for the new company. The name 'Canberra' was chosen (more or less randomly from world renowned place names and narrowed down to capital cities) to avoid any technical connotation that would restrict its future interests, but fortuitously 'CI' was also a play on 'TI', the successful company that the owners hoped to emulate. 'CI' therefore featured prominently in their original logo. In 1968, Canberra moved to nearby Meriden, and the German nuclear chemist, Hans Fiedler (see Section 11.12), joined them and started manufacturing Ge(Li) detectors. In 1972, Greer resigned as president and Olcott took over. Also at this time, Bob Lothrop, who had made Si(Li) detectors in Fred Goulding's facility at Berkeley, joined Canberra to start Si(Li) detector production. This was planned to feed the embryonic Canberra Analytical x-ray business; however, this business did not thrive and Canberra refocused on gamma-ray spectroscopy. Canberra bought the assets of the short-lived successor company to TMC, which had a multichannel analyzer (MCA) product line and more importantly, Harvey Roberts, the engineer who became known as the father of the Canberra MCA, leading the design group through several generations of hard-wired MCAs.

In 1972, Fiedler, who had previously been at McMaster University and Nuclear Diodes, left Canberra and returned to Germany where he established a Ge(Li) detector manufacturing facility, GETAC, in Mainz, and began representing Kevex in Germany in the analytical x-ray field. He then founded Eumex, an EDXMA company, in Germany with Garry Baerwaldt (ex-EDAX) in 1992. Orren Tench, who was general manager from 1969, assumed Fiedler's responsibilities and continued as general manager of the detector division. In 1977, K.M. (Mike) Yocum joined Canberra to develop HPGe detectors, which were very successful. He also resurrected Si(Li) production and continues to lead the Canberra detector R&D effort in the United States. In 1980, Canberra invested in Berkeley Germanium, the company founded by

University of California Berkeley personnel, including George Mraz and Scott Hubbard, to grow HPGe crystals. They funded this unsuccessful effort for several years, but around 1981 formed a joint venture with Walter Schoenmaekers, the HPGe Crystal specialist at Hoboken, to make HPGe detectors for the European market. This company was known as Canberra Detectors N.V. and was located in Olen, Belgium, near the Hoboken plant.

Emery Olcott remained CEO for many years, during which the company grew, and purchased Packard Instruments in 1986, Radiomatic in 1988, and Nuclear Data in 1989. Orren Tench became its detector expert in the early 1980s and became VP and general manager of Detector Products Group in 1992. Canberra purchased Tennelec-Nucleus from Oxford Instruments in 1994, thus giving them (like ORTEC) their own source of HPGe crystals with Larry Darken. Canberra became a subsidiary of Packard Bioscience and was eventually sold (in 2001) to the partly state-owned French company, Cogema (now Areva), which already owned Eurisys Mesures. Canberra is now the trade name of the various instrument companies now owned by Areva.

Technical Measurements Corp. In 1964, TMC (Newhaven, CT) had a base in San Mateo, CA, with Dick Frankel as the salesman. Ed Woo (see Section 11.12) was the technician making their Si(Li) particle detectors, which were cooled by liquid nitrogen. Woo started as a technician at Berkeley under Glenn Seaborg, the Nobel laureate, measuring nuclear radiation with Ge(Li) detectors. By 1966, TMC had an XRF Spectrometer (see Figure 11.49) with better than 1-keV resolution using Si(Li) crystals with areas from 80 to 1000 mm^2 and a drifted depth up to 5 mm.

TMC wanted to get out of the detector business, and Frankel and Woo left to set up Kevex Corporation in nearby Burlingame in 1967. TMC ceased trading around 1970. Two years later, Gary Kramer (from NEC), Peter Lum, and Rolf Woldseth (a Norwegian) (see Figure 11.50), joined Kevex.

Figure 11.49 Early TMC XRF spectrometer.

Figure 11.50 Ed Woo and Rolf Woldseth at Kevex.

In 1970, Aitken (Stanford University) and Woo gave a paper on Si(Li) detectors at the Toronto meeting [335], showing interfaces for an EM and a SEM (see Figure 11.51).

Seaborg kept in close touch with Frankel and Woo and became chairman of Kevex in 1978. By then, Dick Cushing had joined and John Colby (ex-Bell Laboratories) was writing analytical software (Quantex). Kevex grew rapidly, manufacturing in San Carlos with its headquarters in Foster City, CA. In 1978, it purchased the x-ray tube division of Watkins-Johnson (Scotts Valley, CA). Dick Frankel's son, Todd Frankel, later managed this company. Woo retired to Hawaii in January 1978. In 1988, Kevex merged with Surface Science Laboratories, a company that designed and marketed electron spectroscopy instrumentation for chemical analysis, and its

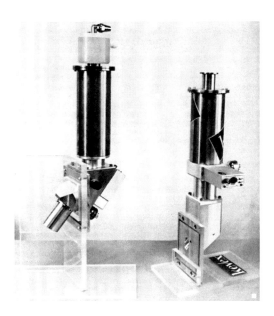

Figure 11.51 Early Kevex detectors interfaced to different microscopes. (From Russ, J.C. (coordinator), Energy dispersive X-ray analysis, ASTM STP-485, 57 Rep. on Symp., Toronto, June 1970. With permission.)

president, Michael Kelly, became president of Kevex. Kelly had previously been with Hewlett-Packard. He left Kevex in 1989, when it was purchased by Vacuum Generators (VG), the UK TEM manufacturer. This was after an attempted purchase by the Link Scientific Group, part of UEI plc. It was later purchased by Fisons (a British chemical company) and eventually became part of Thermo Scientific Co (see below) that by then also included NSI (Tracor-Northern, NORAN) (see below).

United Scientific Corporation (USC) was the parent company of Nuclear Semiconductor Inc. (NSI) (by which name the detectors were traded in the 1970s). The company was a spinoff from Stanford University. H.R. (Hans) Zulliger and L.M. Middleman had worked with D.W. Aitken at Stanford, and W.E. Drummond also came from Stanford, where he had worked on HPGe detectors. They were later joined by Louis Wang (who had worked at SSR). Their Si(Li) crystals were of the 'top hat' geometry developed at Berkeley and Stanford. The company used the name 'Spectrace' (the name of an unrelated long-established laboratory instrument company; see below) and 'Autotrace' for their spectrometers. Jim Stewart (software) joined around 1975. In the 1970s and 1980s NSI was the main competitor to Kevex, EDAX and PGT in EDXRF and EDXMA. Figure 11.52 shows personnel from a USC 1980 advertisement.

In 1973, Tracor, the large defence instrumentation company based in Texas, bought Northern Scientific, a nucleonics and MCA company based in Middleton, WI, thus forming Tracor-Northern. In 1970, the VP of Northern Scientific was Garry Williams. In 1984, NSI was acquired by Tracor-Northern, becoming NORAN, and then sold to Baker-Hughes in 1990. Jon McCarthy (who became business unit president of NORAN) had started with Northern Scientific. He, with Fred

Figure 11.52 USC personnel in 1980. Bill Drummond is at extreme left.

Schamber, introduced the first MCA with its own operating system (NS-880) in 1977. NORAN eventually became part of Thermo Scientific Co. (see below).

Nuclear Enterprises (NE) was first set up in 1956 in Winnipeg, Canada, by Robert Pringle. He was an expert on scintillators and chairman of the Nuclear Physics Department at the University of Manitoba. In the same year, he returned to Edinburgh (his birthplace) and formed what was to become the headquarters of NE. In 1967, a U.S. company (probably EG&G) attempted to take over the nucleonic subsidiary of UK-based Elliott-Automation [350]. Isotope Developments, which specialised in low-cost nucleonics and medical applications, and Baldwin Instruments, which specialised in industrial process control, were already then part of the Elliott Nucleonics Division. The British government realised the importance of having a nucleonics base and (as the Industrial Reorganisation Corporation) decided to step in. Nuclear Enterprises was the largest nucleonics company in the UK and, in the rationalisation, took over these interests of Elliott and EMI. As part of the strategy, a Nuclear Enterprises subsidiary making semiconductor detectors for the U.S. market, Nuclear Equipment Corp. (see below) was set up in San Carlos, Silicon Valley. The detector technology was led by J.D. (David) Ridley. He had started growing crystals at the Royal Radar Establishment, Malvern, UK, in 1964 and had already worked on Ge(Li) detectors at the University of Oxford. He was working on Ge(Li) detectors at NE in Winnipeg and later in Edinburgh in 1968. He then extended his interest to semiconductor x-ray detectors [351]. Albert Muggleton joined NE from AWRE, Aldermaston, around 1970 with a view to manufacturing Ge(Li) and Si(Li) detectors in the UK, which they did for a short time [352]. However, during all the structural changes, the manufacturing of semiconductor detectors was abandoned and an opportunity was lost. Muggleton moved to work with Tavendale at ANU, Australia. In 1987, a management buyout formed NE Technologies in Reading, UK. In

1995, NE Technologies was purchased by Saint Germain (they had previously purchased Bicron and Engelhard (Harshaw) in 1990, mainly for their scintillator technology).

NEC (San Carlos, CA, USA) was manufacturing Si(Li) x-ray detectors in the mid 1960s [353]. The company was acquired from NE (see above) by Gus Klein and included G. Kramer, P. Werp, and H.P. Hotz (senior scientist). Gary Kramer later moved to Kevex as mentioned above. Palmer Hotz was originally working with Paul Hurley (who went to Quanta Matrix) on Ge(Li) detectors at NRDL in San Francisco and after this closed down in 1969 moved to NEC and later the Finnigan Corporation. In 1972, NEC introduced the product 'Speediffrax' (see Figure 11.53). This allowed diffraction peaks to be acquired simultaneously with EDXRF spectra for element analysis. It used an air-cooled x-ray tube, sample chamber, Si(Li) detector, an MCA, and strip chart recorder. Spectrum stripping was used for unfolding complex diffraction spectra.

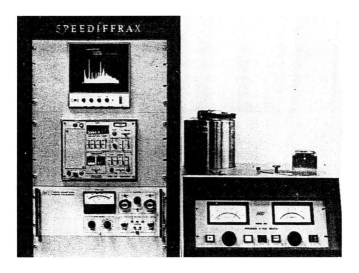

Figure 11.53 Combined EDXRF and diffraction from NEC.

NEC was ultimately sold to Envirotech Corporation, which was taken over by Baker International in 1982, and in 1987 Baker merged with the Hughes Tool Co. to form Baker Hughes Inc. The Si(Li) detector manufacturing side did not survive these changes. Envirotech was acquired by Thermo Scientific (see below) in 1994.

This company is included because it ultimately acquired many of the original companies as we have seen. Thermo Scientific is one of many subsidiaries of Thermo Electron Corp. that was founded by George Hatsopoulos in 1956 [354]. Hatsopoulos was born in Athens, Greece, to a wealthy family in 1927 and after the war went to study at MIT with a particular interest in thermodynamics and its practical applications. While still at MIT, he negotiated a loan to found Thermo Electron with his brother as the only employee. With many patents, the company survived for several years on research grants and work for other businesses, eventually gaining a reputation for industrial power generation and heating. In 1961 Martin Marietta, which was interested in the application of thermodynamics to aerospace ventures, offered to buy a 51% interest in Thermo Electron, but the offer was refused. Soon, the U.S. government also recognized the value of the company, awarding them several contracts related to the space program. In 1968, the Ford Motor Company established a joint venture with Thermo Electron to apply their technology to automobiles and in 1971 Hatsopoulos predicted the ratification of the Clean Air Act with a demand for environmental monitoring devices. The key to the company's success, the *New York Times* noted, is that Thermo Electron 'gambles on innovative technology to attack potentially huge markets in socially important areas'. They have consolidated their position with a number of acquisitions, for example, Gammametrics, Spectra Physics, Nicolet (1993), Envirotech, NORAN, TN-Technologies (all from Baker-Hughes 1994), Baird Analytical, Bennett X-Ray Corp (1995), Fisons Scientific (including Kevex) (1996), Scintag (1998), Niton (2005), and merged with Fisher Scientific (2006).

Simtec was a spinoff from RCA Victor, Montreal, by R.L. Williams and C. Wilburn in 1964. Simtec was based in Montreal, and one of its early products was a room temperature Si(Li) detector with a diffused junction window using an oxide passivation (see Section 11.11). Simtec also employed several ex-Chalk River physicists (M. Martini, I.L. Fowler) and Jim Mayer (ex-Hughes). Mario Martini was with them from 1971 to 1973 before joining Nuclear Radiation Development (NRD), which was part of Electronic Associates of Canada Ltd., Toronto. In 1975, Martini left NRD with some of his technical group to join ORTEC. K.R. Zanio and H.L. Montano, whose chief interests had been in CdTe at Hughes Laboratories, were also at Simtec in 1972. Simtec did not enter the EDXMA market and specialised in particle detectors for nuclear physics. Ultimately, they could not compete with ORTEC. Colin Wilburn now runs Micron Semiconductors, UK, makers of Si detectors mainly for HEP.

APTEC ENGINEERING LIMITED

4251 STEELES AVE. WEST, DOWNSVIEW, ONTARIO M3N 1V7   TELEPHONE (416) 661 9722   TELEX 065 27210

Aptec Engineering (Applied Technology), in Downsview, Toronto, Canada, was formed around 1975 from the NRD division of Electronic Associates of Canada. The detector technology was brought in by Howard Malm (ex Chalk River). One of its first products was a HPGe particle detector (PHYGE). The company was owned and run by Ed Zieba and specialised in HPGe γ-ray detectors. In 1999, it merged with NRC (Nuclear Research Corp), and Canberra acquired Aptec-NRC in 2001.

ENERTEC LASCO INTERTECHNIQUE EURISYS MESURES

The history of these French companies is rather convoluted. Some details were given by Henck [355]. Intertechnique was founded in 1951 with the French Aircraft manufacturer Marcel Dassault as the main shareholder. Its role was to supply the industry with equipment at an international level. It developed experience in nuclear safety and online process control and by the 1960s Intertechnique was an established French company. It made general portable α, β, γ, and neutron monitors as well as MCAs and acquired Lasco (Lab. des SemiConducteurs), which was founded in 1967 by Roland Henck and Paul Siffert of Centre National de la Recherche Scientifique (CNRS). Paul Siffert remained at CNRS, leading one of the major groups in Europe working on semiconductor detectors. He was also the founder and long-time president of the European Materials Research Society.

Lasco manufactured Ge(Li) detectors for the Société d'Applications Industrielles de la Physique (SAIP) located in Bagneux. It was soon joined by Paul Burger and Marc Lefèvre. For the contributions of Burger to silicon detector development, see Heijne [356]. SAIP itself acquired the small detector concern of SEAVOM (Organisation for the Study and Application of Vacuum to Mechanical Optics), Franconville, which was making silicon surface barrier and Si(Li) detectors. In 1972, the giant Schlumberger acquired SAIP and a year later, Lasco. In 1977, Schlumberger reorganised these and other concerns under the name Enertec in Lingolsheim near Strasbourg. It made the first commercial surface-barrier microstrip detectors in 1980 but never entered the EDXMA market. In 1987, the activities at Lingolsheim were acquired by Intertechnique, constituting their Nuclear Department. In 1990, Eurisys Mesures, Strasbourg, was created, by merging with other small businesses. Cogema acquired Eurisys Mesures in 1993 and around 2001 merged them into the Canberra organisation (see above).

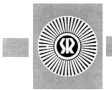

SEFORAD – APPLIED RADIATION LTD.

Turning to the smaller companies outside North America, Seforad (Emek Hayarden, Israel) was founded in about 1970.

The company made Si(Li), Ge(Li), and HPGe detectors and occasionally supplied to SEM manufacturers for EDXMA. It became Jordan Valley Applied Radiation Ltd. in 1980. In 2008, it acquired Bede Scientific (a spinoff from Durham University, UK, founded in 1978) and its EDXRF division became Xenemetrix.

 LINK SCIENTIFIC GROUP

Link Systems (High Wycombe, UK) was founded by Peter Millward, who was previously working as an experimental officer at Manchester University Nuclear Physics Department (together with Brian Sharp, another Link founder). He was representing AEP International, the agent for Nuclear Diodes in the UK in the early 1970s. The company name was chosen for its marketing benefits such as 'your *link* with the future' although, on Link entering the U.S. market later, EDAX was (in rather flattering recognition) to turn it into 'don't let your EDS be the weakest *link* in the chain'. One of the present authors (RAS) was another cofounder and established the Detector Group in 1974. Rob Sareen had worked on semiconductor detectors for nuclear physics at the University of Liverpool in the mid 1960s. He joined ORTEC in Oak Ridge, TN, in 1969. However, he was keen to exploit the pulse processing and FET technology being developed by Kandiah's group at UKAEA [357] and returned to the UK in 1973 to form Nuclan Ltd., with a view to manufacturing Si(Li) detectors and pulse processors. With a government grant, this was merged with Peter Millward's company (which was manufacturing MCAs) a year later. They worked closely with Kandiah's research group at nearby Harwell using their FET (see Section 11.19.3). Malcolm Patterson (from Liverpool University) and Peter Statham (via Cambridge University, Harwell and Berkeley) joined in 1975, and Barrie Lowe (Liverpool, Southampton and the University of Science, Malaysia, working with Ge(Li) detectors) in 1977. They were later joined by Graham White and Chris Cox (both from Kandiah's group at Harwell) and later by Tawfic Nashashibi (from Thorn-EMI) in 1987. Link

Systems was purchased by United Engineering Industries (UEI) in 1978 that then merged with Peter Michael's company Micro Consultants Ltd. in 1981. UEI purchased the x-ray tube company X-Tech in California, run by Bill Hershyn (formerly Watkins-Johnson and Kevex) in 1987 and Tennelec-Nucleus (see above), mainly for their germanium crystal production, in 1988. After an abortive attempt to merge UEI with Oxford Instruments in 1987, Carlton Communications/ITV (which had acquired Micro Consultants/UEI for their video software) finally sold off Link as a separate business to Oxford Instruments in 1989.

Oxford Instruments (OI) was founded as the first successful spinoff from Oxford University, as an electromagnet manufacturer in 1959 by Martin Wood and his wife, Audrey. Martin was working under Prof. Nicholas Kurti, who gave him much help and encouragement in establishing OI in the early days. The company also gained from the adoption of the world recognized and prestigious 'Oxford' name (with the stylised magnet 'O' in the logos). A detailed account of their history has been published by Audrey Wood [358]. The company grew by manufacturing superconducting magnets, mainly for NMR, and also by acquisitions, one of which was Link Systems (then Link Scientific). Continuing as OI Microanalysis Group, Tawfic Nashashibi and Graham White developed PentaFET, and Chris Cox, Rob Sareen, and Barrie Lowe developed Si(Li) and HPGe x-ray detectors. In 1993, the British Government, realising a fundamental problem of transfer of research into industry, published a White Paper, 'Realising our Potential—a Strategy for Science and Technology'. There followed a number of initiatives, one of which involved OI together with University of Leicester, the Rutherford-Appleton Government Laboratory (RAL) and e2v Ltd, in developing new semiconductor x-ray detectors for spectroscopy. These were based on PXDs (RAL) and CCDs (e2v). At the time there was very limited expertise in SDDs in the UK. The results of the PXD work and the CCD work (SCD) were reported (see Sections 7.3.5

and 8.3). They achieved technical specification but were not adopted by OI mainly on the basis of cost and x-ray stopping power. Chris Cox, with Pat Sangsingkeow (ex ORTEC), later managed detector manufacturing at the Tennelec (OI Nuclear Measurements Group) site in Oak Ridge. OI later sold this off, and Sangsingkeow moved back to ORTEC and Cox moved to PGT (becoming CEO in 2002). OI Nuclear Measurements Group, including the HPGe Crystal Growing Facility, under Larry Darken, was sold to Canberra in 1994. It did, however, remain in Oak Ridge. OI purchased the Finnish company Metorex in 2004.

Rob Sareen left OI in 1992 and after a year at Manchester University set up Gresham Scientific Instruments and was soon joined by Tawfic Nashashibi (ex-OI and Moxtek) and Barrie Lowe (ex-OI) in 2000. Gresham was bought by e2v Ltd. (manufacturer of Scientific CCDs) in 2005, forming e2v Scientific Instruments. Greg Bale manages detector manufacturing at e2v Scientific Instruments and Bob Daniel, formerly of Link Systems and OI, leads the technical management.

RNIIRP was established in 1966 in Riga, Latvia. It was a large-volume supplier of semiconductor detectors for all applications for most of the Soviet Bloc countries. From this, there was a spinoff company, Ritec, which was established in 1992, and later Baltic Scientific Instruments, established in 1995, both as private companies based in Riga. Ritec primarily produced CdZnTe detectors (Victor Ivanov). In 2003, BSI was acquired by Bruker AXS, Germany, and is now known as Bruker Baltic. Bruker also acquired the Berlin-based detector company, Rontec (established in 1989) in 2005.

11.18.1 Companies Specialising in WBGSs

Although work on WBGSs had been progressing in parallel with silicon and germanium, the detectors did not begin to be commercialised until relatively late. The following companies tended to specialize in them.

Texas Nuclear Corporation was a spinoff from the University of Texas in Austin. It was founded by I.L. Morgan and three associates from the University in 1956. Ira Lon Morgan was VP and director of research for the company until 1961 and was one of the pioneers of accelerators for nuclear physics. In 1965, Nuclear-Chicago (founded 1946 as a spinoff from the Manhattan Project work at the University of Chicago) bought Texas Nuclear and Morgan joined Nuclear-Chicago as VP until 1968. Then he returned to academia and became a professor of physics and director of the Centre for Nuclear Studies at the University of Texas. In addition to Texas Nuclear Corp., he founded and was the president of Columbia Scientific Corp. The technical operations included J.R. Rhodes and P.F. Berry. The latter was involved with making the HgI_2 detectors [359], using the vapour phase ampoule technique for their handheld XRF analyzer 'Metallurgist-XR'. In 1989, the company moved to Round Rock, Texas and became TN-Technologies and in 1994 was bought by Thermo-Electron.

Radiation Monitoring Devices (RMD) (Watertown, MA) was founded by Gerald Entine in 1974 for commercial applications of CdTe. It was a spinoff from Mobil Tyco Labs Inc. (Waltham, MA). Tyco Semiconductors itself had been founded in 1960 by Arthur Rosenberg to do experimental research on government research contracts and changed to

Tyco Labs 1965. Its focus was on high technology materials, crystal growth, and energy conversion (solar cells), and it was the CdTe technology, started around 1968 [360], which was transferred to the commercial outlet RMD [361]. Michael Squillante became director of research in 1983, and the company has employed many notable names in room temperature detector research such as Jim Lund and Kanai Shah. It continues as one of the world leaders in this area.

The company II-VI Inc. was established in 1971 for optical and electro-optical applications of CdTe, CdZnTe, ZnS, and ZnSe. By 1990, it was using high-pressure Bridgman crystal growth and starting to research room temperature detectors. In 1992, the company purchased eV Products (Saxonburg, PA). This had been founded in 1989 as a spinoff from BNL, making low noise electronics for radiation detectors, and it was soon using Si p–i–n diodes and CdTe for detectors. By 1994, eV Products was making large-volume (up to 6 cm^3) CdZnTe detectors. In 1997, it developed the first application-specific integrated circuit (ASIC) for CdZnTe in collaboration with BNL. In 2009, it was acquired by Endicott Interconnect and, as EI Detection and Imaging, continues to concentrate on large-volume CdZnTe detectors especially for homeland security applications.

Aurora Technologies was founded by Jack Butler, Clinton Lindgren, and others in San Diego in 1988. The firm, along with eV Products, was a pioneer of the HP Bridgman growth of CdZnTe [362]. In 1991, it became Digirad. This company has now moved into solid-state cameras for nuclear medicine, focusing on imaging, and no longer supplies CdZnTe material.

Xsirius Takes New Name: Advanced Detectors Inc.

11.18.2 Advanced Photonics, API

Besides EG&G and Texas Nuclear, a number of companies were involved with mercuric iodide usually using government funded contracts and grants. Xsirius (Camarillo, CA) was founded in 1985 with Andrzej J. Dabrowski, as chairman and CEO. Gerald Huth was a cofounder. Dabrowski had worked on compound semiconductors at the Institute of Nuclear Research (INR), Swierk, Poland, before moving to the University of Southern California in 1979. He was joined by his former student, Jan Iwanczyk, who had researched CdTe x-ray detectors at INR in 1979. They collaborated with Tracor-Northern to produce a HgI_2 EDX product in 1982 [363,364]. In 1989, Iwanczyk joined XSirius, which became Advanced Detectors (1996). Iwanczyk had also worked as a consultant to Texas Nuclear and, with Bradley Patt, helped to establish the subsidiary Advanced Photonix Inc. (Camarillo, CA) in 1991. This company manufactured large-area avalanche photodetectors. Iwanczyk also cofounded Photon Imaging, Inc., with Bradley Patt. This company is now involved with SDD manufacture (see Section 11.27). Radiant Detector Technologies, which was the commercial outlet for Photon Imaging's industrial and scientific technology, was bought by the Japanese giant Seiko Instruments Inc. (SII) in 2005 to form SII Nanotechnology USA.

After a career in metallurgy and crystal growth (Cominco, Amistar), Bob Redden founded Redlen Technologies in Sidney, Victoria, Canada, in 1999. Its specialty is CdZnTe manufactured by THM.

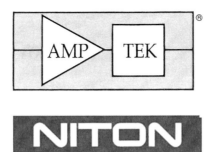

A number of companies exploited the capabilities of HPSi photodiodes (at first the commercially available ones) without being involved in lithium drifting. Using Peltier coolers, it was then relatively straightforward for them to expand into compound semiconductor detectors.

Amptek was founded in 1977 by Alan Huber and John Pantazis originally to fulfil the need for light compact x-ray detectors for space instrumentation. They were joined by Valentin Jordanov (electronics) and later Robert Redus (chief scientist). Their detectors flew on satellites and deep space probes, and the XR-100 was used on the Pathfinder Mission to Mars in 1997. Amptek remains a world leader for manufacturing portable x-ray and γ-ray detectors as well as MCAs.

Niton was founded by Lee Grodzins in 1987. He had graduated from Purdue in 1954 and worked as a nuclear physicist at Brookhaven, eventually becoming professor of physics at MIT in 1969. There he was interested in PIXE and EDXRF, devising a method of measuring heavy metals, such as U and Pu as contaminants and Pb in paint. Niton received an SBIR (Small Business Innovative Research) grant from the Department of Defence in 1990 and developed the Niton 'XL' portable analyzer. He remained at MIT as a consultant to Niton with Ethel Romm, and later his son, Hal Grodzins, who stepped in to run the company. It grew rapidly and became a world leader for portable XRF instruments by 2000. After some tragic events to key personnel and the departure of others to start a rival company in 2000, Niton was acquired by Thermo-Scientific in 2005.

11.19 1970s and 1980s: Evolution in Commercial Environment

There are companies that, because of lack of space, have not been included above, but we have attempted to include most of those that played a major part in semiconductor x-ray detector evolution. The EDXMA application was a turning point. The EM was a powerful tool across all the sciences: materials analysis, biology, medicine, forensic science, the list is endless. All researchers wanted EMs and most wanted EDXMA interfaced to them. The semiconductor x-ray detector industry changed forever. On the downside, not only were small companies stuck with a difficult technology and 'cottage industry production processes' not lending themselves to large-scale production, but they now needed detector designs engineered for every SEM and TEM on the market and a variety of the access ports on them. It was a time for clever engineering. The detectors had to fit into confined spaces (this is particularly true of TEMs, where a maximum active solid angle was demanded) and sometimes, a hostile radiation and electromagnetic environment [365]. At this stage, the big boys and some of the smaller ones (SSR, TMC, and later Canberra and ORTEC, and all but Link outside the United States) dropped out. During the early 1970s, alliances with the major EM and EDXRF manufacturers were being sought. Because many of these were Japanese (JEOL, Hitachi, Shimadzu, Rigaku), there was promotional interest in the Sixth International Conference on X-ray Optics and Microanalysis held in Osaka in 1971. ORTEC and EDAX were represented, and a favourable evaluation of a Kevex detector installed on a Hitachi SEM in Japan was reported [366]. As we shall see, the main contenders—Kevex, Tracor-Northern, EDAX, PGT, and Link—took their own evolutionary paths in the engineering of the detectors.

11.19.1 Si(Li) Crystals

Companies used basically either grooved or top-hat crystal geometries. Like surface passivation, companies were free to vary them at will as they were not in the position of having to tool-up for a manufacturing process or geared to produce vast numbers of identical components, as was the case for the microchip companies. Much variation and 'R&D on the fly' found its way into production at that time. Thus, there was a mix in each company but in general the following companies favoured grooved geometries: Kevex, ORTEC, PGT, Link, and the following favoured 'top hat': NSI, EDAX. Figure 11.54 shows a sample of crystals from various manufacturers.

Shown from left to right is Tracor-Northern (80 mm^2), Gresham (50 mm^2), Link (10 mm^2), EDAX (10 mm^2), Kevex (10 mm^2; guard ring), and Kevex (10 mm^2; standard). The grooves could be easily drilled using a diamond drill or a mild steel tool using a slurry of carborundum (SiC). The periphery, which was left after drilling, also made the handling and labelling easier. The grooved crystals usually had a major part of the outer

Figure 11.54 (**See colour insert.**) Detector geometries salvaged from redundant equipment.

wall ground off, producing the inverted 'T' geometry used to good effect by Kevex (Figure 11.53). The top hat and inverted T geometries gave a slightly lower electrical capacitance and assisted in the inspection of crystal wall surfaces. As a rough rule of thumb, the capacitance could be calculated by assuming a simple parallel plate ($C \sim \varepsilon\varepsilon_0 A/d$) giving 0.03 pF/mm^2 and adding 0.2–0.3 pF for stray capacitance from the periphery. The top-hat shape could be achieved by modifying a groove or rotating the crystal against a small circular diamond saw, but ultrasonic drilling could give any desired shape. Handling and labelling were more difficult but vacuum pencils could be used on the edges, and NSI managed to etch serial numbers on the front rim of the crystal. EDAX produced smaller overall diameter crystals and developed the art of applying a flat on one edge. This was possible, even into the active volume, if the damage was completely removed by etching. It facilitated the achievement of large solid angles with angled crystals in tightly restricted areas such as between the pole pieces of a TEM. This is illustrated in Figure 11.55.

Another approach to increasing the solid angle of x-ray collection, patented by EDAX [367], was to use oblong—rather than circular—crystals, which were drilled ultrasonically. Kevex was the master of miniaturisation. The company's

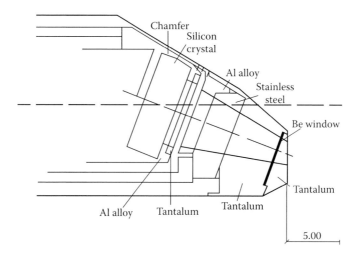

Figure 11.55 Extremely compact geometry needed for an angled TEM detector.

5-mm-diameter crystals (Figure 11.53) were the smallest of all. Together with a very short FET package (Figure 11.56), it often enabled them to tilt the crystal by tilting the whole cryo-head package. The cryo-head packages are discussed below.

In the very early crystals, the front was merely finely lapped and etched right across the whole area. This is still done for some planar HPGe crystals, where the junction is ion implanted. However, the surface barrier contact on a Si(Li) is very fragile. Any scratch or brush will cause electron injection, requiring a re-etch to remove the damage. As discussed in Section 11.12, a slight recess afforded some protection particularly when crystals were placed face down during processing. The active area or 'window' of the crystal could be delineated by copper staining and the insensitive uncompensated rim around it masked with wax for etching the recess. A mirror finish etched surface without any blemishes was the aim. Figure 11.57 shows a beautifully etched window of an early Kevex crystal.

Figure 11.56 Compact Kevex FET package.

Figure 11.57 (**See colour insert.**) Front face of a Kevex crystal illustrating their excellent etching technology.

The engineering of the cryo-head that housed the Si(Li) crystal, the FET chip, the charge–restore LED, heater resistor and all the connections, electrical and thermal, was a complex, many-parameter, compromise. The design constraints were discussed in Section 1.16.4 where some examples were also given (Figure 1.61a and b shows schematically the early approach taken by Kevex and Link, respectively).

11.19.2 Cryostat

Most companies used an oxygen-free copper cold finger connected to the boss on the inner vessel of a liquid nitrogen dewar using a short copper braid. The latter accommodated the relative movements during cryostat evacuation and cooling and dampened external vibration. Since 1976, most companies used dewars from the U.S. company Kadel (Danville, IN, USA), which were welded aluminium and cosmetically smart. The company supplied them in any desired colour and was open to any mechanical variations requested. NSI sometimes used an aluminium cold finger in the form of a tube used as a wiring conduit. Although the thermal conductivity is about half of that of copper, the emissivity is higher and being less dense, less energy was stored in any vibrations and the microphony was consequently reduced. Often, the copper cold fingers were nickel plated or wrapped in aluminium foil to raise the surface emissivity.

11.19.3 Cryo-Head

The cryo-head assembly was usually machined from aluminium alloy and screwed or clamped onto the copper cold finger, taking advantage of the greater contraction of the aluminium on cooling. For microanalysis, streamlining of the cryo-head was of the essence. The challenge was to insert it into a 14- or 12-mm-diameter stainless steel tube without giving rise to any heat shunt. As cold fingers tended to move on pumping and cooling, some sort of spacer was necessary either on the cold finger or on the cryo-head itself. The design had to be firm while not providing a path for heat or vibration. Some forms of spacers used are illustrated in Figure 11.58.

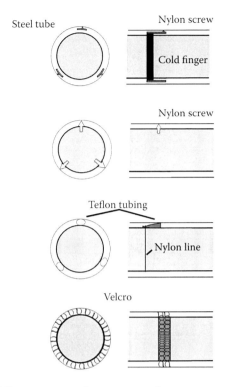

Figure 11.58 Different types of spacers used to support cold finger.

Figure 11.59 shows a schematic diagram of the Harwell arrangement that Link attempted to implement in the 1970s.

The PTFE shrinks on cooling, and no glue was used for the FET gate contacts but the PTFE was etched to glue the copper components. The Ni wire was long enough to sufficiently decouple the Si(Li) crystal thermally and optically from the FET.

Besides Link, several other companies, such as PGT, also started with PTFE packages.

Figure 11.60 shows an early example from PGT.

Although the nature of the electrical contact to the FET had to be nonresistive and solid (to avoid resistive noise, stray capacitance, dielectric noise, and microphony), the *thermal* contact had to be poor in order to maintain the temperature differential (20–25 K) that had to exist between the FET chip and the Si(Li) crystal. This was an engineering challenge. The contact could be maintained by an insulating strap or disk

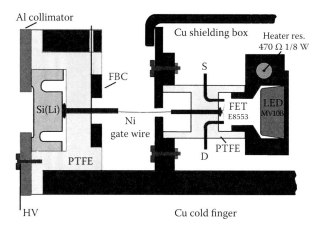

Figure 11.59 Schematic layout of AERE Harwell system.

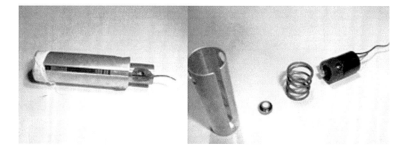

Figure 11.60 Early PTFE front–end design.

(such as a Mylar strip or PTFE bar) clamping a pin or screw against the crystal mesa. Such insulators contract considerably on cooling and therefore tightened the contact. A fine nickel wire could be soldered to the pin for the connection to the FET. This was used in early EDAX and Link detectors. NSI in its EDXRF detectors clamped a ball of the 25-μm gold wire against the crystal mesa. Figures 11.61 and 11.62 show EDAX examples where the soldered wire to the BN FET package is just visible as is the Velcro spacer and PTFE disk:

The FBC was a pin mounted in the BN package close to the gate pin. This was also used by NSI, but Kevex and Link used a cirlclip around the package. To avoid added components and noise, the FET package was often pushed against the mesa

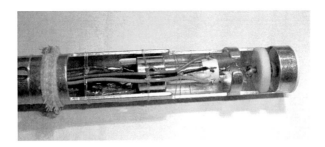

Figure 11.61 An early EDAX front–end.

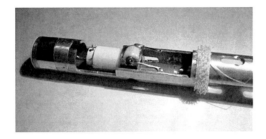

Figure 11.62 **(See colour insert.)** Early EDAX front–end.

using the FET gate pin directly. These pins were kovar and could be sharpened to a point contact. This is shown in Figure 11.63, an early Link example.

The FET gate pin compressed a crinkle washer behind the collimator at the front. If the thermal isolation of the FET package was not good, the crystal leakage current increased with the FET package heater resistor current. Besides using a point contact, the isolation could be improved by increasing the thermal path length for example with a spring arrangement to the FET gate pin itself. A stiff nickel wire was used in early

Figure 11.63 Early Link front–end.

EDAX cryo-heads and a phosphor-bronze strip, which could be tensioned with an insulating crossbar, was used by NSI. The latter also worked well with angled crystals although it did add some stray capacitance.

In 1967, TMC was using 2N3823 FETs cooled and mounted in the cryostat. Later, most companies were using the 2N4416 made by Union Carbide (UC 150) or TI, Dallas. Up to the early 1980s, all Si(Li) detector manufacturing companies were selecting FETs from those bought commercially. They were supplied direct on pins or decanned and remounted in the BN package. When the manufacture of these at TI (then the only supplier) looked threatened, EDAX purchased all the remaining wafers. This prompted a spinoff company, InterFET, led by Michael Hoye in 1982, to carry on a line in specialist FETs. Link had been using a modified 2N4416 (the E8553) manufactured at TI Bedford UK for AERE Harwell and later by EMI, Hayes, UK. The modification had been made under the contact pads at Kandiah's request in order to reduce the long-term effects of the restore light pulse and reduce high rate degradation effects. The work was originally stimulated by the problem of analyzing radioactive specimens where all spectral detail could be destroyed. These rate effects were caused by surface states and were often seen in the conventional 2N4416s manufactured by TI, Dallas and Motorola. Unfortunately, Kandia's 'optical modification' also increased the input capacitance significantly. However, in 1981, Link gained exclusive right of use for a Tetrode FET (E8552) [368], again originally manufactured for AERE Harwell. This had had quite a different architecture from the 2N4416 and E8553, as shown in Figure 11.64.

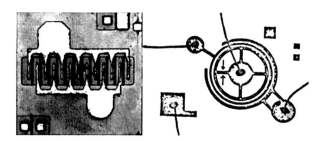

Figure 11.64 Two successive FET designs by Kandiah AERE, Harwell.

The operational gate (top gate) was a small area ring of low capacitance (~1 pF compared to ~4 pF for the interdigital 2N4416) and quite separate from the chip substrate and the mounting pin. As a result of this geometry, a higher intensity of LED was required for restore with the added problem of maintaining a light tight package. The Tetrode retained the optical modification but had redressed the capacitance problem. Also, by varying the substrate (bottom gate) potential on the mounting pin, the channel could be moved to 'tune out' noise producing traps at the edges of the channel (see Section 1.13). The respective noise curves are shown in Figure 11.65 [369].

An added bonus of the Tetrode was that the FET top gate was no longer connected electrically to the kovar pin, as mentioned, on which FET chips were normally mounted. This meant that it need not necessarily be *thermally* connected to the Si(Li) crystal. A 25-µm gold wire bonded to the FET top gate could be used to support the considerable temperature differential required. The concept of a separate 'FET package' could be abandoned

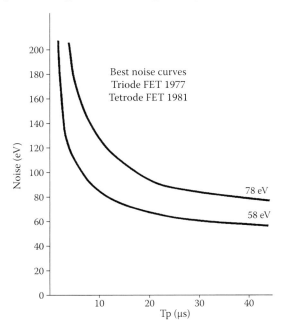

Figure 11.65 Noise curves obtained using two different FETs. (From Statham, P.J., Assoc. Nat. de la Res. Techn. (ANRT), Microanalyse et Microscopie Electr. A Balayage, Paris, Nov. 1985. With permission.)

Figure 11.66 Example of a package that took advantage of four-terminal FET with isolated gates.

with the FET chip and LED mounted transversely on the cryo-head body itself. Thermal isolation was achieved by mounting the kovar pin in a PTFE platform, but this was later changed to a PCB board as shown in Figure 11.66.

The threaded brass ring allowed the Si(Li) crystal mounted in an aluminium cup to be pressed tightly against the central pin at the same temperature. The HV bias could be applied through this brass ring. The FBC was originally a pin mounted in the PTFE platform alongside the kovar pin.

11.19.4 Case of 'Self-Counting' Detectors and 'Ghost Peaks'

Self-counting was briefly mentioned in Section 3.6. During the 1980s, a strange phenomenon appeared in some detectors interfaced to HV TEMs. Users complained that the detectors were counting x-rays even when the electron beam was turned off! Furthermore, the effect did not seem to diminish much with time. The authors first observed this on an NSI TEM detector but, although Kevex seemed to be immune, it ultimately also occurred in a Link detector. The customers gave assurance that they had not used any radioactive specimens, but investigations nevertheless proceeded with some trepidation. The signals looked like x-rays and this was confirmed by long counting times in which the fluorescence of the Al Kα peak from the cryo-head was revealed. After warming the detector to room temperature, this self-counting at first increased alarmingly and then disappeared. On recooling, the phenomenon did not reoccur.

This was discussed at the Aussois Workshop [370] in France in January 1988. J.-P. Chavalier believed that BN could be 'charged up' by the TEM beam and subsequently discharge electrons into the vacuum after the beam was turned off, causing Bremsstrahlung and fluorescence of the surrounding materials. It became apparent that electrons could flood the detector even through a Be window under certain circumstances, including low magnification, electron 'showering' (to clean the specimen chamber) and scattering when the beam hit a grid bar. HTO angle detectors were particularly vulnerable. In the cases of companies (e.g., as NSI, EDAX, and Link) that used BN at the front of the cryo-head, these electrons were responsible for this charging and being at low temperature in vacuum this charge resided for extremely long periods (many weeks or months) causing self-counting. A redesign was necessary, screening the BN or making the whole Si(Li) crystal cup of aluminium.

Ghost peaks were discussed in detail in Section 3.4. They are well-defined peaks associated with larger parent peaks but are actually Artefacts of the semiconductor crystal. They were first noted in the early years of semiconductor detector development (around 1960) in diffused junction (see Section 11.13.1) and surface barrier particle detectors when they were referred to as 'multiple peaks' [76]. It was noted that the peaks were remarkably narrow and were localised to specific regions of the crystal, being influenced by collimation. Some detectors were worst at low bias. The problem was again discussed in the book by Dearnaley and Northop [61] in 1966 and at the Gatlinburg Conference [137] 1969, where they were explained in Ge(Li) detectors as uncompensated 'pea pockets'. Irradiation of surface barrier detectors with α particles produced increased leakage and multiple peaking. The effect increased as the bias was decreased. Workers proposed that the problem was attributable to charge loss when the depletion depth was comparable to the range of the ionising radiation.

Later in the 1980s, it was noticed that p–i–n diodes that were not fully depleted (such as the Hamamatsu S-series used by some companies) gave rise to 'artifact spectral peaks', which moved with bias. It was thought that they were due to charge diffusion from the undepleted back region and could sometimes be collimated out [371]. Since then, some studies have been made of artifact peaks in Si(Li) detectors [372–374]. Campbell also noted that collimation removed the artifact. At very low bias (<20 V), some Si(Li) detectors will display well-defined ghost peaks with an Fe-55 source [375] but at full bias x-ray energies of >15 keV are usually required in order to detect them, if they are present. Very large 'anomalous' peaks have also been observed in PXDs when irradiated on the patterned side [376]. These were thought to be due to positive fixed charge beneath the unmetallised oxide regions that may have resulted in an undepleted layer of the high resistivity (>5 kΩ cm) n-type silicon. Ghost peaks from an α-particle source have also been observed in some configurations of SDDs [377]. These were thought to be associated with hole accumulation around the guard rings. Tests using a high-energy source (such as Cd-109) need to be carried out on Si(Li) detectors to ensure these Artefacts are absent.

11.20 1970s and 1980s: Low-Energy EDXMA

We have already seen that O Kα x-rays were detected by Nuclear Diodes Inc. in 1969 (albeit almost certainly internally generated) on an EMP. Systems termed 'double-window systems' where the beryllium window could be replaced by a self supporting ultrathin window [UTW; but not atmospheric pressure supporting (ATW)] of Al or polymer using an inline gate valve for EDXMA were discussed by Aitken et al. in 1970 [378]. Manny Elad and Dale Gedcke of ORTEC also speculated on the requirements that had to be met for light element analysis [379]. We also noted in Section 11.17 that the practicality of windowless analysis was demonstrated at Berkeley in 1970 by Jaklevic and Goulding [380] using an electron gun especially designed to produce soft x-rays. This was mounted on a vacuum chamber shared with a Si(Li) detector having the window replaced with a gate valve as shown in Figure 11.67.

With this setup, they were able to unambiguously detect elements down to C Kα (277 eV), despite displaying some nonlinearity, as shown in Figure 11.68.

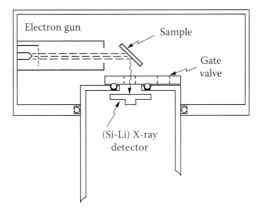

Figure 11.67 Early gate valve assembly enabling the detector to be operated in a 'windowless' mode. (From Jaklevic, J.M., Goulding, F.S., *IEEE* NS-18, 1, 187, 1971. With permission.)

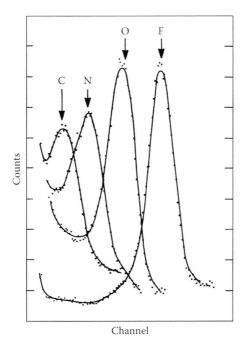

Figure 11.68 Spectra taken with an early windowless Si(Li) detector. (From Jaklevic, J.M., Goulding, F.S., *IEEE* NS-18, 1, 187, 1971. With permission.)

Somewhat preempting the later Kevex Quantum Window (see Section 11.21), Aitken et al. [378] suggested that alternatives to Be foil for windows should be considered. For example, they pointed out that a 3.75-μm mylar foil was vacuum-tight (but as we now realise, not impervious to water vapour). A 2.5-μm graphite ATW [381] was demonstrated at ORNL in 1974 but not taken up commercially.

11.20.1 'Windowless' EDXMA Detectors

The first commercial windowless detector was the 'ECON I', which was offered by EDAX around 1974 [382]. The name referenced the fact that EDAX could now offer 'semiquantitative' detection of the elements C, O, and N. The ECON was bulky and the solid angle was poor. The window mechanism swung to the side and had to clear the specimen stage and relied on the operator's skill and strength to secure a positive closure before venting the SEM chamber. ORTEC had a prototype rotating turret detector working in 1973 [383] and introduced their version, EEDS-I, for windowless operation on SEMs around 1975.

EDAX started work on a rotating turret design that allowed three different window choices (usually Be, UTW, and open), around 1974 and launched ECON II in 1976 [384]. This was more compact (50-mm-diameter turret) and the solid angle improved (0.0082 steradian for a sample 3 mm from the turret end). The Be window was sealed by a concentric O-ring by retracting the outer tube by ~1 mm. The mechanism was then locked. An example is shown in Figure 11.69.

The first detectors produced the O Kα photopeak with a large low energy tail (see Figure 11.70a) [384] and the C Kα photopeak was barely detectable even for a pure graphite sample (see Figure 11.70b) [385].

Nonlinearity was mentioned as an issue but only in the context that the N Kα peak was 'high' because it was 'on the tail of the noise'. In spite of these problems, the transverse sliding O-ring in the rotating turret design worked surprisingly well and was copied by most of the other companies in due time. They also encountered similar spectral problems.

Figure 11.69 Example of an EDAX ECON 11 multiple window cryostat.

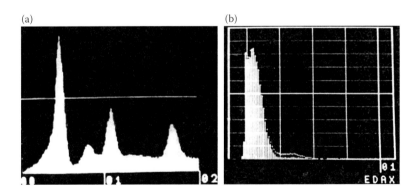

Figure 11.70 (a) Large peak on the left is an oxygen peak with a pronounced low energy tail. (From Coates, D.G., *Electron Microsc.*, 3, 26, 1980. With permission.) (b) Carbon peak almost buried in noise. (From Russ, J.C. et al., *XRS*, 5, 212, 1976. With permission.)

Kevex chose to go down the gate valve route with a design by Musket and Bauer. Ron Musket [343,344] was also the first to detect the B Kα photopeak in 1974 using PIXE. After joining Kevex, he developed 'Alpha-X' (see Figure 11.71), an elegant product that incorporated an α-source for excitation but it never really took off commercially, mainly because of the complications caused by the very high activity sources required for practical excitation.

In 1979, Kevex launched their Kevex/6 UTW detector (see Figure 11.72).

Like many of the windowless detectors of the time, a small ion pump backed up the vacuum. A welded metal bellows was used for retraction and an Al-coated 0.2-μm paralene UTW (which was not accessible to the user) to shield the Si(Li) crystal from infrared (IR) radiation. Because of the space required

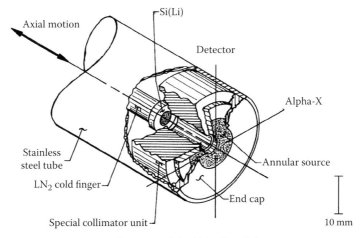

Geometry of the Alpha-X probe

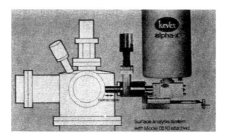

Figure 11.71 Alpha-X, a source-excited product launched by Kevex.

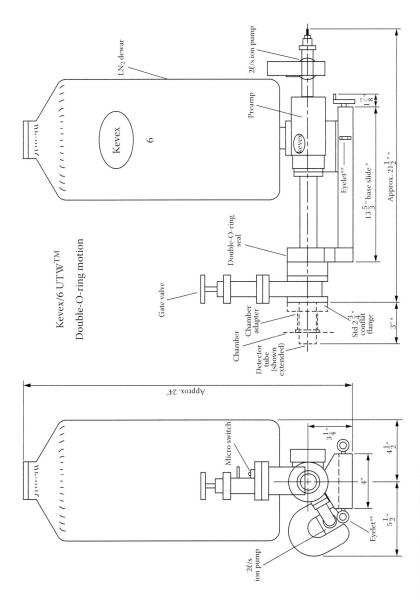

Figure 11.72 A Kevex/6 UTW detector.

by the ceramic magnet electron trap located in front of the Si(Li) crystal, the 10-mm^2 crystal was collimated to a 2-mm-diameter aperture on their narrowest tube version. The resolution offered (145 eV) was the best for the time. Their flyer gave the true peak position for C Kα as 277 eV (which is correct) when most of the x-ray tabulations and 'slide rules' were giving it as 283 eV. The latter figure seems to have been a mistranslation from the wavelength tables but was making the achievement of linearity by EDX companies all the more difficult! This error persists even into recent times [386,387].

At the NBS Workshop on EDXS in Gaithersburg, MD, USA, in April 1979, it was clear that the C Kα peak shift was attributable to ICC at the surface of the Si(Li) crystal window: for example, the P escape peak at 275 eV was in the correct position, whereas the C Kα peak from the same detector under identical conditions could be 10 eV or more below it. At this workshop, Ron Musket [388], then with Kevex, showed that the C Kα peak could shift up 35 eV as the bias was increased from 200 to 1000 V. He also discussed the effect of ice on the crystal window, that is, a relatively large O Kα peak in all spectra with a paucity of counts just above it (reaching beyond 1 keV) due to the O Kα absorption edge from the ice. Indeed, a survey (unpublished) by the present authors carried out on many different commercial detectors at the time all showed these symptoms. The other major players, Tracor-Northern, PGT, and Link, soon followed EDAX with windowless detectors. Link was the last but caused quite a stir in claiming for the first time the detection of B Kα x-rays on an SEM. The name for their detector, LZ-5 (due to Peter Milward), succinctly indicated this. This had a 32-mm-diameter turret and was demonstrated 'in-house' in 1981 with three major advantages. First of all, it worked with *guaranteed* B Kα x-ray detection (Be Kα x-rays had also been detected occasionally by 1983). The other two advantages both originated from the work of 'Wrangy' (he had been a Cambridge mathematics Wrangler) Kandiah and his team at Harwell, namely, the long-established 'Harwell' Pulse Processor [389] and the recently designed and manufactured low-noise Tetrode FET mentioned earlier.

Kathirkamathamby Kandiah (also referred to as 'KK') was born in Jaffna, Sri Lanka, in 1916 and won a scholarship to London University. He received an MA from Cambridge University in 1941 and stayed on at the Cavendish Laboratory throughout the war years designing instrumentation. He moved to AERE Harwell in 1946, where he set up their first instrumentation team that designed a modular system, the 'Harwell 2000' Series, which sat side by side on a single 19-in rack. The width of the modules was chosen to accommodate a 'CV 138' thermionic vacuum tube lying horizontally across it. In the late 1950s, he had in his team a young Birmingham University graduate, Fred Goulding. Goulding moved to Chalk River, Canada, and in the early 1960s to Berkeley, CA, where he ran their pioneering nuclear instrumentation group. A friendly rivalry between Kandiah and Goulding remained throughout their careers. Kandiah died in 2002.

Link Systems adopted the 2000 series pulse processor (as the '2010') in 1974 and started their business by purchasing Kevex and NSI Si(Li) detectors and improving their resolution by adapting them to the 2010 Pulse Processor. Most pulse processors at the time set a noise discriminator in the processing channel and processed both signals and any noise exceeding the discriminator in the same way. Naturally, because of the random nature of the noise, the discriminator could never eliminate the Gaussian tail of the noise altogether. Thereby lay their problem. Getting good efficiency at low energies involved lowering the discriminator and swamping the processing channel with noise. The spectra, for example, Figure 11.69b, showed peaks sitting on the enormous tail of the noise. Not only was the photopeak not well resolved, but one had no idea how it was being distorted by the underlying noise. The Harwell processor used a time-variant filter with a 'truncated cusp' shaper. As it turned out, this was superior to the Gaussian shaper in the presence of dielectric and $1/f$ noise. Another of its features was the division of the signal into three channels as shown in Figure 11.73.

Besides the processing channel with its delay line, a series switch (designed to eliminate effects of slow charge) and time-variant filter, there were two 'recognition' channels both with their independent discriminators. Their purpose was to recognize *bone fide* signals above the noise in the respective channel and to open the processing channel to accept them. The

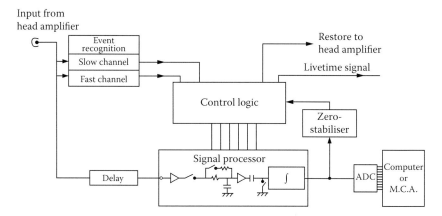

Figure 11.73 Harwell pulse processor. (From Statham, P.J., *NBS* 604, 127, Heinrich, K. et al. (Eds.) (Gaithersburg Meeting April 1979), 1981. With permission.)

time constants were much shorter than the processing channel and could therefore also recognize and reject events that would normally 'pile up' in the time-variant filter. The 'Fast' channel (Figure 11.72) had a time constant of 0.1 μs that led to a noise FWHM of about 900 eV. This meant that pileup inspection could be carried out for about 2.7 keV x-rays and above. The Slow channel had a time constant of about 1 μs. This meant that pileup inspection could be extended down to about 900 eV x-rays and above. This was adequate for Be window detectors that were not used for analysis below ~1 keV. Thus, only events triggered by the quite separate recognition channels appeared in the processing channel. If there were no x-ray signals present, only those noise excursions that crept over these discriminators triggered the processing channels to open. Furthermore, their *rate* could be set (usually to ~50 Hz for analysis) by the discriminators in the recognition channels. For the true noise Gaussian peak, there was a series of pulses that opened the processing channel at regular intervals providing there were no x-ray signals recognized. This 'strobe' could be displayed and the noise FWHM therefore accurately measured and monitored. With the advent of the LZ-5, it was necessary to recognize x-rays down to ~100 eV. This could be done (at the cost of reduced pileup rejection) by reducing the noise in the recognition channels. This was achieved by making the time constant in the 'Slow' channel 10 μs and in the 'Fast' 0.5 μs. In the 'windowless' mode, the strobe could be switched off and the low-energy efficiency increased by lowering the discriminators in the recognition channels without swamping the spectrum with noise. Later, three recognition channels were used to give improved pileup protection for a wide range of energies.

The second advantage the LZ-5 had was Link's exclusive use of the Tetrode FET (E8552). The importance of low noise at all processing times for soft x-ray spectrometry cannot be overemphasised. Not only are peaks better separated from each other and the noise tail, but the detection efficiency threshold can be lowered by lowering the recognition channel discriminators into the reduced noise tail.

The third advantage was the earlier launch of the 'Series E' detectors by Link in 1980. In these, the background count level at low energies (<3 keV) had been reduced by a factor of 3 or more by using a thin Ni contact to the Si(Li) crystal window instead of Au (see Section 2.8.2). In the study of alternative contacts, they were significantly aided by the diagnostic programs of Peter Statham and his software team. These characterised the relatively pure spectrum of x-rays from a Fe-55 source, and the characteristic of the detector (or specifically the ICC) could for the first time be *quantified* (see Figure 1.53 in Section 1.14.3). Two metals stood out as giving very low background around 1 keV, Ni and Co. At the time the reason was not appreciated (the formation of metal silicides at the interface was suspected and the idea of using silicides was suggested later by David Joy [390]), but eventually the true reason became clear (J.L. Campbell, private communication, 1998). The Au M-shell photoelectrons (0–3 keV) from the contact were spraying into the active volume of the silicon crystal. By 1984, spectra unequivocally showing B Kα and Be Kα were published. Figure 11.74 shows a Be spectrum [391].

Although the turret design was made more and more compact, for example, the LZ-5 Mk 3 (1982) had a 25-mm-diameter turret, they were too large for use on TEMs. For this application, there was no alternative to the retractable inline gate valve design of Kevex. For example, the Link LZ-5 Mk 6 (1985) was retractable behind one or two gate valves. The latter

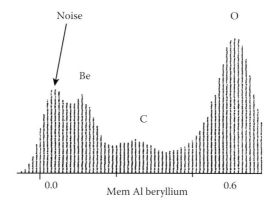

Figure 11.74 Spectrum showing the detection of the Be x-ray, 1984. (From Statham, P.J., *J. Phys.* 45, 2, pC2-175, 1984. With permission.)

enabled the UTW and collimator to be inspected or changed and the TEM chamber to be baked without venting the TEM or removing the detector. This was as the result of collaboration with VG Microscopes and was ultrahigh vacuum (UHV)–compatible and commonly used on the VG HB 501 TEM.

11.20.2 Question of Nonlinearity

Nonlinearity in Si(Li) detectors was first discussed by Zulliger and Aitken [392] in 1967 using a TMC Si(Li) crystal in a modified SSR cryostat at Stanford University. They found field-dependent nonlinearity of a few percent right up to 75 keV but they identified the field dependent component as due to trapping (poor compensation) and the residual (~1% at 4.9 keV) as due to crystal window effects. Three years later, improved drifting had removed most of the bulk trapping, and Jaklevic and Goulding [380,393] remarked that even the C Kα and N Kα peaks were only 0.7 and 0.9, respectively, from their expected positions. They also attributed this nonlinearity to ICC in the Si(Li) crystal window. Commercial companies were more reticent about admitting any such 'defects' in their crystals. To admit to peak *distortion* was one thing, but to admit to the peaks being in the *wrong place* was quite another. As mentioned above, Ron Musket—working with an early Kevex detector while still at Sandia Laboratories in 1973 [343]—had shown that nonlinearity at 277 eV could be a problem but published a plot showing good (to 2%) linearity down to B Kα. Nonlinearity was discussed at the NBS Workshop on EDXS, Gaithersburg, 1979 reported in 1981. As we have seen, Musket showed results from the Alpha-X [388]. The linearity was field-dependent (see Figure 11.75) and therefore presumed a *crystal* artifact. This was backed by a plot that showed different detectors having different nonlinearity but similar field dependence. Incidentally, this field dependency depends on the processing of the front contact (see Section 4.6), and his results were not generally observed.

As Musket admitted, 'typically, complete commercial detector systems have resolved C Kα and O Kα pulse heights of about 0.9 and 0.97 of the values expected from linearity considerations', but he was careful not to necessarily blame the Si(Li) crystal. There was also an apparent inconsistency in the fact that the O Kα peak (525 eV) always had a large tail, whereas the C Kα peak was fairly symmetric (see Figure 11.76). He suggested that an analysis of gaseous oxygen might show this anomaly to be a 'solid state effect on the true x-ray spectrum', again avoiding blaming the crystal.

At the same workshop, Peter Statham (Link) [394] suggested that the tail on the C Kα peak might be suppressed as a result of the rapid falloff in electronic efficiency below 250 eV. John Russ (EDAX) admitted that low energy peaks 'do not

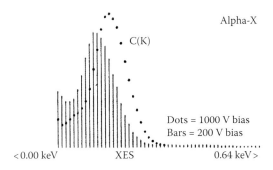

Figure 11.75 Shift in C peak position with applied bias using Kevex, Alpha-X product, 1979. (From Musket, R.G., *NBS* 604, 97, Heinrich, K. et al. (Eds.) (Gaithersburg Meeting April 1979), 1981. With permission.)

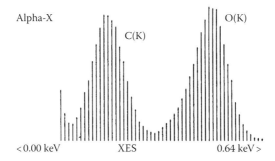

Figure 11.76 Visible tail on oxygen peak. (From Musket, R.G., *NBS* 604, 97, Heinrich, K. et al. (Eds.) (Gaithersburg Meeting April 1979), 1981. With permission.)

always align well with the energy markers'. At this time, the main reason for these effects (ICC in the crystal) was not identified. The 'nonlinearity topic' was clearly a sensitive issue for all companies, including EDAX, which carefully avoided the topic in its subsequent report on the meeting [395].

Bill Drummond, working at NSI, did much work in the late 1970s and early 1980s to improve the crystal window processing including minimising the native oxide (W.L. Drummond, private communication, 1980). The major reduction of ICC at the front contact, however, came with contact engineering from the Link group in the mid 1980s that largely solved both the tailing and nonlinearity problems. Historically, a-Ge was first applied to a Si(Li) crystal in the hope of increasing the carrier yield at low energies. This also worked well on HPGe. Later SiO was used, since it was a recognized passivation. This improved the ICC still further. Theoretical explanations have been discussed in Section 4.6.

11.20.3 Icing

As mentioned previously, ice on the crystal window was first identified by Musket at the Gaithersburg meeting in 1979. He showed that an O Kα absorption edge could even be detected in a freshly cooled detector. Layers of a few 100 nm of ice were sufficient to cause the observed effects. Ice 'build-up on ORTEC Be window detectors was further researched by David Cohen

[396] working at Lucas Heights, Australia, in a series of papers from 1982. The ice layer could be removed by warming up the cryostat for sufficient time to achieve thermal equilibrium and subsequently recooling. There were generally no adverse effects to this and the low energy efficiency was restored for a time. However, it was inconvenient and detector parts could work loose. The liquid nitrogen had to be boiled off and then the detector left to warm up, preferably overnight. After drying out the dewar, it then required an hour or so to be cooled down again. Musket pointed out a more serious problem. Oil could not be removed in this way and 'oil may render the crystal unusable for high-resolution purposes. This excludes the use of windowless detectors on diffusion pumped sample chambers'. The authors found that the oil from a rotary backing pump was even more reactive on the front contact and affected stability. To answer some of the inconveniences of windowless detectors, Link developed [397] an internal heater that they termed a 'Conditioner'. It was discovered that a heater resistor in good thermal contact with the cryo-head would warm the Si(Li) crystal sufficient to cryo-pump the ice film away in a matter of minutes without removing the liquid nitrogen from the dewar or the detector from the EM. The low-energy x-ray sensitivity (usually gauged by the Ni L/K line ratio at 20 kV electron excitation) was brought back to a maximum. The conditioner was engineered into the electronics to go through its cycle safely and automatically, removing the detector bias, which might damage the FET, at the simple pressing of a button. An added bonus was that often the noise also reduced, because water in the permeable BN part was also removed.

The maximum detection efficiency could only be obtained by operating completely windowless or with an ultrathin membrane (UTW) to restrict contamination. Target specimens could sublime onto the Si(Li) crystal or the specimen itself could be contaminated (e.g., by carbon 'build-up) by the activated charcoal in the detector cryostat. Isolating the dewar vacuum, either with metal bellows or by closing a bypass valve, considerably eased the latter problem. Companies worked hard for solutions on an individual basis but they were costly and the operation was often complex. An alternative came from an unexpected quarter.

11.21 1987: Kevex Quantum Window: Convenience versus Performance

There was a rumour around the industry that Kevex were about to launch a significant new product related to the 'front–end' at the Pittsburgh Conference in 1987 (we believed the code name was 'Buba'). The 'Quantum Window' [398] took all of the competitors by surprise. The Be window was simply replaced by an alternative: a vacuum sealing atmosphere supporting window (ATW) of a secret new material and construction. Compared to the windowless detectors available, this gave an inferior solid angle, poorer x-ray transmission efficiency, and was less robust. But it was a cheap and convenient solution. Elements from carbon upward were detectable with all the convenience of the standard Be window detector to which all users were now well accustomed. No more retraction, gate valves, rotating turrets, or interlocks. The speculation was that it was made of amorphous diamond, but in fact it consisted of an amorphous BN membrane supported on a Si lattice structure. The BN was deposited onto a Si wafer that was then masked and etched away to form a pattern of BN membranes. The Si formed ribs between the panes and acted as a support structure and it could be made large enough for a 10-mm^2 Si(Li) detector. The technology itself had come from Hewlett Packard [399] in Palo Alto via Mike Kelly and Surface Science, which Kevex had purchased. Despite the window rupturing at the exhibition (and competitors being accused of sabotage [400] with the FBI involved), the idea caught on and all EDX companies were suddenly frantically trying to catch up with a technology they were not familiar with. Thin polymer windows for soft x-rays had been used for many years on gas proportional detectors and these were the first port-of-call. By laying one thin layer of, for example, formvar, floated in solution, onto successive layers and aluminising each one in turn, strong windows could be made. They held vacuum, but there was always a concern about their imperviousness to water vapour. Link learned that a group at King's College London [401] was working on silicon nitride membranes for x-ray lithography and a project was set up with Peter Anastasi [402]. Window alternatives have been reviewed [403] and companies such as Moxtek [404] specialize in fabricating thin polymer windows that are sourced by most EDX companies.

11.22 High-Purity Germanium and Silicon

There was a great incentive to produce HPGe (or 'intrinsic' Ge as it is sometimes referred to). This would deplete without the need for any lithium drifting and hence was stable even if accidentally warmed to room temperature. From a manufacturing point of view, it could be both stored and shipped at room temperature. It was found that exposing n-type germanium crystals to penetrating radiation produced levels in the forbidden gap that acted as acceptors and hence partially compensated the material ('transmutation doping') increasing its resistivity sufficiently to deplete up to 5 mm. Such detectors were manufactured in 1966 by the Russians [405] at the Ioffe Institute and later at ORNL [406]. But not everyone wanted to experiment with neutrons or with 200 Ci of cobalt-60! Pioneering work on germanium crystal growth was being carried out by Rob Hall at GE Schenectady under a contract from the US AEC. He was using Hoboken germanium as the starting material for purification. At the NRC Conference on Semiconductor Materials for Gamma Ray Detectors in New York in 1966, he presented evidence that HPGe might soon be produced at impurity levels $<10^{12}/cm^3$. The main problem was copper, which diffuses readily at temperatures above 300°C. The possibility of using HPGe for detectors was also discussed at the Second Gatlinburg Conference in 1967 [407], and in 1968 Hall took out patents. He thought then that the cost of HPGe material might ultimately be an issue. Finally, with R.D. Baertsch [408], he reported the manufacture of such material resulting in the first HPGe γ-ray detector in 1970 [409]. These crystals were grown by T.J. Soltys [410]. About the same time, Tavendale [411] at Chalk River also made planar surface barrier detectors using the GE n-type HPGe. The following year, Llacer [412] was using GE material at Brookhaven.

Rob Hall, whose name repeatedly crops up in connection with germanium crystal growth, was in fact a rare example of an all-round inventor. He retired from GE in 1987 having 43 U.S. patents to his name and having invented a type of magnetron (1945) commonly used in microwave ovens and the semiconductor laser (1962). His HPGe was a major breakthrough for gamma-ray detector manufacture. For example, detector arrays (separate crystals in a single cryostat), would have been impossible with Ge(Li) detectors. HPGe made it possible and by 1979 both PGT and ORTEC had produced such arrays. In 1971 Eugene Haller, who had worked on probing the surfaces of large Ge(Li) detectors using Hoboken material for his Ph.D. thesis at Basel University, Switzerland [413], joined the Berkeley group. Bill Hansen was already there and they quickly became a team with Hansen [414] growing HPGe crystals and Haller characterising them using photothermal ionisation spectroscopy (PTIS). From then on, the Berkeley group tended to dominate HPGe work. Soon Bill Drummond [415] was making detectors at Stanford University using material from NPC Metals Co. Los Angeles. Detectors were commercially available from GE and by 1973 also from PGT. With the advent of HPGe the problem of making surface barrier contacts also went away. Coaxial Ge(Li) detector production was eventually wound down but they were still being sold in the United States up to around 1983.

HPSi did not take a similar route. The reasons for this include the stability of Si(Li) material even at high temperatures (allowing for example vacuum baking up to ~50°C) and the lack of availability of HPSi on a scale sufficient to base an industry on it. In 1966 Meyer and Langmann [416] working in Karlsruhe, Germany, used Wacker material to manufacture 3-mm-deep particle detectors. In 1982, the Japanese company, Komatsu, was producing small amounts of high-resistivity p-type silicon (up to 80 kΩ cm) from monosilane gas. This could also be depleted to 3 mm and surface-barrier detectors were manufactured at the Rikkyo University [417], Nagasaki. Later in the 1980s, the Japanese detector company Horiba (a subsidiary of Hitachi) was basing silicon x-ray detectors on this material. The Danish company Topsil could also supply small amounts, but a total switch from Si(Li) detectors to thick HPSi detectors did not arrive.

11.23 1990s: HPGe X-Ray Detectors

In 1970, Aitken (NSI and San Jose College), Woo (Kevex), and Russ (EDAX) [418] compared the performance of Si(Li) and Ge(Li) detectors at the EDX Meeting in Toronto. They observed that the best resolution on Ge(Li) detectors was ~175 eV compared to ~153 eV on Si(Li) detectors. As they remarked: 'although ω is less in germanium the resolution advantage has not been realised. This situation undoubtedly can be improved, but in our opinion it will require some major solid-state surgery'. In discussing the low-energy performance of germanium detectors, the world authority on the subject, the Berkeley Group, had also made some rather damning observations. In 1971, Goulding and Jaklevic [298] compared the performance of Si(Li) detectors and Ge(Li) detectors for XRF. It was concluded that trace elements were obscured by 'various Artefacts' in the Ge(Li) spectra caused by the detector itself and 'it appears that germanium detectors have little value for trace element analysis'. HPGe fared no better. In 1977, in what was in fact the classic paper on window effects in semiconductor detectors, Jorge Llacer et al. [419], stated: 'the use of (HP) germanium for a low-energy x-ray detector does not seem possible'. The reason for this conclusion was that the thick effective 'window dead layer' gives rise to extensive tailing particularly below the Ge L absorption edges (1.2 to 1.4 keV) (see Figure 4.6 in Chapter 4).

Llacer's conclusion was reaffirmed by both Fink [420] and by Barbi et al. [421] at the NBS Workshop on EDXS in Gaithersburg in 1979. The former stated that the efficiency of the HPGe detectors with gold surface barrier contacts purchased back in 1971 had declined over time but expected that 'ion-implanted contacts should be better'. The latter paper describes a 5-mm-deep HPGe and 3-mm-deep Si(Li), both manufactured by PGT, with resolutions ~160 eV on an SEM. He concluded 'the tailing below 3 keV precludes HPGe from any analytical work'. Figure 11.77 shows the Si Kα for both these detectors [421].

The first suggestion that things might not be so bad came in 1982 with reports using ion-implanted contacts. These had already been used on n-type HPSi and HPGe detectors giving effective dead layers as thin as 20 nm [422,423]. Mieczyslaw Slapa [424], originally from (and now back at) the Institute of Nuclear Research, Swierk, near Warsaw, was then working with Gerald Huth at the University of California, Los Angeles. He used a Ga-implanted contact on HPGe that gave reasonable line shapes down to about 1.25 keV. However, there was still marked tailing below this energy. In 1986, E.B. Steel [425] reported on an experimental PGT HPGe detector on a JEOL 200CX AEM, at the NBS Gaithersburg Workshop. The resolution was good (~133 eV) and line shapes acceptable down to about 2 keV (see Figure 11.78).

The report concluded that the HPGe detector was useful for the analysis of high-energy K lines (see Figures 1.2 and 1.36) for elements where there could be significant overlap of the L lines. This was despite a considerable falloff in intensity of K-lines because of the decreasing ionisation cross section for atomic numbers 50–96.

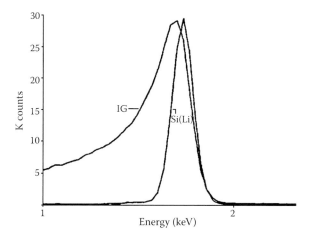

Figure 11.77 Comparison between a Si(Li) and a HPGe detector at 1.74 keV, 1981. (From Barbi, N.C. et al., *EDX Spectrometry*, Heinrich, K.F.J. et al. (Eds.), *NBS* 604, 35 (Gaithersburg Meeting April 1979), 1981. With permission.)

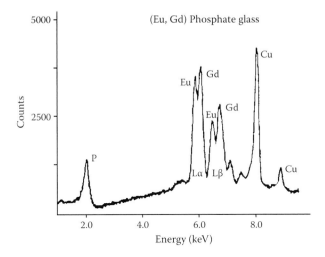

Figure 11.78 Results from HPGe detector on an SEM, 1986. (From Steel, E.B., MAS-21, 439, 1986. With permission.)

The present authors were working at Link Analytical in the UK in the mid 1980s to improve the performance at low energy of the LZ5 Si(Li) x-ray detectors. They applied similar techniques to small planar p-type HPGe crystals from Hoboken in a government-funded 'germanium for microanalysis' (GEM) project. The results [426] were presented at the IEEE meeting in Orlando, FL, in 1988 (see Figure 11.79, and also Figure 4.10 in Chapter 4).

The B Kα (183 eV) peak is clearly separated from the noise in the HPGe detector and the linearity is good. The crystals were stable and even the tailing on higher energy peaks was improved. Link was able to produce them commercially (LGEM) and to place a good premium on them. Unfortunately, however, from the production point of view, this biased the customer orders toward the high end of the market, in the form of windowless retractable gate valve detectors on the high-resolution, high-voltage, and high-vacuum TEMs, the most complex of all combinations. Nevertheless there were quickly ~20 such units shipped worldwide and by 1997 [427], more than 170 LGEM detectors had been installed.

In 1990, Canberra introduced an ultralow energy HPGe detector [428] that was used in a current mode at the KEK synchrotron, Japan. NORAN [429], in reply to the Link GEM,

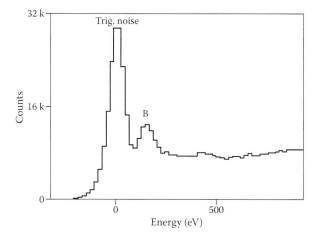

Figure 11.79 Boron measured with HPGe detector, 1988. (From Cox, C.E. et al., *IEEE* NS-35, 28, 1988. With permission.)

collaborated with Canberra and combined their ATW technology ('diamond window'), achieving results in TXRF in Vienna [430] that approached those of LGEM detectors (see Figure 11.80).

This spectrum was taken with a 0.4-µm 'diamond window' detector with a resolution of 125 eV. Soon Link was achieving resolutions of ~109 eV and occasionally detection down to Be Kα. The escape peaks that could have been troublesome above 11

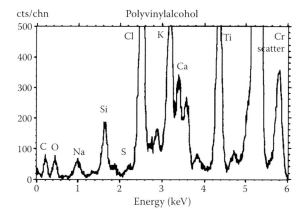

Figure 11.80 Low-energy spectrum using HPGe detector, 1993. (From Streli, C. et al., *NIM A*, 334, 425, 1993. With permission.)

keV could be handled by the software. The real problem was in sustaining the difficult production technology with the demanding high specifications. With detection levels of ~30 electrons, often in quite noisy environments, and failures due to instabilities in the contact in high radiation environments, the yield was not as high as for Si(Li) detectors. Furthermore, germanium being extra sensitive to IR, required an intermediate delicate self-supporting aluminium UTW at a temperature of ~100 K. Depositing the aluminium directly onto the crystal contact was not successful, presumably because of diffusion or pin holes.

There were thoughts at that time (late 1980s) that the whole EDX market could be switched from Si(Li) detectors to HPGe detectors. The germanium was available and it had the advantage that it did not need to be drifted. Apart from some extra cooling and an IR shield, it could be treated very much as a Si(Li) detector. But in the end, the technological difficulties won over and EDX detector companies have since fallen back on the old 'workhorse' of the Si(Li) detector and lately the SDD, although they may be persuaded to manufacture HPGe as 'specials'. The importance of germanium detectors, particularly segmented detectors, for the next generation of synchrotrons may revive the technology but it is likely to come from those companies that specialize in germanium such as Canberra.

11.24 1990s: Bespoke FETs

We have remarked that most EDX companies were using commercial FETs (2N4416) from TI, Motorola or (later) InterFET. Link was using a modified version of this. The gate capacitance of all these FETs was 3–4 pF, which was far from the perfect match for the Si(Li) transducer (~0.5 and 1.2 pF for a 10- and 30-mm^2 active area crystal, respectively). After 1981, Link used the Tetrode (E8552) that, as we remarked above, was a big improvement. In 1987 Link employed a semiconductor specialist (Tawfic Nashashibi from Thorn-EMI) with a view to designing a bespoke FET with improved performance for EDX. Thorn-EMI had produced the last batch of Tetrode FETs for Harwell (and used under license by Link) using the 'optical modification' invented by Wrangy Kandiah. The culmination of this work was the PentaFET [431] in 1990. This was based on the tetrode architecture and had a capacitance of ~1 pF but had further innovations such as an on-chip electronic restore and an on-chip FBC. Nashashibi later developed similar FETs for Moxtek, Gresham, and Semifab. Moving away from pulsed optical restore meant that the FET package no longer had to be light-tight, and dielectric noise and stray capacitance could be reduced (see Figure 1.61c).

11.25 Convenience versus Performance: A New Approach

In a sense, the Si(Li) detector itself was always a choice of 'convenience'. WDS has always had superior resolution and sensitivity. But its inconvenience in the form of bulk, low acceptance solid angle, and slow sequential elemental analysis, etc., was against it. Similarly, the liquid helium-based 'cryogenic' detectors that evolved in the 1990s [432] had the advantage of excellent resolution in addition to simultaneous elemental analysis under exactly the same analytical conditions but at present the technology is complex and expensive, very low temperatures are required, and in most cases the solid angle is very restricted. We have also seen that although HPGe detectors *can* be superior to Si(Li) detectors, for EDX the technology is difficult. Si(Li) detectors themselves have now also recently succumbed to the notion of inconvenience. The requirement of liquid nitrogen makes them relatively bulky and inconvenient particularly for portable and industrial applications where liquid nitrogen is not on hand. They are also still relatively expensive.

11.25.1 1990s and 2000s: Enter Lithography and Planar Technology

In Chapter 7, we discussed lithographic techniques as applied to x-ray detectors. This started with the work in the early

1970s at the TUM [433] on p–i–n diodes for medical applications and culminated in the work on PXDs, CCDs (Chapters 7 and 8), and most significantly, SDDs (Chapter 9). Much of the historical progress on these devices was outlined there. Here, we will concentrate on the significant impact made by the p–i–n diode and SDDs on EDXRF and EDXMA markets and the companies involved.

Amptek launched the XR-100 in 1994 offering a resolution of 250 eV using a Peltier cooler. On July 4, 1997, an XR-100 was landed on Mars as part of the Pathfinder Mission and successfully relayed EDXRF spectra from Martian rocks. Following Amptek, many companies introduced products based on p–i–n diodes. These included Aracor, Moxtek, Niton, Jordan Valley, Metorex, and Oxford Instruments (an added channel to their 'MDX' WDS product). The fragile surface barrier technology for the window contact was never developed for lithographic techniques and the handling of wafers of thickness greater than the standard 270 μm was difficult. Thus, low- and high-energy x-ray efficiency of such devices could not compete with the 3-mm-deep Si(Li) crystal with a surface barrier window contact. However, the enormous advantage of the stable low leakage currents at elevated temperatures, because of their small volumes, and the excellent thermal SiO_2 passivation techniques available in wafer processing, was recognized. Not only does this obviate the need for the liquid nitrogen cryogen, but there are advantages (at least in theory) resulting from reduced ω, enhancing resolution and electron thermalisation. The capacitance per unit area for wafer technology will always be a disadvantage, however. The p–i–n diodes are still regarded as adequate for many less demanding EDXRF applications in industry today.

11.25.2 Peltier Cooled Detectors (Thermo-Electric Coolers) and Other Cooling Engines

The great advantage of the thin lithographic detectors lies in the fact that they can operate with only moderate cooling (−30°C to −90°C). Thus, a three- to four-stage Peltier Cooler [also known as Thermo-Electric Cooler (TEC)] can be used to replace the liquid nitrogen cryostat. TECs are purely electrical, have no moving parts, and are relatively cheap. The main problem is supplying them with a sufficiently efficient heat sink. A copper block with fins for air-cooling or internal piping with water or oil cooling is typically used. They were pioneered by Abraham Ioffe and other Russians in the 1950s [434] and subsequently used to cool p–i–n diodes, at first as IR detectors, mainly for the military.

After the success of TECs for p–i–n diodes for EDXRF, companies turned to their use for Si(Li) detectors for EDXMA. The freedom from liquid nitrogen would allow their use for special applications such as in clean rooms. The first to achieve this was the Berkeley group in 1979 [435]. This showed the way, and Kevex introduced a TEC Si(Li) for EDXRF ('No-LN') 10 years later, with a resolution <200 eV, and Tracor ('Spectrace 6000') in the same year. Rolf Woldseth, at Kevex, did pioneering work using a five-stage water-cooled TEC to produce a commercial EDXMA product (Kevex 'Supercool') in 1991 [436]. Then, as an independent consultant, he was instrumental in the production of Peltier cooled detectors for Tracor and Scintag [437] ('ARX001') soon after. All these products used a guard ring Si(Li) crystal, as had the Berkeley group. Scintag, for x-ray diffractometry, used one of oval geometry. Baltic Scientific also later offered a Peltier cooled Si(Li) for EDXRF [438]. Although the TEC itself was free from the effects of vibration, there was often an issue with the water cooling and image distortion on an EM. An alternative was to couple the TEC to another cooling device such as a heat pipe [439].

Alternative cooling engines had been developed vigorously for military applications and these could be the subject of a whole book. In general, they have greater cooling power than TECs and can easily cool to 100 K. Hence, they do not require special guard ring Si(Li) crystals. However, this comes at the expense of greater complexity and cost. Here we will briefly mention some which have been used for S(Li) and HPGe detectors with lesser or greater amount of success. Joule-Thomson (JT) devices are attractive because they can cool quickly and liquefy high-pressure nitrogen by expansion. A copper braid or multicomponent strip (or in the case of IR detectors, a 'fuzz washer') was used to decouple any vibration. Such devices as the MAC series from Hymatic [440] have been used by Laben [441] and OI. In the case of the latter the requirement of 300 atm nitrogen (from a gas cylinder or compressor) was deemed to make them unsuitable for EDXRF. APD Cryogenics [442] introduced a product (Cryotiger) based on JT cooler technology [443] that was originally developed in Russia [444]. This used various low and high boiling point gas mixtures, together with oil in a closed cycle at a lower pressure and therefore could use a cheaper compressor. This technology was used by a number of companies including OI (Ultracool), EDAX (CyoX) [445], NORAN (CryoCooled) [446], and PGT (JT-Cool). Some of the gases (such as propane) were inflammable and problematic for air transport.

Stirling and Solvay cycle engines using He gas as a working medium have also been used effectively but more often for HPGe rather than Si(Li) detectors. The introduction of HPGe allowed systems to be cooled on demand in remote regions. The Solvay cycle was preferred as the compressor was physically separate from the cold-finger. As early as 1973, GE, Philadelphia [447] had evaluated such engines for a small HPGe detector with acceptable results and ORTEC [448] did the same for both HPGe and Si(Li) detectors in 1986. In the case of the latter the cooling engine added ~10 eV to the energy resolution. PGT claimed to have the first Stirling engine–cooled HPGe detector on the market (PGT 77K) in 1987. The present authors have demonstrated excellent results with a Si(Li) detector cooled by a Stirling engine designed at Oxford University [449]. Microphony was not a problem providing the cryohead was held rigidly relative to the detector body and isolated mechanically by a short flexible copper braid from the cold finger.

Orifice pulse tube coolers, with no moving parts other than the oscillating helium gas medium, have been developed and improved since their invention in Russia [450] in 1984. The first commercially available cooler was introduced by the Japanese company Iwatani Plantech [451] and was further developed by Sumitomo [452]. OI used this technology [453] for their EDXMA product DryCool in 2003. The German Company VeriCold Technologies used their low vibration pulse tube cooled cryostat [454] with a microcalorimeter sensor developed by LLNL USA to develop a gamma ray detector (UltraSpec) that won the R&D Award in 2006. OI acquired VeriCold Technologies GmbH in 2007.

As far as semiconductor x-ray detectors go, the TEC-cooled SDDs have largely negated the demand for mechanically cooled Si(Li) detectors as indicated by their product names (NORAN UltraDry and PGT Sahara), but for Home Security and spacecraft applications there has been a renewed interest in TECs for WBGS detectors and the Stirling engine for HPGe detectors [455,456].

11.26 SDD: Influence of Nuclear Physics Again

The nuclear physics community in the meantime were seeking better position-sensitive detectors (PSDs) for particle tracking to supersede the likes of emulsions, scintillation counter telescopes, bubble chambers, spark chambers, and multiwire gas (Charpak) chambers. It was also useful to measure the energy of the particle as well as its position of interaction. This was first achieved with semiconductors by incorporating a charge dividing layer [457] into the back contact of a surface barrier detector with contacts either end of this layer. Using amplifiers connected to the ends of the charge-dividing layer, the ratio of the signals measured gave the position of interaction of the particle. Its full energy was obtained by taking a signal from the other surface of the detector in the normal manner. These detectors were manufactured at Bell Laboratories on strips of 'web silicon' (thin silicon strips supported by dendrites).

The discovery of long-lived charmed particles [458] in 1976 provided a major boost for the future of tracking detectors in high-energy physics. In the United States, a study of the application of planar lithography technology for nuclear physics detectors had been published by BNL [459] in 1977. Also, discussions by the Isabelle hadron collider workers at BNL had taken place with ORTEC in December 1979 (T. Ludlam, BNL private communication, given by Heijne [357]). Photolithography could be used to create the necessary complex geometries and structures. The low density of interface states for SiO_2 resulted in extremely low leakages even at room temperature and the high accuracy achieved by step-and-repeat lithography led to mass production and low cost of manufacture. Not only was material now available but also so was the high-technology planar processing foundry equipment necessary for production. Furthermore, the production of devices did not require any huge infrastructure. The semiconductor equivalent of the multiwire gas drift detector led to silicon strip detectors [460], effectively an array of p–i–n diodes in the form of parallel strips. The first practical strip detectors using photolithography came from Kraner and others at BNL in 1983 [461] who were influenced by the work of Kemmer in Germany (see below). These now form the basis of most of the large 'vertex detectors' used in HEP.

The history of the SDD has been summarised recently for its 25th anniversary at the 11th European Conference on Semiconductor Detectors, Wildbad, Bavaria [462,463]. Emilio Gatti had been working in Milan on low noise front–end amplifiers. He was interested in the possibility of using fully depleted silicon for use as high value resistors. One way of achieving full depletion efficiently is to deplete a wafer from both faces. For example, an n-type silicon wafer can be fully depleted from p⁺ contacts on both faces with one quarter the bias needed to bias from one face only (see Chapter 9). Meanwhile, Pavel Rehak was working at BNL with Veljko Radeka on PSDs [464]. The motivation was to limit the number of strips on a conventional detector and the problems associated with having an amplifier connected to each strip. They were particularly intrigued with the idea of a semiconductor version of a gas drift chamber. The aim was a simplified semiconductor PSD combining position and energy resolution. Using a transverse drift field, it should be possible to measure the position of an event by looking at the time of arrival of electrons at an anode (e.g., after a trigger signal from a scintillation counter) as well as the magnitude of the charge package produced in the ionisation event. Thus, one PSD detector with its measuring amplifier could replace a large array of strips and their need for individual associated electronic channels. The saving in space, power, and cost would be significant.

In one of Gatti's annual visits (1982) as a guest scientist to BNL, he discussed over a few days the possibilities with Rehak. They realised that fully depleting an n-type silicon wafer from both faces resulted in a potential minimum at the center that would act as a natural collection channel for electrons produced by ionising radiation. It was the idea of Rehak [463] to superpose an 'external' transverse field onto the depletion field. This would have the effect of 'tilting' the collection channel so that the electrons would drift along it much as the ions do in the gas chamber version. It was proposed to do this using 'drift field electrodes' or parallel strips on both faces. These p+ electrodes could also act as the depletion electrodes. The collecting 'anode' would be at one edge at the highest positive bias. The concept was reported at the Workshop on Collider Detectors held at LBL in the early spring of 1983 [465] in a short paper without any detailed analysis. They called the proposed detector a 'semiconductor drift chamber', in analogy to a conventional gas drift chamber. A test device 300 μm thick and ~9 mm diameter was built in 2 days by R.H. Beutenmuller (BNL) using 10 kΩ cm resistivity Komatsu n-Si; the material was supplied by Kramer (BNL). It was tested in the Physics Laboratories at Berkeley, California. Full depletion was indicated as a sudden drop in capacitance (see Figure 9.2). The position resolution was estimated to be better than 2–5 μm and was better than could be achieved by emulsions. The results were presented at the Second Pisa Conference on Advanced Detectors [466] in Grosetto, Italy, in the summer of 1983.

11.27 SDDs as X-Ray Detectors

This proposal also postulated how the design could be applicable to 'ultra low capacity large area x-ray spectrometers and photodiodes and also fully depleted thick CCDs'; this was foresight indeed. Gerhard Lutz (MPI, Munich), for example, showed an active interest. The concept was brought to a wider public in the following year [467]. In parallel with these presentations, the first patent assigned to the U.S. Department of Energy was filed on February 24, 1984 [468]. The first true SDD, as distinct from a test device, was made by Jack Walton at LBL, California, and the results reported at the Third European Symposium on Semiconductor Detectors at Munich in 1983 [469]. The first SDD built in Germany by Joseph Kemmer was tested at LBL and reported in 1985 [470]. Figure 11.81 shows a spectrum of the 88 keV γ of Cd-109 taken at room temperature from this device.

If the energy deposited was only of interest, that is, position information abandoned and not used as a PSD, better resolution than the conventional p–i–n detector should be possible. Low-noise photodiodes [471,472] were tackled next with an interest from the Aerospace company Messerschmitt-Bolkow-Blohm. The low noise also has relevance if they are to be usefully coupled to a scintillator, for example. Since the drift field only needs symmetry normal to the wafer plane, the entrance window could be a uniform implant on one side with a concentric ring electrode structure on the other (see Figure 1.65). Furthermore, there was the possibility of the JFET being located inside the small central anode ring. It was also possible to integrate the JFET [473] onto the SDD wafer

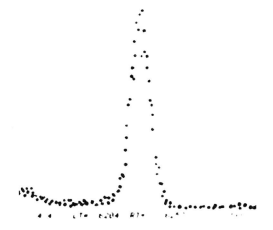

Figure 11.81 An 88-keV γ-ray as measured on the first SDD showing its response to Cd-109 at room temperature, 1985. (From Rehak, P. et al., *NIM*, 235, 224, 1985. With permission.)

using ion implantation doping to reduce the input capacitance still further. The use for EDXRS [474] was just an extension of these ideas. As we have seen (Chapter 9), these detectors have shown remarkable and rather unexpected superior performance in the 100 eV to 10 keV x-ray energy range. In 2004, the NASA Martian 'Rover' explorers *Spirit* and *Opportunity* carried α-excited EDXRF detectors using 10-mm^2 SDDs. They gave superior performance to the previous p–i–n diode detectors. The first use as a HEP tracking detector, as first intended, was at CERN 1992 [475].

We have seen (Section 11.12) that the idea for a 'transverse field' in a silicon slice had already been tried out for different reasons by Kuhn [195] and Kraner et al. [196] by lithium drifting an open-ended coaxial geometry crystal 20 years earlier. Progress in planar processing including double-sided wafer processing and innovators such as Josef Kemmer now provided the possibility of controlling this field in high-purity silicon from Topsil, Wacker, and Komatsu with linear or concentric circular electrodes using photolithography. The latter geometry also gave a low capacitance central anode for charge collection, which is essential for x-ray detection. Gatti and Rehak had shown the way.

Emilio Gatti was born in Italy in 1922 and has a long history of innovation in electronics for radiation detectors. He was a professor of Electronics and Physics at the Milan Polytechnic from 1957 to 1992 and is still professor there (without tenure). He was guest senior scientist at Brookhaven from 1973 to 1995. He worked on low-noise charge-sensitive amplifiers in the 1950s, a sliding scale digital converter, the 'Gatti Wobbler' (1963), and the theory of optimum signal processing (1955–1995). Pavel Rehak was born in Czechoslovakia and earned a doctorate at the University of Prague in 1969 and a second Ph.D. at the University of Pisa, Italy, in 1972. In 1974, he went to Yale University in the United States and was assigned to CERN, Geneva. He went on to BNL in 1976 and joined their Instrumentation Division working under Radeka in 1994.

Rehak himself [476] reviewed the work of Joseph Kemmer in a memoriam paper in 2008. As we saw earlier, Kemmer was working on semiconductor detectors at Heidelberg in the mid 1960s and also attended the MPI in Heidelberg, where he had made his own detectors, including a Ge(Li) with an FET amplifier in 1966 [477] and an ion-implanted dE/E counter. In the late 1970s, like Keil and Lindner, at TUM, he started manufacturing oxide passivated p–i–n diodes by lithographic processes specifically as nuclear radiation detectors [478]. Although his Ph.D. (TUM 1970), like many detector pioneers, was in nuclear physics, he was well experienced with semiconductor detector manufacture and experimentation. Furthermore, his detectors had very abrupt ion implanted contacts and in 1980 he patented [479] the process. Up to this time, the best silicon surface barrier α-particle detectors gave ~15 keV resolution for 5 MeV α particles at room temperature. Kemmer's new detectors gave ~10 keV resolution and noise of 1.5 keV. Leakage currents went from ~120 nA/cm^2 to less than ~5 nA/cm^2 at 100 V bias. He worked closely with Paul Berger (then working for Lasco), who started to make detectors based on the planar technology on an industrial scale in their new Lingolsheim factory. They later collaborated on improving the oxidation process. The work at Enertec-Schlumberger backed and enhanced the pioneering work of Kemmer at TUM, and by 1982 they were manufacturing ion-implanted strip PSDs [480]. Later, while running his own company (Ketek, see below), he was also supervising the MPI semiconductor laboratory in Munich and was instrumental in the design and construction of the pn-CCD x-ray detector flown on the XMM satellite [481] in 1999. He died in 2007.

11.28 SDD EDXRS Companies

All the major EDXRS companies now offer SDDs, and for many applications they have replaced Si(Li) detectors. SSDs have the advantages, as discussed in Chapter 9, of not requiring liquid nitrogen and of having high rate capability. The falloff in their high-energy efficiency (>10 keV) is the only drawback at present. SDDs have been available commercially for EDXMA since about 1996 [482]. We will briefly look at the major companies involved in the manufacture of SDDs.

As we have noted, Joseph Kemmer was quick to manufacture SDDs in Germany from 1985 and founded his own company, Ketek, Haimhausen, Munich in 1989. Also involved in this company were Tobias Eggert, Gerhard Lutz, Susanne Krisch, Peter Lechner, Nicola Findeis, Gerard Forcinal, and Peter Holl, and they worked closely with MPI, Munich and Rontec GmbH, Berlin. For many years, they were the only source of commercial SDDs and had delivered a thousand units by 2002. In 2006, they were integrating SDDs into OI's INCA product retaining the PentaFET technology. Ketek used an ion implant to form a p–n junction on n-type silicon with an aluminium coating (~100 nm) [483,484] to reduce light sensitivity. Later, they introduced an integrated JFET [485]. Their recent product is a complete Peltier cooled package (see Figure 11.82).

In 2010, Ketek claimed to have achieved a resolution of 121 eV on Fe-55 (46 eV on C Kα), very close to the Fano limit.

PNSens•r

In 2002, a spinoff company from MPI Munich and Ketek was set up as PNSensors. The company continued to use the

Figure 11.82 **(See colour insert.)** SDD assembly from Ketek.

laboratories of MPI. Those involved were Gerhard Lutz, Heike Soltau (the wife of Lothar Struder (MPI)), Rouven Eckhard, Andreas Pahlke (ex-Wacker Siltronic), Peter Lechner (from Ketek), Robert Hartmann, and Peter Holl (from Ketek). They currently use an integrated FET offset in a 'droplet' (SD3) geometry (see Figure 9.17). PNSensors also manufacture pn-CCDs and DEPFET detectors. Ketek and PNSensors supply devices to the majority of EDXRS companies worldwide.

As discussed in Section 11.18, Jan Iwanczyk and Bradley Patt founded Photon Imaging, Inc. (Northridge, CA). Radiant Detector Technologies was the commercial outlet for Photon Imaging's industrial and scientific technology. Caroline Tull (ex-Berkeley), J.D. Segal (ex-XSirius), and Jan Iwanczyk (ex-XSirius) worked on SDDs using Stanford University facilities [486]. They were joined later by Shaul Barkan (ex-Kevex). By 2005, they were a supplier of SDDs [487] and were then bought by the Japanese giant, Seiko Instruments Inc. (SII) to form SII Nanotechnology USA [488]. Their product goes by the name Vortex. Vortex-EM is a 50-mm^2 SDD for EDXMA. In 2009, they introduced an XRF analyzer using the Vortex SDD.

XGLab (Milan, Italy) was founded in 2009 as a spinoff from Milan Polytechnic, fittingly the home of Prof. Gatti, one of the co-inventors of the SDD. XGLab has as an industrial partner, GNR. Srl, Novara (Analytical Instruments). The founders consisted of Frizzi Tommaso (CEO), Antonio Longoni, Carlo Fiorini, Roberto Alberti, and Luca Bombelli, who all worked on radiation detectors at Milan Polytechnic and Michele Gironda from GNR. They use integrated FETs with 'multidroplet' architecture up to 100 mm^2 active area. One product is

Figure 11.83 (**See colour insert.**) SDD from XGLab.

XGL-SPCM-8100. Figure 11.83 shows one of their SDDs. The company is not limited to SDDs.

In general, these companies have remained SDD package (including TEC) suppliers rather than complete detector or system manufacturers. Examples of the SDD products available commercially are as follows: Thermo Scientific (UltraDry System 6), PGT (Sahara II), EDAX (XLT Apollo), e2v Scientific (SiriusSD), Rontec (XFlash), OI (X-Max 80), IXRF Systems (Vortex -EM, SiriusSD), and Amptek (XR-100SDD).

11.29 Future

The demand for x-ray spectrometers will continue to increase for applications in pure research, industrial, medical, space, security, etc. The question is, what will be the form of these spectrometers? It has taken us half a century to get where we are, and there is no reason to suppose that we have yet arrived. There is no indication that the SDD is the perfect solution. The falloff in efficiency above ~10 keV and spectral contaminations are limitations. Also, for sub-120 eV resolution, HPGe is required. Possibly, it will be just a question of 'horses for courses' and diversification. There have been a few surprises in the semiconductor x-ray detector industry in the past. These can mainly be summed up in the move from ultimate technical performance toward lower cost, greater convenience, and reliability. One can think of the survival of the Si(Li) detectors (rather than thick HPSi or HPGe) detectors, the persistence of the contoured geometry for thick detectors, the option for ATW x-ray windows (rather than true windowless), and the acceptance by the industry of thin lithographically manufactured devices such as p–i–n diodes and SDDs. The ultimate surprise, however, has been how far the quality of all these devices has been pushed. The performance of Si(Li) detectors, thin windows, SDDs, CCDs, etc., has far exceeded what was first expected of them. Much work is still being done to achieve a better theoretical understanding of the ionisation processes and ICC. It may be hoped that realistic calculations of ω and F might lead to further material and contact engineering improvements. Despite the success of SDDs, other new detector concepts might emerge to take advantage of progress made in electronics. WBGS have so far failed to make a major impact on XRS, but are there any materials still waiting in the wings? An innovative vibration free, compact, cheap cooling engine capable of rapidly cooling to ~100 K would dramatically change the situation. Nanotechnology may hold the key to this. The superiority of HPGe might then come to the fore, providing that a hold on the technology could be affected. Should such a cooling device be capable of very low temperatures, the cryogenic detectors (bolometers, transition edge devices, etc.) could completely take over from semiconductor detectors. However, commercial pragmatism will always prevail. It is not always the best technology that succeeds but the most reliable, easily delivered, economic solution that is fit for purpose.

References

1. John O., *The Story of Semiconductors*, Oxford University Press, 2004. ISBN 0 19 853083 8.
2. Braun F., *Ann. Phys. Chem.* 2, 153, 556, 1875.
3. Bose J.C., Detector for Electrical Disturbances, US Patent 755840, 1901.
4. Garratt G.R.M., *Early History of Radio from Faraday to Marconi, IEEE History of Tech. Series*, 1994.
5. Joffé A., *Z. Phys.* 33, 393, 1932.
6. Hecht K., *Z. Phys.* 77, 235, 1932.

7. Mott N.F., *Proc. Camb. Philos. Soc.* 34, 568, 1938.
8. Schottky W., *Z., Phys.* 113, 367, 1939.
9. Davydov B., *Acad. Sci. USSR* 20, 283, 1938.
10. Ramo S., *Proc. IRE*, 584, 1939.
11. Shockley W., *J. Appl. Phys.* 9, 635, 1938.
12. Jen C.K., *Proc. IRE* 29, 345, 1941.
13. Cavalleri G. et al., *NIM* 21,177, 1963.
14. Cavalleri G. et al., *NIM* 92, 137, 1971.
15. Fano U., *Phys. Rev.* 63, 222, 1943.
16. Fano U., *Phys. Rev.* 72, 26, 1947.
17. Seitz F., *Modern Theory of Solids*, McGraw-Hill, 1940.
18. Seitz F. and Einspruch N.G., *Electronic Genie—The Tangled History of Silicon*, Univ. of Illinois Press, Urbana, GA, 1998.
19. Riordan M., How Europe Missed the Transistor, www.spectrum. ieee.org/nov05/2155/3, 2005.
20. Van Heerden P.J., Ph.D. thesis, Univ. of Utrecht N.V. Noord Hollandishe, 1944.
21. Van Heerden P.J. et al., *Physica* 16, 505, 1950.
22. Freriches P.R., *Phys. Rev.* 72, 594, 1947.
23. Ahearn A.J., *Phys. Rev.* 73, 524, 1948.
24. McKay K.G., *Phys. Rev.* 74, 1606, 1948.
25. AEA & Office of Naval Research, First Conference on Scintillation and Crystal Counters, Rochester, July 22–23 (Proceedings were not published), 1948.
26. Morton G.A., *IEEE* NS-15, 3, 7, 1968.
27. Seitz F., ORNL, Conf. on Scintillation Counting, 1949.
28. Hofstadter R., Crystal counters, *Nucleonics* 4, 2, 1949.
29. Hofstadter R., Crystal counters, *IRE* NS-38, 726, 1950.
30. Price, W.J., *Nuclear Radiation Detectors* (1st ed.), McGraw-Hill, 1958.
31. Taylor J.M., *Semiconductor Particle Detectors*, Butterworth, 1963.
32. Davies W.D., *J. Appl. Phys.* 29, 231, 1958.
33. Van Putten J.D. et al., *IRE* NS-8, 1, 124, 1961.
34. Mayer J.W., *IRE* NS-9, 3, 124, 1963.
35. Riordan M. and Hoddeson L., *Crystal Fire*, W.W. Norton & Co., 1997.
36. Shurkin J.N., *Broken Genius: The Rise and Fall of William Shockley, Creator of the Electronic Age*, Palgrave Macmillan, 2006.
37. Lilienfeld J.E., US Patent 1745175, 1926.
38. Morris P.R., *A History of the World Semiconductor Industry*, UK Inst. of Eng. & Tech., 1989.

39. Lilienfeld J.E., US Patent 1,900,018, 1926.
40. Heil O.E., British Patent 439457, 1935.
41. Shockley W., *Electrons and Holes in Semiconductors*, Van Nostrand, 1950.
42. Shockley W. and Pearson G.L., *Phys. Rev.* 74, 232, 1948.
43. Shockley W., *IRE* 40, 1365, 1952.
44. Dacey G.C. and Ross I.M., *IRE* 41, 8, 970, 1953.
45. Hyde F.J., *Semiconductors*, Mcdonald & Co., 1965.
46. Benzer S., US Patent 2504627, 1946.
47. Benzer S., *Phys. Rev.* 70, 105, 1946.
48. Bardeen J. and Brattain W.H., *Phys. Rev.* 74, 230, 1948.
49. McKay K.G., *Phys. Rev.* 76, 1537, 1949.
50. Orman C. et al., *Phys. Rev.* 78, 5, 646, 1950.
51. McKay K.G., *Phys. Rev.* 84, 833, 1951.
52. McKay K.G., *Phys. Rev.* 84, 829, 1951.
53. Brattain W.H. and Pearson G.L., *Phys. Rev.* 80, 846, 1950.
54. McKay K.G. and McAfee K.B., *Phys. Rev.* 91, 1079, 1953.
55. Ryvkin S.M. et al., *Zhur. Tekh. Fiz.* 27, 7, 1599, 1957.
56. Vavilov V.S. et al., *Kernenergie (E. Germany)* 1, 279, 1958.
57. Walter F.J. et al., ORNL Rep. CF58-11-99, 1958.
58. Dearnaley G. and Whitehead A.B., AERE-3278, 1960.
59. Baily N.A. et al., *Rev. Sci. Instrum.* 32, 865, 1961.
60. Sharpe J., 1964, *Nuclear Radiation Detectors*, Methuen, 1964.
61. Dearnaley G. and Northop D.C., *Semiconductor Counters for Nuclear Radiation*, Spon, 1966.
62. Teal G.K. and Little J., *Phys. Rev.* 78, 5, 647, 1950.
63. Pfann W.G., *Trans. Am. Inst. Mining Metall. Eng.* 194, 747, 1952.
64. Hall R.N., GE Report 55-RL-1316, 1955.
65. Theuerer H.C., US Patent 3060123, 1952.
66. Cressell I.G. and Powell J.A., *Progress in Semiconductors 2*, 138, Heywood, 1957.
67. Powell A.R., *J. Appl. Chem.* 1, 541, 1951.
68. Billig E., *Proc. R. Soc. A* 229, 1178, 346, 1955.
69. Freck D.V. and Wakefield J., *Nature* 193, 669, 1962.
70. Dearnaley G., *IEEE* NS-17, 3, 282, 1970.
71. Pell E., *Phys. Rev.* 119, 1014, 1960.
72. Adda L.P. et al., *IEEE* NS-15, 347, 1968.
73. de Wit R.C. and McKenzie J.M., *IEEE* NS-15, 352, 1968.
74. Webb P.P. et al., *NAS Publ.* 1592, 138, 1969.
75. Hansen W.L., *NAS Publ.* 1592, 189, 1969.
76. Dabbs J.W.T. and Walter F.J. (Eds.), *Semiconductor Nuclear Particle Detectors*. (Asheville Conf. 1960), NAS-871, Washington, D.C., 1961.

77. Fuller C.S. and Allison H.W., *J. Appl. Phys.* 35, 1227, 1964.
78. Arkad'eva E.N. et al., *Sov. Phys.-Tech. Phys.* 11, 847, 1966.
79. Arkad'eva E.N. et al., *Sov. Phys.-Semicond.* 1, 1967.
80. Mayer J.W., *NAS Publ.* 1592, 377, 1968.
81. Willig W.R., *NIM* 96, 615, 1971.
82. Swierkowski S.P. et al., *IEEE* NS-21, 1, 302, 1974.
83. Moseley H.G.J., *Phys. Mag.* 26, 1024, 1913.
84. Mayer J.W., *IRE* NS-7, 2–3, 178, 1960.
85. Blankenship J.L. and Borkowski C.J., *IRE* NS-7, 2, 3, 190, 1960.
86. 7th Ann. Meeting on Semiconducting Rad. Dets., Gatlinburg (1960), IRE NS-8; 1, 1961.
87. Bertolini G. and Coche A., *Semiconductor Detectors*, Wiley Interscience, 1968.
88. Friedland S.S. et al., *IRE* NS-9, 3, 391, 1962.
89. West H.I. et al., *Rev. Sci. Instrum.* 33, 380, 1962.
90. Buck T.M. et al., *IEEE* NS-11, 3, 294, 1964.
91. Fertin J., *Instrum. Nucl. (Paris)* 29, 1967.
92. Hayashi I. et al., *IEEE* NS-13, 214, 1966.
93. Wilkes J.G., *Proc. IEE* 106B, 866, 1959.
94. Dearnaley G. et al., *IEEE* NS-17, 3, 282, 1970.
95. Donovan P.F. et al., *Bull. Am. Phys. Soc.* 5, 5, 355, 1960.
96. Koch L. et al., *Comp. Rend.* 252, 74, 1960.
97. Blankenship J.L. and Borkowski C.J., *IRE* NS-7, 2, 3, 190, 1960.
98. Mayer J.W., *J. Appl. Phys.* 30, 1937, 1959.
99. Shockley W., *Proc. Intl. Conf. on Semiconductor Phys.*, Prague Aug. 29–Sept. 2, 1960.
100. Mann H.M. et al., *IRE* NS-9, 4, 43, 1962.
101. Gruhn C.R. et al., *Phys. Lett.* 24B, 266, 1967.
102. Harvey N.P. and Bernou J.P., *IEEE* NS-17, 3, 306, 1970.
103. Baily N.A. et al., *Rev. Sci. Instrum.* 32, 865, 1961.
104. Mayer J.W. et al., *Nucl. Electron.* 1, 567, 1962.
105. Mann H.M. and Haslett J.W., *US AEC ANL* 6455, 23, 1961.
106. Mann H.M. et al., *IRE* NS-9, 4, 43, 1962.
107. Nybakken T.W. and Vali A., *NIM* 26, 182, 1964.
108. Webb P.P. and Williams R.L., *NIM* 22, 361, 1963.
109. Tavendale A.J. and Ewan G.T., *NIM* 25, 185, 1963.
110. Fox R.J. et al., *NIM* 35, 331, 1965.
111. Hofker W.K., *Philips Tech. Rev.* 27, 323, 1966.
112. Elad E. and Nakamura M., *NIM* 41, 161, 1966.
113. Elad E. and Nakamura M., *NIM* 42, 315, 1966.
114. Tamm I.E., *Phys. Z. Sov. Union* 1, 733, 1932.
115. Shockley W., *Phys. Rev.* 56, 317, 1939.

116. Bardeen J., *Phys. Rev.* 71, 717, 1947.
117. Brown W.L., *Phys. Rev.* 91, 518, 1953.
118. Schrieffer J.R., *Phys. Rev.* 97, 641, 1955.
119. Miller G.L. et al., *IRE* NS-7, No.2–3, 185, 1960.
120. Miller G.L. and Gibson W.M., *BNL* 5391, 35, 1961.
121. Frosch C.J. and Derick L., *J. Electrochem. Soc.* 104, 547, 1957.
122. Kahng D. and Atalla M.M., IRE SSDR Conf. Carnegie Inst., Pittsburgh, 1960.
123. Miller G.L. et al., *IRE* NS-7, No.2–3, 185, 1960.
124. Gibson W.M., in Asheville Conference (see ref. [76]), 1960.
125. Hansen W.L. et al., *J. Appl. Phys.* 34, 1570, 1963.
126. Madden T.C. and Gibson W.M., *Rev. Sci. Instrum.* 34, 50, 1963.
127. Hansen W.L. and Goulding F.S., *NIM* 29, 345, 1964.
128. Buck T.M. and McKim F.S., *J. Elecrochem. Soc.* 105, 709, 1958.
129. Statz H. et al., *Phys Rev.* 101, 1272, 1956.
130. McKenzie J.M., *NIM A* 162, 49, 1979.
131. Dearnaley G. and Whitehead A.B., *Nucleonics (USA)* 19, 1, 72, 1961.
132. Dearnaley G. and Whitehead A.B., *NIM* 12, 205, 1961.
133. Yeh L.-S.R., US Patent, 5268578, 1993, to North American Philips, 1993.
134. Muggleton A.H.F., *NIM* 37, 345, 1965.
135. Buck T.M. and McKim F.S., *J. Electrochem. Soc.* 105, 1, 1958.
136. Buck T.M. et al., *IEEE* NS-11, 3, 294, 1964.
137. Brown W.L. (Ed.), *Proc. Conf. Semiconductor Nuclear Particle Detectors*, Gatlinburg (1969), NAS 1593, Washington, 1969.
138. Wilburn C.D. and Mallamo R., *IEEE* NS-14, 569, 1967.
139. Grainger R.J., Asheville Conference (see ref. [76]), 116, 1960.
140. Sandor J.E., US Patent 3158505 to Fairchild, 1962.
141. Jenny D.A., US Patent 3397449 to Hughes Aircraft, 1965.
142. Goulding F.S., *NIM* 43, 21, 1966.
143. Fox R.J. et al., *IRE* NS-9, 3, 213, 1962.
144. Lothrop R.P. et al., UCRL-16190, 1965.
145. Lothrop R.P., UCRL-19413, 1969.
146. Walton J.T., *IEEE* NS-31, 1, 331, 1984.
147. Sher A.H. et al., *NIM* 53, 341, 1967.
148. de Witt R.C. and McKenzie J.M., *IEEE* NS-15, 352, 1968.
149. Hansen W.L. and Jarret B., *NIM* 31, 301, 1964.
150. Palms J.M. and Greenwood A.H., *Rev. Sci. Instrum.* 36, 1209, 1965.
151. Stab L. et al., *NIM* 37, 113, 1965.
152. Adams F., *NIM* 48, 338, 1967.
153. Dinger R.J., *IEEE* NS-22, 135, 1975.

154. Dinger R.J., *J. Electrochem. Soc.* 123, 9, 1398, 1976.
155. Herdt W. et al., *Thin Solid Films* 82, 293, 1981.
156. Hansen W.L. et al., *IEEE* NS-27, 247, 1980.
157. Buck T.M. and McKim F.S., *J. Electrochem. Soc.* 105, 709, 1958.
158. Benton J.L. et al., *Appl. Phys. Lett.* 36, 8, 670, 1980.
159. Chabal Y.J. et al., *Phys. Rev. Lett.* 50, 1850, 1983.
160. Ubara H. et al., *Solid State Commun.* 50, 673, 1984.
161. Kern W. (Ed.), *Hand Book of Semiconductor Wafer Cleaning Tech*, Noyes, 1993.
162. Shockley W., US Patent 2672528, 1954.
163. Amsel G. et al., *NIM* 8, 92, 1960.
164. Mann H.M. et al., *IRE* NS-9, 4, 43, 1962.
165. Huth G.C. et al., *Rev. Sci. Instrum.* 34, 1283, 1963.
166. Huth G.C. et al., US Patent 3491272, 1963.
167. Huth G.C. et al., NYO Tech. Report 3246-8, 1966.
168. Huth G.C. et al., *IEEE* NS-12, 275, 1965.
169. Locker R.J. and Huth G.C., *Appl. Phys. Lett.* 9, 227, 1966.
170. Coleman J.A. and Rogers J.W., *IEEE* NS-11, 3, 213, 1964.
171. Llacer J., *IEEE* NS-11, 221, 1964.
172. Llacer J., BNL-9535, 1965.
173. Llacer J., *IEEE* NS-13, 1, 93, 1966.
174. Hayashi I. et al., *IEEE* NS-13, 214, 1966.
175. Hansen W.L. and Goulding F.S., UCRL 7436, 1960.
176. Goulding F.S. and Hansen W., *NIM* 12, 249, 1961.
177. Hansen W.L. et al., *J. Appl. Phys.* 34, 1570, 1963.
178. Lothrop R.P. and Smith H.E., UCRL-16190, 1965.
179. Lothrop R.P. and Goulding F.S., UCRL-16718, 1966.
180. Goulding F.S. and Lothrop R.P., UCRL-17557, 1967.
181. Goulding F.S. and Lothrop R.P., US Patent 3413529, Jan. 1968.
182. Lothrop R.P. and Goulding F.S., UCLR-17299 L, 349 *Nucl. Chem. Ann. Report*, 1966.
183. Woldseth R., *X-ray Energy Spectrometry*, Kevex Corp, p. 23, 1973.
184. Mayer J.W., *J. Appl. Phys.* 33, 2894, 1962.
185. Fox R.J. and Borkowski C.J., *IRE* NS-9, 3, 213, 1962.
186. Campbell J.L. et al., *NIM* 92, 237, 1971.
187. Walter F.J., *IEEE* NS-17, 3, 196, 1970.
188. Miller G.L. et al., *IEEE* NS-10, 1, 220, 1963.
189. Malm H.L. et al., *Can. J. Phys.* 43, 1173, 1965.
190. Tavendale A.J., *IEEE* NS-12, 1, 255, 1965.
191. Tavendale A.J., US Patent 3374124, 1968.
192. Fiedler H.J., *NIM* 40, 229, 1966.

193. de Bruin M. and Hoekstra W., *NIM* 51, 353, 1967.
194. Tavendale A.J., *IEEE* NS-13, 3, 315, 1966.
195. Kuhn E.W. et al., *Proc. 1st European Symposium on Semiconductor Detectors for Nuclear Radiation*, Eichinger P. and Keil G. (Eds.), 127, TUM, Munich, Elsevier, 1970.
196. Kraner H.W. et al., *IEEE* NS-17, 3, 215, 1970.
197. Rossington C.S. et al., *IEEE* NS-40, 354, 1993.
198. Hall R.N., US Patent 3825759 to GE, 1972.
199. Ohkawa S. et al., *MRS* 16, 185, 1984.
200. Ohkawa S. et al., *NIM* 226, 122, 1984.
201. Hall R.N. and Dunlap W.C., *Phys. Rev.* 80, 467, 1950.
202. Alferov B.M. et al., *J. Tech. Phys. (Moscow)* 25, 11, 1955.
203. Konovalenko B.M. et al., *J. Tech. Phys. (Moscow)* 25, 18, 1955.
204. Goetzberger A. et al., *Bull. Am. Phys. Soc.* 4, 455, 1959.
205. Miller G.L. et al., *IRE* NS-7, 2–3, 185, 1960.
206. Goulding F.S., UCRL-16231, 87, 1965.
207. Goulding F.S., *NIM* 43, 1, 1966.
208. Benzer S., US Patent 2622117, 1950.
209. Moore A.R. and Herman F., US Patent US 2691076, 1951.
210. Pantchechnikov J.I., *Rev. Sci. Instrum.* 23, 135, 1952.
211. Simon E., Ph.D. thesis, Purdue University, 1955.
212. Mayer J.W. and Gossick B., *Rev. Sci. Instrum.* 27, 407, 1956.
213. Walter F.J. et al., ORNL Rep. CF-58-11-99, 1958.
214. McKenzie J.M., *NIM* A162, 49, 1979.
215. Andreyeva V.V. and Shishakov N.A., *Zh. Fiz. Khim.* 35, 1351, 1961.
216. Siffert P. and Coche A., *IEEE* NS-11, 3, 244, 1964.
217. Ryvkin S.M. et al., *Zhur. Tekh. Fiz.* 27, 7, 1599, 1957.
218. McKenzie J.M. and Bromley D.A., *Phys. Rev. Lett.* 2, 303, 1959.
219. McKenzie J., *Bull. Am. Phys. Soc. Ser.* 2, 5, 355, 1960.
220. McKenzie J.M., *Bull. Am. Phys. Soc.* 4, 422, 1959.
221. Nordberg E., *Bull. Am. Phys. Soc.* 4, 457, 1959.
222. Mayer J.W., *J. Appl. Phys.* 30, 1937, 1959.
223. Walter F.J. et al., *IRE* NS, 8, 79, 1961.
224. Andrews P.T., *Proc. Symp. on Nucl. Instr.*, Birks J.B. (Ed.), Harwell 1, 1961.
225. Inskeep C.N., Thesis, University of Tennessee, 1963.
226. Inskeep C.N. and Eidson W.W., *Nucl. Electr. OECD* 163 (Paris), 1963.
227. Mead C.A., *Bull. Am. Phys. Soc. Ser.* 2, 5, 355, 1960.
228. Bramley A., *Bull. Am. Phys. Soc. Ser.* 2, 5, 355, 1960.
229. Caywood J.M. et al., *NIM* 79, 329, 1970.
230. Amsel G., *Rev. Sci. Instrum.* 32, 1253, 1961.

231. England J.B.A. and Hammer V.W., *NIM* 96, 81, 1971.
232. Hansen W.L. and Haller E.E., *IEEE* 24, 1, 61, 1977.
233. Ewins J. and Llacer J., *IEEE* NS-21, 1, 370, 1974.
234. Ishizuka Y., Husimi K. et al., *IEEE* NS-33, 1, 326, 1986.
235. Kingsbury E.F. and Ohl R.S., *Bell Syst. Tech. J.* 31, 802, 1952.
236. Ohl R., *Bell Syst. Tech. J.* 104, 1952.
237. Heinz O. et al., *Rev. Sci. Instrum.* 27, 1, 43, 1956.
238. Alvager T. and Hansen N.J., *Rev. Sci. Instrum.* 33, 567, 1962.
239. Ferber R.R., *IEEE* NS-10, 2, 15, 1963.
240. Martin F.W. et al., *IEEE* NS-11, 3, 280, 1964.
241. Laegsgaard E. et al., *IEEE* NS-15, 3, 239, 1968.
242. Fuller C.S. and Ditzenberger J.A., *Phys. Rev.* 91, 193, 1953.
243. Reiss H., Morin F.J. et al., *Bell Syst.Tech. J.* 35, 3, 535, 1956.
244. Pell E.M., *J. Appl. Phys.* 31, 291, 1960.
245. Pell E.M., *J. Appl. Phys.* 31, 1675, 1960.
246. Whoriskey P.J., *J. Appl. Phys.* 29, 867, 1958.
247. Mann H.M. et al., *IRE* NS-9, 4, 43, 1962.
248. Mayer J.W. et al., IAEA, Belgrade, May 15–20, p 35, 1961.
249. Mayer J.W., US Patent 3225198 (filed May 16, 1961), 1961.
250. Baily N.A. et al., *Rev. Sci. Instrum.* 32, 865, 1961.
251. Mayer J.W. and Dunlap H.L., *Bull. Am. Phys. Soc. Ser.* 2, 6, 107, 1961.
252. Gibbons P.E., *NIM* 16, 284, 1962.
253. Elliott J.H., UCRL-9538, 1961.
254. Elliott J.H., *NIM* 12, 60, 1961.
255. Mann H.M. et al., *IRE* NS-9, 4, 43, 1962.
256. Blankenship J.L. and Borkowski C.J., *IRE* NS-9, 3, 181, 1962.
257. Freck D.V. et al., GB Patent 1058753, 1967.
258. Tavendale A.J., *IEEE* NS-11, 3, 191, 1964.
259. Davies D.E. and Webb P.P., *IEEE* NS-13, 3, 78, 1966.
260. Mann H.M. and Yntema J.L., *IEEE* NS-11, 3, 201, 1964.
261. Landis D.A. et al., *IEEE* NS-36, 185, 1989.
262. Dearnaley G. and Lewis J.C., *NIM* 25, 237, 1964.
263. Siffert P. and Coche A., *Comp. Rend.* 256, 3277, 1963.
264. Mayer J.W., *J. Appl. Phys.* 33, 2894, 1962.
265. Lothrop R.P., UCRL-19413, 1969.
266. Kuchly J.M. et al., *NIM* 47, 148, 1967.
267. Goulding F. and Hansen W.L., *IEEE* NS-11, 3, 2861, 1964.
268. Goulding F. and Hansen W.L., UCRL-11261, 1964.
269. Landis D.A. et al., *IEEE* NS-36, 185, 1989.
270. Cappellani F. et al., *NIM* 37, 352, 1965.
271. Lauber A., *NIM* 75, 297, 1969.
272. Chasman C. and Allen J., *NIM* 24, 235, 1963.

273. Bernt H. and Eichinger P., *Proc. 1st European Symposium on Semiconductor Detectors for Nucl. Rad.*, Eichinger P. and Keil G. (Eds.), 70, TUM, Munich, Elsevier, 1970.
274. DeLyser H. et al., *IEEE* NS-12, 265, 1965.
275. Hansen W.L. and Jarrett B.V., UCRL-11589, 1964.
276. Hansen W.L. and Jarrett B.V., *NIM* 31, 301, 1964.
277. Miner C.E., UCRL-11946, 1965.
278. Miner C.E., UCRL-17201, 1966.
279. Miner C.E., *NIM* 55, 125, 1967.
280. Bowman H.R. and Hyde E.K. et al., *Science* 151, 562, 1966.
281. Radeka V., BNL-6953, 1962.
282. Blalock T.V., *IEEE* NS-11, 3, 365, 1964.
283. Blalock T.V., ORNL-TM-1055, 1965.
284. Blalock T.V., *IEEE* NS-13, 3, 457, 1966.
285. Elad E. and Nakamura M., *NIM* 41, 161, 1966.
286. Elad E. and Nakamura M., *NIM* 42, 315, 1966.
287. Elad E. and Nakamura M., *IEEE* NS-14, 523, 1967.
288. Harris R.J. and Shuler W.B., *NIM* 51, 341, 1967.
289. Kandiah K., *Proc. Conf. Rad. Meas. in Nucl. Power*, 420, IOP London, 1966.
290. Kandiah K. and Stirling A., in ref. [137], Gatlinburg NAS 1593, 495, 1969.
291. Goulding F.S. et al., *NIM* 71, 273, 1969.
292. Landis D.A. et al., *IEEE* NS-18, 1, 115, 1971.
293. Goulding F.S. et al., US Patent 3611173, Nov. 1, 1969.
294. Goulding F.S. et al., *IEEE* NS-17, 1, 218, 1970.
295. Radeka V., *Proc. Int. Symp. on Nucl. Electronics*, Versailles, France, Vol. 1, 61, 1968.
296. Elad E., in *Energy Dispersive X-Ray Analysis*, Russ J. (Ed.), ASTM STP-485, 57 Report on Symp. on EDXA, Toronto, June 1970.
297. Landis D.A. et al., *IEEE* NS-18, 1,115, 1971.
298. Goulding F.S. and Jaklevic J.M., UCRL-20625, 1971.
299. Goulding F.S. et al., UCRL 19377, 1969.
300. Kern H.E. and McKenzie J.M., *IEEE* NS-17, 3, 425, 1970.
301. Radeka V., *IEEE* NS-17, 3, 433, 1970.
302. Elad E., *IEEE* NS-19, 1, 403, 1972.
303. Ryan R.D., *NIM* 93, 2, 241, 1971.
304. McKenzie J.M. and Witt L.J., *IEEE* NS-21, 1, 794, 1974.
305. Beckhoff B. et al., *Handbook of Practical X-Ray Fluorescence Analysis*, Springer, 2006.
306. Giauque R.D. et al., *Anal. Chem.* 40, 3, 2075, 1968.
307. Goulding F.S. and Jaklevic J.M., UCRL-20625, 1971.

308. Jaklevic J.M. and Goulding F.S., *IEEE* NS-18, 1, 187, 1971.
309. Ansell K.H. and Hall E.G., in Ziegler C.A. (Ed), *Recent Developments in Low Energy Photon Sources in Applications of X- and Gamma Rays*, Gordon & Breach, 1970.
310. IEEE standard 759 (Standard approved 1981), *Standard Test Proc. for Semiconductor X-Ray Detectors*, IEEE, Std. 759, 1984.
311. Rhodes J.R., ORNL-IIC-10, 442, 1967.
312. Rhodes J.R., *Special Tech. Pub. ASTM* 485, 243, 1971.
313. Cranston F.P. and Anspaugh L.R., UCRL 50569, 1969.
314. Locker R.J. and Huth G.C., *Appl. Phys. Lett*. 9, 227, 1966.
315. Langheinrich A.P. and Forster J.W., *Adv. XRA* 11, 275, 1968.
316. Carr-Brion K.G. and Payne K.W., *The Analyst* 95, 977, 1970.
317. Frankel R.S., *Analytical Methods for EDX Spectroscopy*, Kevex, 1969.
318. Jaklevic J.M. et al., *Adv. XRA* 15, 266, 1972.
319. Hurley J.P. et al., *NIM* 57, 109, 1967.
320. van Roosbroeck W., *Phys. Rev* 139, 5a, A 1702, 1965.
321. Meyer O. and Langmann H.J., *NIM* 39, 119, 1966.
322. Mann H.M. and Bilger H.R., *IEEE* NS-13, 252, 1966.
323. Yamamoto S., *Anal. Chem*. 41, 337, 1969.
324. Giauque R.D. et al., *Anal. Chem*. 45, 671, 1973.
325. Yoneda Y. et al., *Rev. Sci. Instrum*. 42, 1069, 1971.
326. Castaing R., Ph.D. thesis, University of Paris, 1951.
327. Duncumb P., *J. Anal. Atomic Spectr. (JAAS)* 14, 357, 1999.
328. Dolby R.M., *Proc. Phys. Soc*. 73, 81, 1959.
329. Duncomb P., *X-ray Optics and X-ray Microanalysis*, Academic Press, 1963.
330. Heinrich K.F.J., *Electron Beam X-Ray Microanalysis*, Van Nostrand Reinhold, 1981.
331. Fitzgerald R. et al., *Science* 159, 528, 1968.
332. Wittry D.B., *Proc. 5th Int. Conf. on X-Ray Optics and Microanalysis*, 206, 1969.
333. Russ J.C. and Kabaya A., *Proc. 2nd SEM Symp*., Chicago, 1969.
334. Elad E. et al., *IEEE* NS-17, 1, 354, 1970.
335. Russ J.C. (coordinator), Energy dispersive X-ray analysis, ASTM STP-485, 57 Rep. on Symp., Toronto, June 1970.
336. Bender S.L. and Duff R.H., *ASTM Tech Publ*. 485, 180, 1971.
337. Beaman D.R. and Isasi J.A., *MRS* 11, 11, 8, 1971.
338. Beaman D.R. and Isasi J.A., *Anal. Chem*. 44, 9, 1598, 1972.
339. Tenny H., *Metallography* 1, 221, 1968.
340. Myklebust R.L. et al., *ASTM Tech Publ*. 485, 232, 1971.
341. Cliff G. and Lorimer G.W., *EMSA / MAS*, Boston, San Francisco Press, 1464, 1992.

342. Chadwick J., *Philos. Mag.* 24, 594, 1912.
343. Musket R.G. and Bauer W., *NIM* 109, 449, 1973.
344. Musket R.G., *NIM* 117, 385, 1974.
345. Johansson T.B. et al., *NIM* 84, 141, 1970.
346. Duggan J.L. et al., *Adv. XRA* 15, 407, 1972.
347. Johansson T.B. and Campbell J.L., *P.I.X.E. A Novel Technique for Elemental Analysis*, Wiley, 1988.
348. Friel J.J. and Mott R.B., *Microsc. Microanal.* 4, 2, 559, 1999.
349. Bronson F., The Evolution of Laboratory Instruments for the Operational Health Physicist, AAHP Special Session, Providence, USA, 2006.
350. *London Times*, Dec. 1967.
351. Cameron J.F. and Ridley J.D., *IEEE* NS-17, 1, 363, 1970.
352. Muggleton A.H.F., *NIM* 101, 1, 113, 1972.
353. Aitken D.W., *IEEE* NS-15, 3, 10, 1968.
354. *International Directory of Company Histories*, St James Press, Vol. 17, 520, 1993.
355. Henck R., *NIM A* 288, 275 1990.
356. Heijne E.H.M., *NIM A* 591, 6, 2008.
357. Kandiah K., *Rad. Meas. Nucl. Power* 420, 1966.
358. Wood A., *Magnetic Venture*, Oxford Univ. Press, 2001.
359. Berry F.P. et al., *Pitcon Today*, 1238, 20, 1989.
360. Bell R.O., *IEEE* NS-17, 3, 241, 1970.
361. Whited R.C. and Schieber M.M., *NIM* 162, 113, 1979.
362. Raiskin E.R. and Butler J.F., *IEEE* NS-35, 1, 81, 1988.
363. Ames L. et al., *Adv. XRA* 26, 325, 1982.
364. Dabrowski A.J. et al., *NIM* 213, 89, 1983.
365. Lowe B.G., *Ultramicroscopy* 28, 150, 1989.
366. Fujiyasu T. et al., *Proc. 6th Int. Conf. X-ray Optics and MA*, 505, 1971.
367. Yeh Ray et al., EDAX patent, US Patent 6437337B1, filed 1997.
368. Kandiah K. and White G., *IEEE* NS-28, 613, 1981.
369. Statham P.J., Assoc. Nat. de la Res. Techn. (ANRT), Microanalyse et Microscopie Electr. A Balayage, Paris, Nov. 1985.
370. Chavalier J.-P., *Proc. NSF/CNRS Workshop*, Aussois, France p. 231, 1988.
371. Gooda P.H. et al., *NIM A* 255, 222, 1987.
372. Campbell J.L. et al., *NIM B* 5, 39, 1984.
373. Selin E. et al., *XRS* 20, 325, 1991.
374. Pfauntsch S.J., MSc thesis, King's College London, 1992.
375. Lowe B.G., *NIM A* 439, 247, 2000.
376. Rossington Tull C. et al., *IEEE* NS-45, 3, 421, 1998.
377. Kemmer J. et al., *NIM A* 253, 378, 1987.

378. Aitken D.W. et al., Future of X-ray detectors (in ref. [335]), 36 (Toronto) 1970.

379. Elad E. and Gedcke D.A., *6th Int. Conf. X-Ray Optics & Micro Analysis*, 1971.

380. Jaklevic J.M. and Goulding F.S., *IEEE* NS-18, 1, 187, 1971.

381. Sparks C.J. and Ogle J.C., US Patent 3916200 1974.

382. Barbi N.C. et al., *Proc. 7th SEM Symp.*, 151, Chicago, 1974.

383. Jones R.A., The Use of Windowless Cryostats in EDX, Presentation, SEM meeting, ORTEC, Gibraltar, Jan 1974.

384. Coates D.G., *Electron Microsc.* 3, 26, 1980.

385. Russ J.C. et al., *XRS* 5, 212, 1976.

386. Ryon R.W., *XRS* 30, 361, 2000.

387. Newbury D.E., *Scanning* 27, 227, 2005.

388. Musket R.G., *NBS* 604, 97, Heinrich K. et al. (Eds.) (Gaithersburg Meeting April 1979), 1981.

389. Kandiah K. et al., *Proc. Int. Symp. on Nucl. Electronics*, Versailles, France, Vol. 1, 69, 1968.

390. Joy D.C., in *X-ray Spectrometry in e-Beam Instruments*, Chuck Fiori Memorial Publication, Williams D.B. et al. (Eds.), 53, Plenum Press, 1995.

391. Statham P.J., *J. Phys.* 45, 2, pC2-175, 1984.

392. Zulliger H.R. and Aitken D.W., *IEEE* NS-14, 1, 563, 1967.

393. Jaklevic J.M. and Goulding F.S., UCRL-20152, 1970.

394. Statham P.J., *NBS* 604, 127, Heinrich K. et al. (Eds.) (Gaithersburg Meeting April 1979), 1981.

395. Russ J.C. and Sandborg A.O., *NBS* 604, 71, Heinrich K. et al. (Eds.) (Gaithersburg Meeting April 1979), 1981.

396. Cohen D.D., *NIM* 193, 15, 1982.

397. Lowe B.G. and Tyrrell G.J., US Patent 4931650, 1988.

398. Aden G. and Isaacs D., *Res. Dev.*, 29, 85–90, 1987.

399. Karnezos M. and Jones R.S., US Patent 4862490, Hewlett Packard, 1988.

400. Frankel R.S., *Res. Dev.*, June, 1987.

401. Charalambous P.S. et al., *SPIE* 2516, 1995.

402. Anastasi P., Silson Ltd, Blisworth, UK.

403. Lund M.W., in ref. [390], *Chuck Fiori Memorial Publication*, 53 Plenum, 1995.

404. Moxtek, Provo, USA, now part of Japanese Company, Polatechno.

405. Ryvkin S.M. et al., *Sov. Phys. Dokl.* 10, 1116, 1966.

406. Borkowski C.J., *NAS Publ.* 1593, 196.

407. Hall R.N., in ref. [137], Gatlinburg, NAS 1593, Washington, 1969.

408. Hall R.N. et al., NYO-3870-1 (GE Reports), 1968.

409. Baertsch R.D. and Hall R.N., *IEEE* NS-17, 3, 235, 1970.

410. Hall R.N. and Soltys T.J., *IEEE* NS-18, 1, 160, 1971.
411. Tavendale A.J., *NIM* 84, 314, 1970.
412. Llacer J., *NIM* 98, 259, 1972.
413. Baldinger E. and Haller E.E., *Helv. Phys. Acta* 43, 833, 1970.
414. Hansen W.L., *NIM* 94, 377, 1971.
415. Drummond W.E., *IEEE* NS-18, 2, 91, 1971.
416. Meyer O. and Langmann H.J., *NIM* 39, 119, 1966.
417. Shiraishi F. and Takumi Y., *NIM* 196, 137, 1982.
418. Aitken D.W. et al., in ref. [335], 36, *Report on Symp. on EDXA*, Toronto, June 1970.
419. Llacer J. et al., *IEEE* NS-24, 53, 977.
420. Fink R.W., *EDX Spectrometry*, Heinrich K.F.J. et al. (Eds.), NBS 604, 5 (Gaithersburg Meeting April 1979), 1981.
421. Barbi N.C. et al., *EDX Spectrometry*, Heinrich K.F.J. et al. (Eds.), *NBS* 604, 35 (Gaithersburg Meeting April 1979), 1981.
422. Baertsch R.D., *IEEE* NS-18, 1, 166, 1971.
423. Herzer H.S. et al., *NIM* 101, 31, 1972.
424. Slapa M. et al., *NIM* 196, 28, 1988.
425. Steel E.B., MAS-21, 439, 1986.
426. Cox C.E. et al., *IEEE* NS-35, 28, 1988.
427. Lowe B.G., *NIM A* 399, 354, 1997.
428. Kohagura J. et al., *Rev. Sci. Instrum.* 66, 2317, 1995.
429. McCarthy J.J. et al., *Proc. 12th Int. Congr. for EM*, 2, 90, 1990.
430. Streli C. et al., *NIM A* 334, 425, 1993.
431. Nashashibi T., *IEEE* NS-37, 452, 1990.
432. Stodolsky L., *Phys. Today* 44, 8, Part 1, 24, 1991.
433. Keil G. and Lindner E., *NIM* 104, 209, 1972.
434. Ioffe A.F., *Semiconductor Thermo Elements and TECs*, Infosearch Ltd., 1957.
435. Madden N.W. et al., *NIM* 159, 337, 1979.
436. Woldseth R., US Patent 5075555, 1990.
437. Scintag Inc., Cupertino, CA, USA, since 1998 part of Thermo-Optek.
438. Loupilov A. et al., *Rad. Phys. Chem.* 61, 463, 2001.
439. Iwanczyk J.S. et al., US Patent US 7129501 B2, 2004.
440. Hymatic Ltd., Redditch, UK.
441. Alberti G. et al., *NIM* 158, 425, 1979.
442. APD Cryogenics, Allentown, PA, USA.
443. Longsworth R.C., US Patent 5337572, 1994.
444. Alfeev V.N. et al., British Patent 11336892, 1973.
445. Gallagher B.W. et al., US Patent 5552608, 1996.
446. Foote S.J. and McCarthy J.J., US Patent 5816052, 1998.
447. Marler J.M. et al., *IEEE* NS-20, 1, 522, 1973.

448. Stone R.E. et al., *IEEE* NS-33, 1, 299, 1986.
449. Davey G., US Patent 4475335, 1984.
450. Mikulin E.I. et al., *Adv. Cryogen. Eng.* 31, 629, 1984.
451. Yanai M. et al., US Patent 5522223, 1996.
452. Li R., US Patent 5927081, 1999.
453. Radley I. et al., US Patent 6573509 B2, 2003.
454. Hohne J., Patent Appl. 0098752 A1, 2008.
455. Becker J.A. et al., *NIM A* 505, 167, 2003.
456. van den Berg L., *AIP Conf.* 632, 71, 2002.
457. Ludwig E.J. et al., *IEEE* NS-12, 1, 247, 1965.
458. Goldhaber G. et al., *Phys. Rev. Lett.* 37, 255, 1976.
459. Kanofsky A., *NIM* 140, 430, 1977.
460. Amondolia S.R. et al., *NIM* 176, 457, 1980.
461. Kraner H.W. et al., *IEEE* NS-30, 405, 1983.
462. Guazzoni G. et al., *11th European Conf. on Semiconductor Detectors*, Wildbad, Bavaria, June 2009.
463. Guazzoni G. et al., *NIM A* 624, 247, 2010.
464. Radeka V. and Rehak P., *IEEE* NS-26, 225, 1979.
465. Gatti E. and Rehak P., LBL Workshop on Collider Detectors, 1983.
466. Gatti E. and Rehak P., 2nd Pisa Meeting on Adv. Dets. Grosetto, Italy, 1983.
467. Gatti E. and Rehak P., *NIM* 225, 608, 1984.
468. Gatti E. and Rehak P., US Patent 4688067, 1984.
469. Gatti E. et al., *NIM* 226, 129, 1984.
470. Rehak P. et al., *NIM* 235, 224, 1985.
471. Kemmer J. and Lutz G., *NIM A* 253, 365, 1987.
472. Kemmer J. et al., *NIM A* 253, 378, 1987.
473. Radeka V. et al., *IEEE Electron. Devices Lett.* ED-10, 91, 1989.
474. Lechner P. et al., *NIM A* 337, 346, 1996.
475. Chen W. et al., *IEEE* NS-39, 619, 1992.
476. Rehak P., *IEEE* NS-55, 1, 570, 2008.
477. Kalbitzer S. et al., *Z. Naturforsch.* 21a, 1178, 1966.
478. Kemmer J., *NIM* 169, 3, 499, 1980.
479. Kemmer J., US Patent 4442592, 1980.
480. Kemmer J. et al., *IEEE* NS-29, 733, 1982.
481. Struder L. et al., *Astronomy Astrophys.* 365, L18, 2001.
482. Friel J.J., *Adv. Mater. Processes*, March, 41, 2008.
483. Kemmer J. et al., *NIM A* 288, 92, 1990.
484. Eggert T., *Adv. X-ray Anal.* 48, 210, 2005.
485. Eggert T. et al. *NIM A* 512, 257, 2003.
486. Iwanczyk J. et al., *IEEE* NS-46, 3, 284, 1999.
487. Gatti E. and Rehak P., *NIM A* 541, 47, 2005.
488. SII Nanotechnology, www.siintusa.com.

Index

Page numbers followed by f and t indicate figures and tables, respectively.

Q

R

S

Z

X

Printed and bound by CPI Group (UK) Ltd, Croydon, CR0 4YY

18/10/2024

01776236-0015